Excellence in BIOLOGY

LEVEL ONE

Excellence in Biology Level One
1st Edition
Martin Hanson

Cover design: Helen Andrewes
Text design: Helen Andrewes
Illustrations: Martin Hanson
Proof reader: Sam Hill
Typesetter: Helen Andrewes
Production controller: Jess Lovell

Acknowledgements
Photographs on pages 7, 11, 20, 39, 43, 52 bottom, 53, 54, 55, 56, 57, 58, 68, 79, 93, 95, 107, 148 and 161 (Shutterstock) and page 47 (US Public Health Image Library). Other photographs by the author.

Front cover: Shutterstock.

National Library of New Zealand Cataloguing-in-Publication Data
National Library of New Zealand Cataloguing-in-Publication Data

Hanson, Martin.
Excellence in biology : level one / Martin Hanson.
Includes Index.
ISBN 978-017019-134-0
1. Biology. 2. Biology—Problems, exercises, etc.
I. Title.
570.76—dc 22

Cengage Learning Australia
Level 7, 80 Dorcas Street
South Melbourne, Victoria Australia 3205

Cengage Learning New Zealand
Unit 4B Rosedale Office Park
331 Rosedale Road, Albany, North Shore 0632, NZ

For learning solutions, visit **cengage.com.au**

Printed in China by China Translation & Printing Services.
1 2 3 4 5 6 7 14 13 12 11 10

Excellence in BIOLOGY

LEVEL ONE

Martin Hanson

Contents

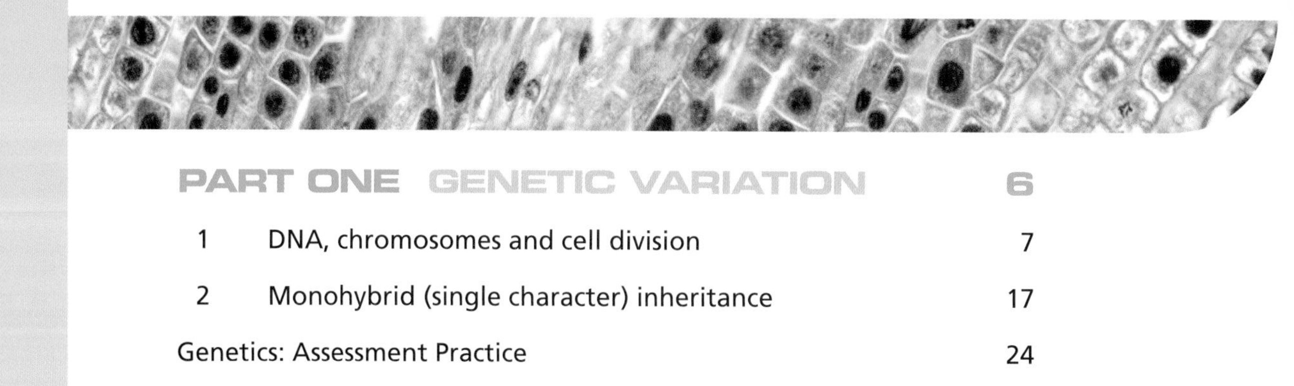

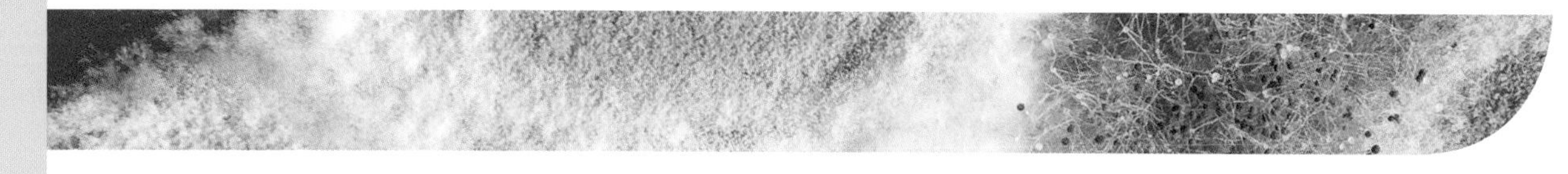

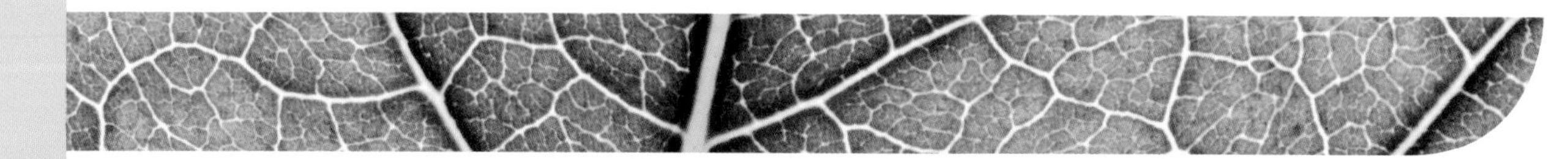

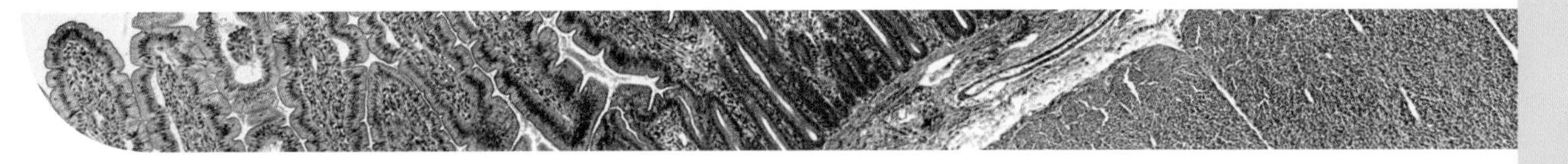

PART FOUR ANIMAL PROCESSES 104

PART FIVE CARBON CYCLING 187

KEY TERMS 207

INDEX 221

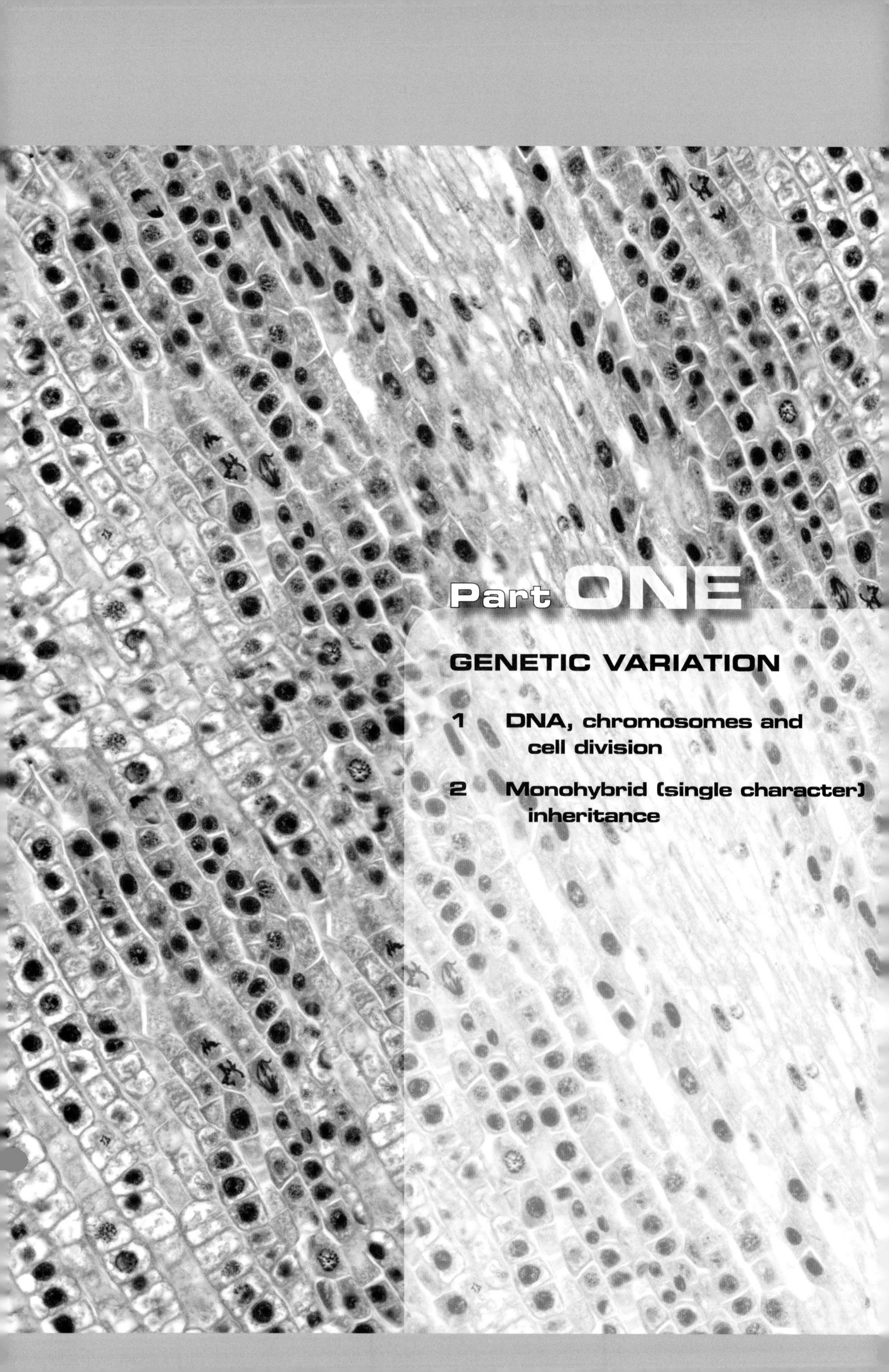

Part ONE

GENETIC VARIATION

1 DNA, chromosomes, and cell division

All sexually reproducing organisms except bacteria begin life as a fertilised egg or **zygote**. This is formed by the joining together of two cells called **gametes** — a large female gamete or **egg**, and a much smaller male gamete or **sperm**. The zygote then divides into two cells, and then each of these cells divides, and so on, eventually producing a body consisting of millions of cells (estimated to be 100 trillion in the case of humans).

GENETIC INFORMATION

For a fertilised egg to grow into an adult, it needs three things:

1. Raw materials.
2. Energy.
3. Information in the form of a 'blueprint' or plan.

The information to grow from fertilised egg to adult is called *genetic* information because it is inherited from the parents. This chapter is concerned with how this information is carried and how it is passed on from generation to generation.

In some cases the genetic information determines whether a characteristic is of one kind or another (e.g. brown or blue eyes). In other cases it influences, rather than determines, development. For example, information that tends to cause a person to grow tall is inherited, but this will only happen if he or she is well-nourished.

This information is inherited from the parents and is carried in threadlike structures called *chromosomes*, located in the cell **nucleus**. When the zygote divides, the two daughter cells each receive a complete set of this information. This is carried in the form of a chemical called **deoxyribonucleic acid**, or **DNA** as it is usually known.

DNA

The information in DNA serves two purposes:

1. It provides the information that cells need to carry out their activities.
2. It enables the DNA to replicate, or make an exact copy of itself, so that the information can be handed down from generation to generation (sometimes the copying is not exact, leading to a *mutation*).

The structure of DNA

Like proteins, starch, and cellulose, DNA is a *polymer*, consisting of large numbers of smaller units called **nucleotides**, joined in a chain. Each chain is a **polynucleotide**.

To make it more complicated, each individual nucleotide consists of three smaller parts (Fig. 1.1):

1. A phosphate group.
2. A pentose (5-carbon) sugar, **deoxyribose**.
3. A nitrogenous (nitrogen-containing) **base**.

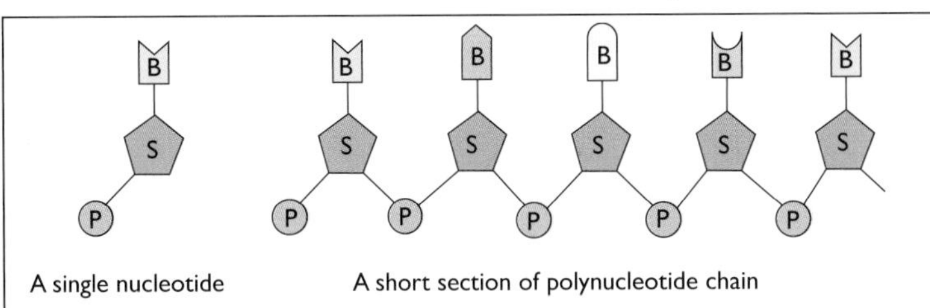

Fig. 1.1
A single nucleotide and a short section of polynucleotide chain

ISBN 9780170191340

Whereas the sugar and the phosphate are the same in all DNA nucleotides, the base can be one of four kinds — **adenine (A), guanine (G), thymine (T)** or **cytosine (C).**

Four kinds of base means there are *four kinds of nucleotide.* It is the *order* of the bases that represents the information for making a protein. The base sequence in DNA is thus a kind of message in code, rather like the order of dots and dashes in the Morse code.

A DNA molecule actually consists of *two* polynucleotide chains, arranged in a double spiral or *helix*, rather like a twisted ladder (Fig. 1.2).

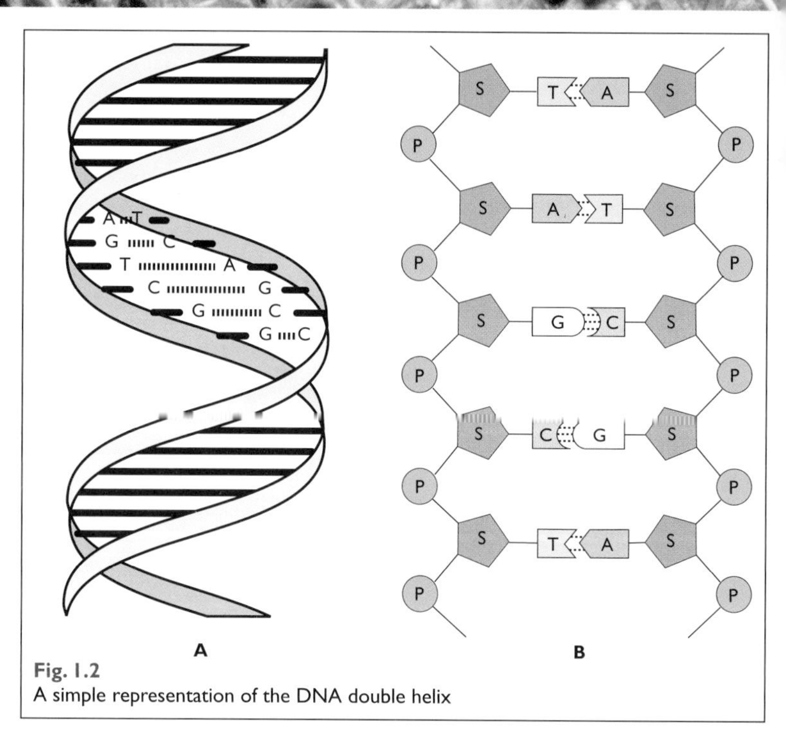

Fig. 1.2
A simple representation of the DNA double helix

The DNA *double helix*, as it became known, has the following important features:

- The two nucleotide chains are cross-linked by *hydrogen bonds* between the bases. The bases are thus in *pairs,* each pair forming a 'rung' of the ladder. Since hydrogen bonds are weak, the two strands can be easily separated.
- The pairing between the bases follows a simple rule: *adenine pairs with thymine,* and *guanine pairs with cytosine.* This arrangement is called **complementary base pairing**. Its significance is that the *order of the bases in one strand determines the order of bases in the other strand.* This enables a DNA molecule to copy itself exactly, because each strand carries the information for building the other.
- Adenine and guanine are larger than thymine and cytosine. Thus a large base always pairs with a small one, so the 'rungs' of the ladder are all the same width, and the two sides are parallel.

The genetic code

A protein consists of one, or sometimes two or more, chains of **amino acids** linked in a particular order or *sequence*. Each chain is called a **polypeptide**. To make a particular polypeptide, a cell must have the information to join the amino acids in the correct sequence.

The length of DNA coding for a polypeptide chain is a **gene**. Thus there is a gene for making the polypeptide hormone insulin. The red oxygen-carrying blood pigment haemoglobin has four polypeptide chains of two kinds, so there are two genes for making haemoglobin.

The trouble is that there are 20 kinds of amino acid in proteins, but DNA contains only four kinds of base. Each kind of base therefore cannot represent one kind of amino acid. A sequence of two bases would give 4 x 4 = 16 possible sequences — still insufficient.

A sequence of three bases would provide more than enough symbols because there are 4 x 4 x 4 = 64 possible orders of three bases.

In fact it has been shown experimentally that each amino acid is indeed represented by a sequence of three bases called a *triplet.* The code is therefore a *triplet code.* A gene for a polypeptide consisting of 250 amino acids would thus contain 250 triplets or 750 base pairs. Most cells make thousands of different proteins, for which thousands of genes are needed.

Since the two strands in the double helix are different, only one of them can be used to make a protein. How the cell 'knows' which strand to use is beyond the scope of Level 1.

DNA replication

In between cell divisions, when the chromosomes are too long and thin to be seen, the DNA molecule in each chromosome makes another copy of itself (Fig. 1.3).

ISBN 9780170191340

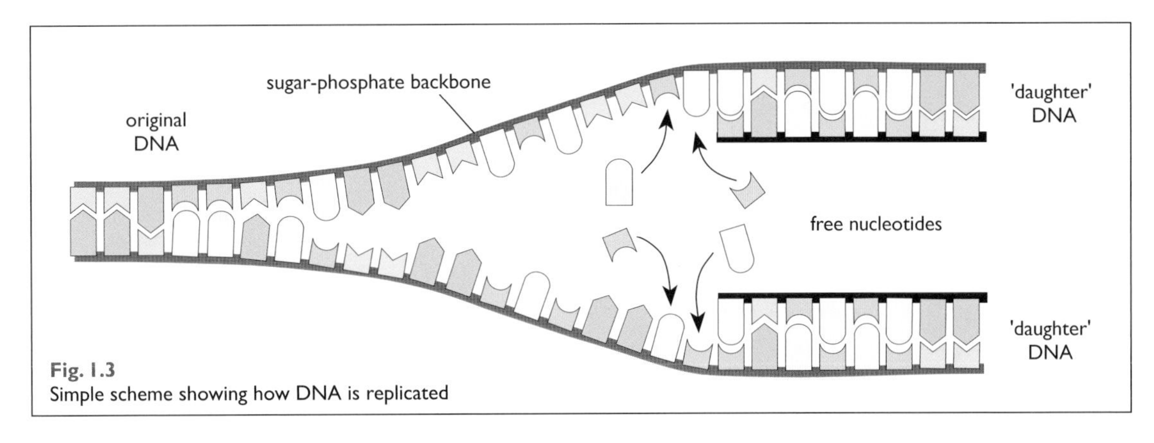

Fig. 1.3
Simple scheme showing how DNA is replicated

First, the two strands separate to expose the bases, rather like the opening of a zip fastener. Under the influence of the enzyme *DNA polymerase*, new nucleotides pair up with the exposed bases of the existing half-strands. The nucleotides are then linked together lengthwise. Because of complementary base pairing, each new half-strand is identical to one of the original half-strands. The result is two identical DNA molecules, each consisting of an 'old' and a 'new' half. Because each replication involves keeping one of the existing strands, DNA replication is said to be **semiconservative** (semi = partial, conservative = keeping), shown in Fig. 1.4.

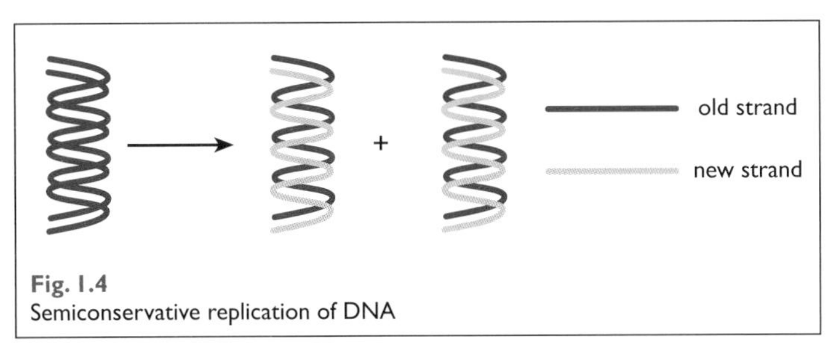

Fig. 1.4
Semiconservative replication of DNA

To replicate its DNA, a cell needs four things:

1. *Raw materials* in the form of new nucleotides.
2. *Energy* supplied by respiration.
3. *Information* in the sequence of bases in each of the two strands.
4. An *enzyme*, DNA polymerase.

Mutation

A gene mutation is a 'mistake' in the replication of a gene. This can change the sequence of amino acids in a polypeptide encoded in the gene, *but the code itself remains unchanged.* Mutations can be of two main kinds:

1. The 'wrong' base is used in the new strand. This changes one triplet, with the result that one of the amino acids in the resulting polypeptide may be of the wrong kind. The effect may be minor, but if it is a critically important amino acid, it may be lethal.
2. An extra base is inserted, or a base is deleted. The effect is to change every triplet 'downstream' of the change, usually with disastrous results.

Most mutations are harmful, but rarely, especially under changed environmental conditions, they can be beneficial.

When a gene mutates, it produces a variant of that gene, called an *allele*. In any population, many genes are present as different allelic forms. For example there are many variants of one of the genes coding for haemoglobin. Some of these alleles are harmless, while others are extremely harmful.

The role of RNA

With the exception of small amounts of DNA in mitochondria and chloroplasts, all the DNA in a cell is in the nucleus. But proteins are produced in the cytoplasm, so the information in the genes must somehow be transferred to the cytoplasm. This is the function of **RNA (ribonucleic acid)**, which is very similar to DNA except:

- The base **uracil** is present instead of thymine.
- RNA is *single-stranded*.
- The sugar is **ribose** instead of deoxyribose.

ISBN 9780170191340

Making a protein involves two main steps (Fig. 1.5):

1. **Transcription** ('copying'). One of the two DNA strands is used to make a *copy* of the gene in the form of **messenger RNA (mRNA)**. The mRNA then passes into the cytoplasm.
2. **Translation**. This is the process in which the information in the mRNA is used to make a polypeptide. Tiny granules called **ribosomes** are used to join amino acids together in the correct sequence.

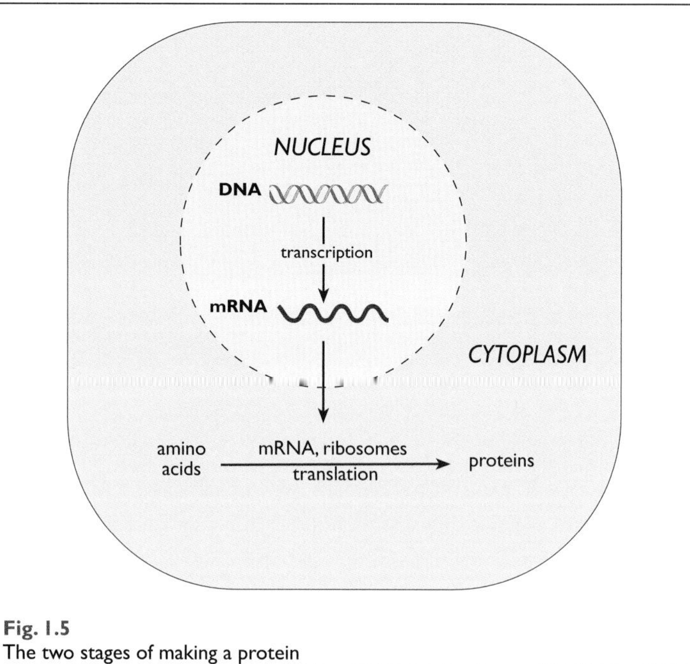

Fig. 1.5
The two stages of making a protein

The spark that ignited a scientific explosion

DNA was first extracted from pus cells in 1869, but it lay on laboratory shelves as a rather uninteresting, grey material for another 80 years before scientists discovered that it was the stuff genes are made of. Soon after, in 1953, James Watson and Francis Crick worked out its structure and were awarded the Nobel Prize for Physiology or Medicine in 1962. Crick and Watson could not have made their discovery without the earlier work of Rosalind Franklin, who used X-rays to work out the arrangement of the atoms in DNA. Franklin could not receive a share of the prize because she had died in 1958. Perhaps in recognition of her vital contribution, the director of her laboratory, Maurice Wilkins, a New Zealander, shared the prize.

Chromosomes

DNA is located in the **chromosomes**, which are threadlike structures in the nucleus. Each chromosome strand is, in fact, a single DNA molecule (though there are proteins attached to it).

Chromosomes only appear as separate structures during cell division, and even then they must first be stained if ordinary microscopes are used (*chromo* means 'colour', and *soma* means 'body').

Divisions between the chromosomes are too long, thin and tangled to be seen. In the years after their discovery in the late 19th century, biologists discovered a number of interesting things about them:

- The number of chromosomes is characteristic of a species. All the body cells of any given species have the same number of chromosomes. For example, human cells have 46 chromosomes, kiwi have 80, mice have 40, kowhai have 18 and rimu have 20.
- For a given species, gametes (eggs and sperm, Fig. 1.6) contain half the number of chromosomes compared with body cells. Gametes are said to have the **haploid** number of chromosomes (n). In humans, $n = 23$.
- In any gamete all the chromosomes differ from each other in length and in the pattern of banding after treatment with special stains. They also differ in the position of the **centromere** (see below). Each gamete has only *one* of each kind of chromosome. Individual chromosomes can thus be distinguished, and each is given a number.

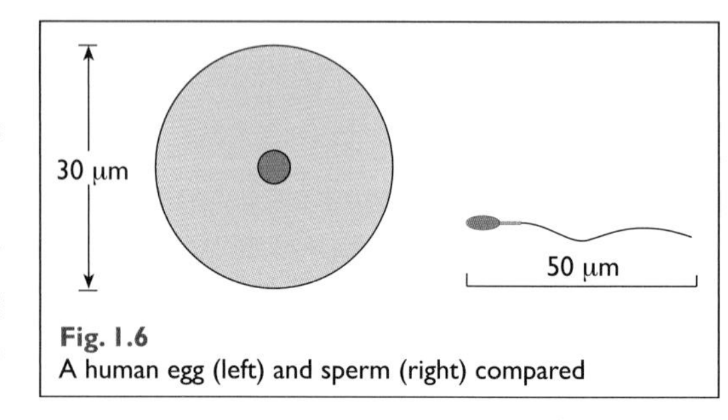

Fig. 1.6
A human egg (left) and sperm (right) compared

ISBN 9780170191340

- In all the body cells of an animal or plant, each chromosome has a 'partner' or *homologue* that looks just like it; the two homologues have the same length and centromere position. There are thus two chromosomes of each kind in the nucleus of a body cell, one member of each pair being inherited from each parent. The two together are called a **homologous** pair. Cells with two of each kind of chromosome have the **diploid** number (2*n*). Chromosomes inherited from the male parent are called *paternal* chromosomes, and those inherited from the female parent are called *maternal* chromosomes. In humans, 23 chromosomes are of maternal origin and 23 are of paternal origin.
- In male mammals, one pair of chromosomes are unlike in appearance, and are called X and Y. They are still homologous, however, because they *behave* like a pair; each gamete can have either an X or a Y, but never normally both. Females have two X chromosomes.

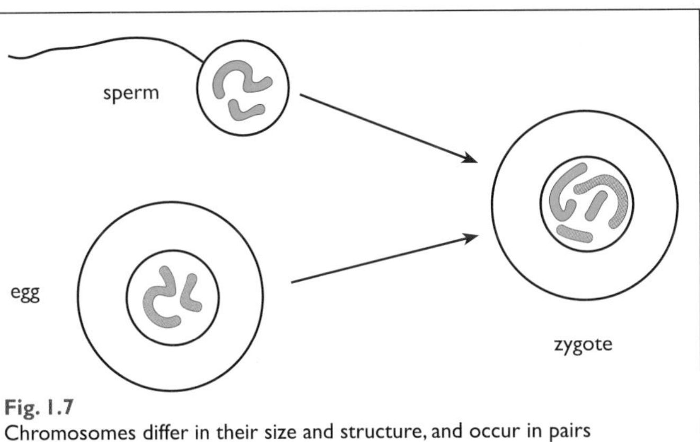

Fig. 1.7
Chromosomes differ in their size and structure, and occur in pairs

Each species can thus be distinguished from every other species by its chromosomes. All these chromosomal characteristics together form an organism's **karyotype**. When photographs of the individual chromosomes are cut out and arranged in order of size, the result is a **karyogram** (Fig. 1.8). Chromosomes are numbered, starting from the largest.

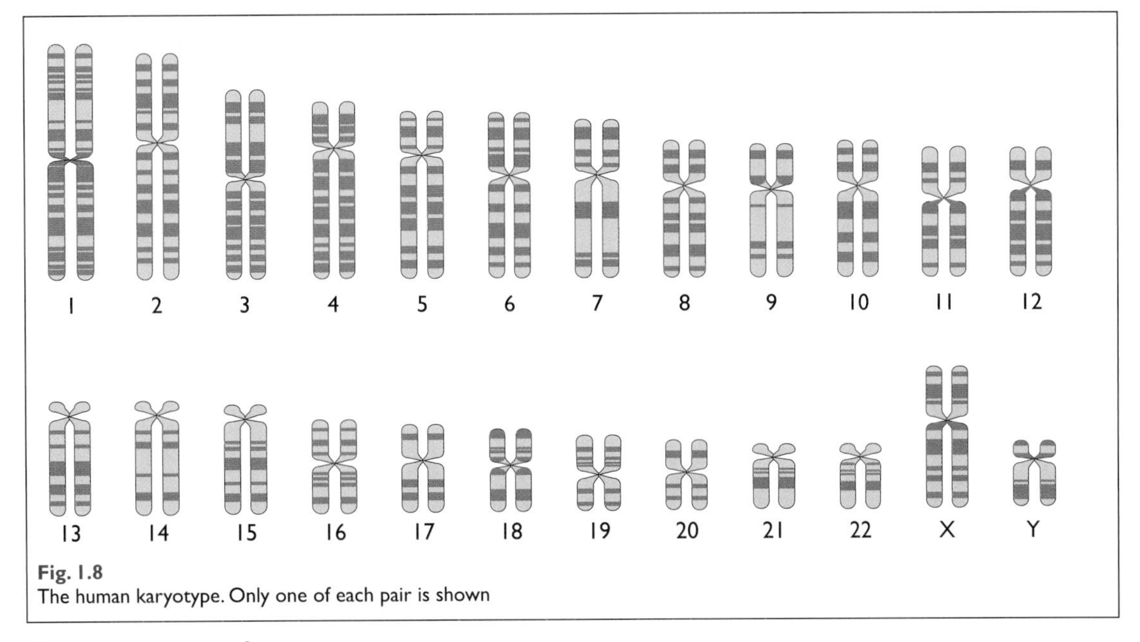

Fig. 1.8
The human karyotype. Only one of each pair is shown

Chromosomes and genes

The complete set of genes in an organism is its **genome**. The human genome contains about 25,000 genes. These are arranged on 23 pairs of chromosomes, so each chromosome has an average of about 1300 genes. Each gene occupies a particular position or **locus** on a particular chromosome. For example, the gene for the ABO blood groups is near one end of chromosome #9.

ISBN 9780170191340

KINDS OF DIVISION

The nucleus of a eukaryotic cell (eukaryotes are all organisms whose cells have nuclei) can divide in two different ways (Fig. 1.9):

1. By **mitosis**. In this kind of division the daughter nuclei are genetically identical to the parent nucleus. Mitosis is concerned solely with increasing the *number* of cells, and in humans it is part of growth ('repair' involves the production of new cells in injured sites and is thus localised growth).
2. By **meiosis**. Meiosis is an essential part of sexual reproduction, and is concerned with producing *genetic variation*. While mitosis occurs throughout life, in most multi-cellular organisms meiosis occurs only in the adult.

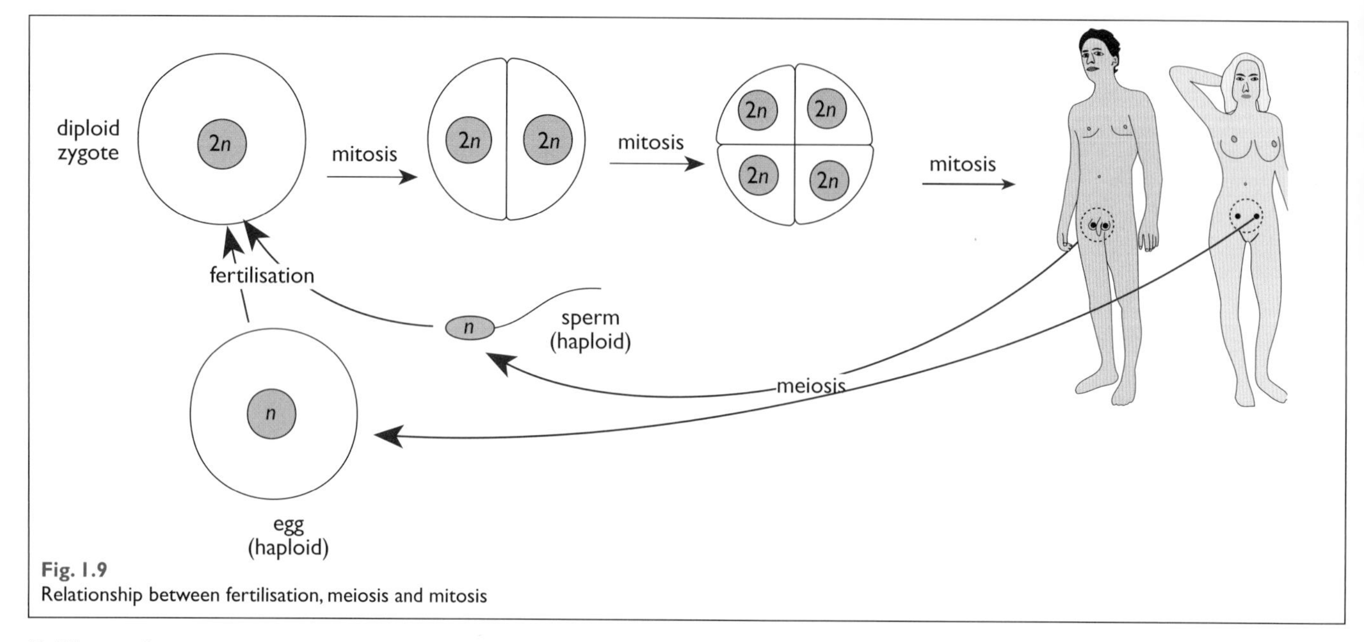

Fig. 1.9
Relationship between fertilisation, meiosis and mitosis

Mitosis

In humans and other multi-cellular organisms mitosis is an important part of growth. In many organisms it is also the basis of *asexual reproduction*. A kind of asexual reproduction that is of commercial importance in plants is *vegetative propagation* (see page 70–71 for examples). Asexual reproduction is the most common form of reproduction in most single-celled organisms.

In animals, mitosis occurs only in diploid cells, but in plants it can occur in diploid or in haploid cells. For example mosses are haploid, and as they grow they increase the number of cells by mitotic division. In plants, gametes are produced by mitotic division of cells that are already haploid.

In the early stages of development of multi-cellular organisms, cell division occurs throughout the entire body. In adults it occurs in certain parts only, such as the skin and bone marrow of mammals and the root tips and shoot tips of flowering plants.

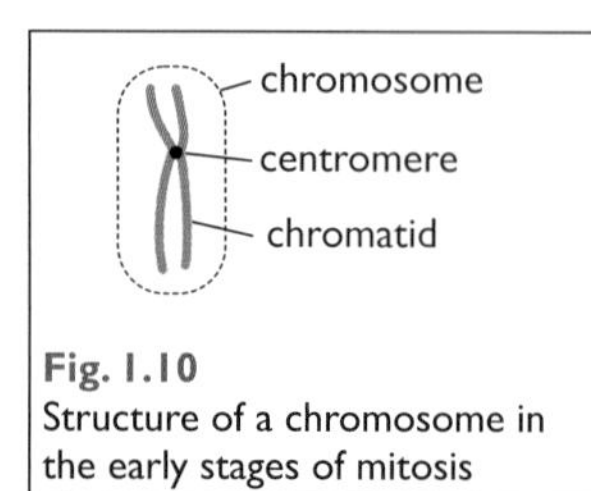

Fig. 1.10
Structure of a chromosome in the early stages of mitosis

Before a cell can divide mitotically, the DNA of each chromosome replicates. The result is two identical threads called **chromatids**, attached at a point called the **centromere** (Fig. 1.10).

As a result of mitosis each of the two daughter nuclei receives one of the two chromatids of each chromosome. If the cell is going to divide again, each chromosome replicates again. During successive cell divisions, therefore, each chromosome goes through a cycle of replication, separation of chromatids, replication, and so on (Fig. 1.11). The essential results of mitosis are illustrated in Fig. 1.12.

ISBN 9780170191340

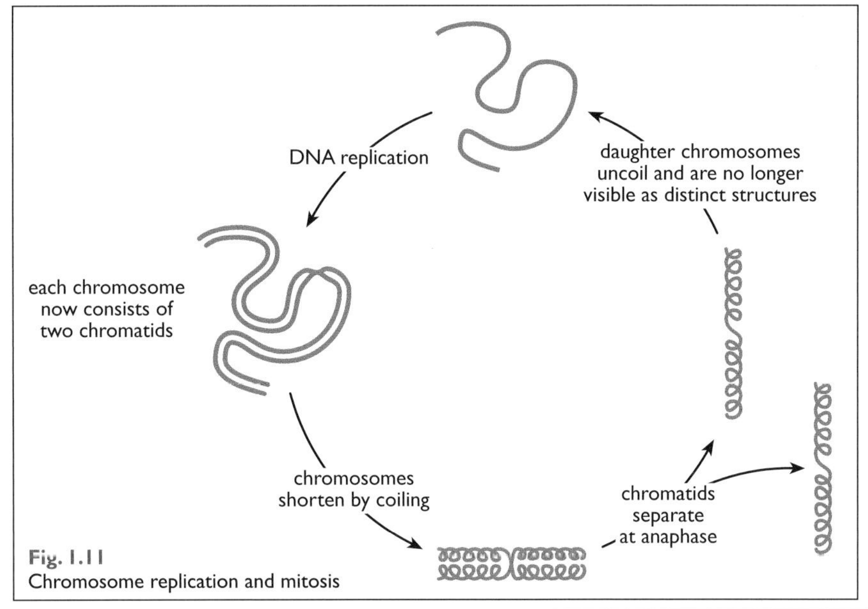

Fig. 1.11
Chromosome replication and mitosis

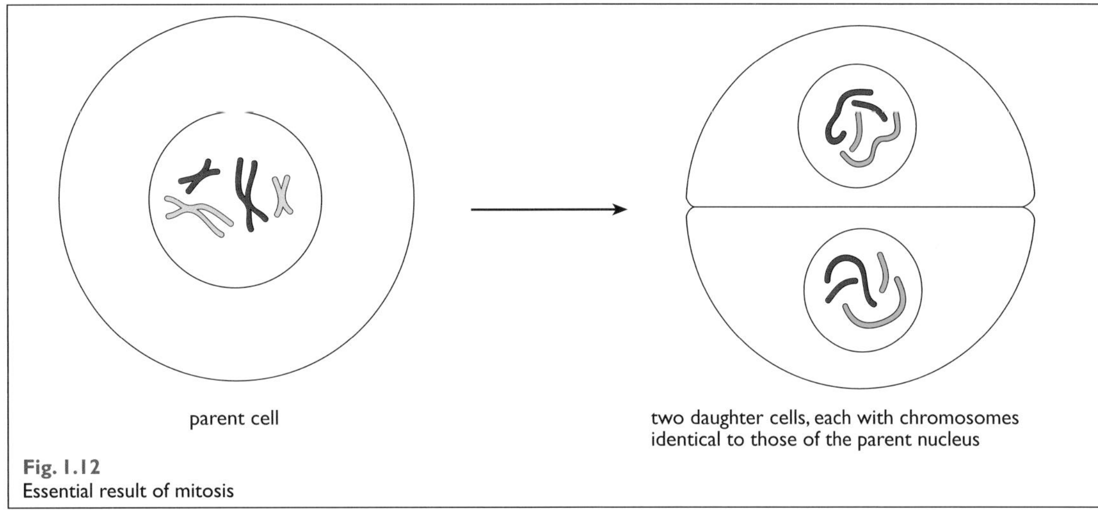

Fig. 1.12
Essential result of mitosis

Now for some more detail. Though mitosis is a continuous process, for convenience it is divided into different stages (Fig. 1.13). (You are not expected to learn these, but it does help with description.) Between divisions (*interphase*) the chromosomes are long and tangled — if it were fully extended, the DNA of the average human chromosome would be about 4 cm long. In this tangled state it would be mechanically impossible for the chromatids to move apart to form two daughter groups. The early stage (*prophase*) of mitosis is concerned with making them short and compact enough for the chromatids to separate.

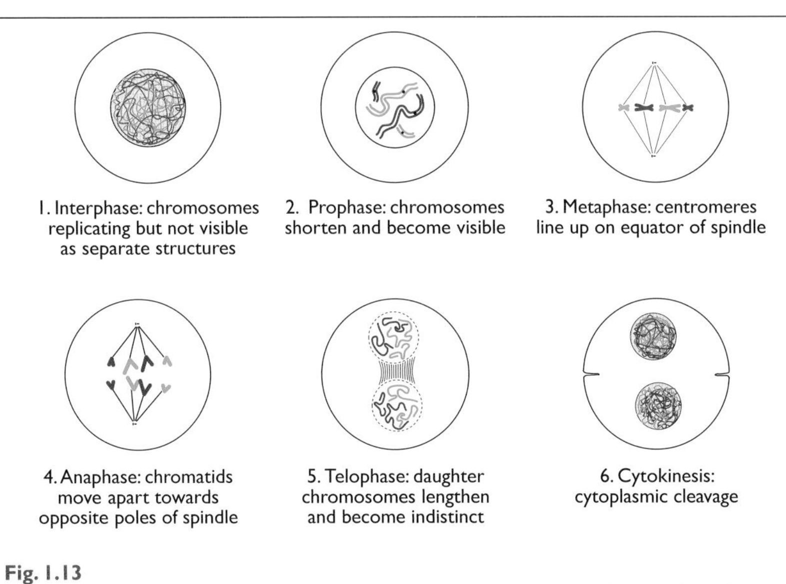

1. Interphase: chromosomes replicating but not visible as separate structures
2. Prophase: chromosomes shorten and become visible
3. Metaphase: centromeres line up on equator of spindle
4. Anaphase: chromatids move apart towards opposite poles of spindle
5. Telophase: daughter chromosomes lengthen and become indistinct
6. Cytokinesis: cytoplasmic cleavage

Fig. 1.13
The main events of mitosis in an animal cell, shown in more detail. Interphase and cytokinesis are not part of mitosis, but are shown for convenience

- **Prophase.** The chromosomes shorten and thicken by coiling up, and become visible when stained. The end of prophase is marked by the formation of the **spindle**, a barrel-shaped system of thousands of protein fibres. At the same time the nuclear envelope breaks down.
- **Metaphase.** Each chromosome becomes attached to the spindle fibres at its centromere. The chromosomes become arranged on the equator of the spindle (the imaginary plane between the two ends or poles).

ISBN 9780170191340

- **Anaphase.** The chromatids of each chromosome move apart to each pole of the spindle.
- **Telophase.** This is like prophase in reverse — the spindle disappears and a nuclear envelope develops around each group of daughter chromosomes, forming two nuclei. The chromosomes elongate and become indistinct again.

The term 'mitosis' refers to division of the *nucleus*, and is followed by division of the cytoplasm.

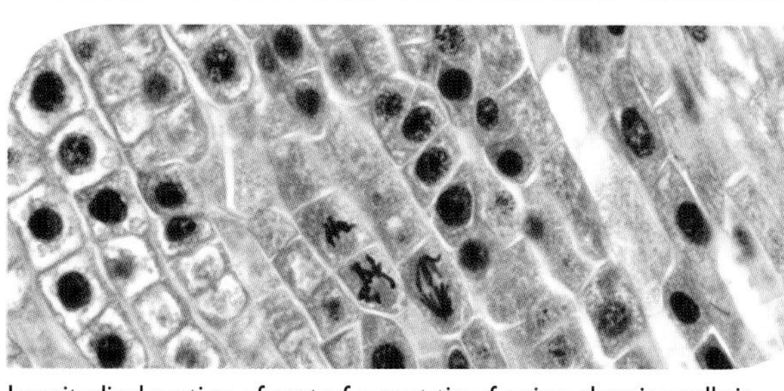
Longitudinal section of part of a root tip of onion, showing cells in various stages of the cell cycle.

The cell cycle

In cells that are repeatedly dividing, mitosis is just the more easily-observed and dramatic period in the process of cell growth and division. Between divisions, when not much *seems* to be happening, the cell is a hive of activity in which mitochondria are dividing, chromosomes are replicating and proteins are being synthesised. The entire sequence of events is called the *cell cycle* (Fig. 1.14).

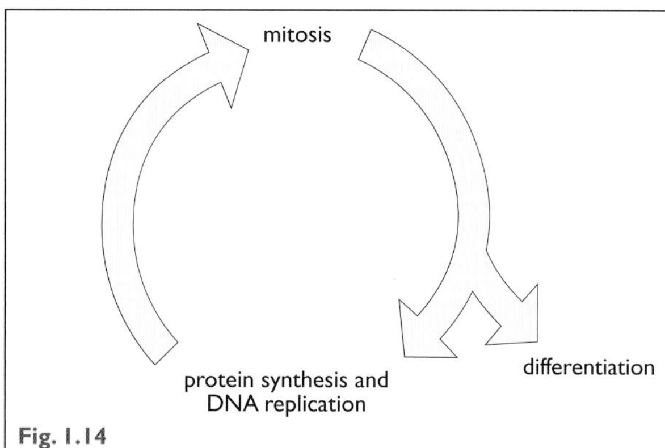

Fig. 1.14
Key events in the cell cycle

Meiosis

Meiosis is concerned with producing *genetic variation*. It is more complicated than mitosis, partly because it involves two divisions. In its *results*, it differs from mitosis in two essential ways:

1. The *number of chromosomes is reduced* from diploid to haploid.
2. The daughter nuclei are genetically *different* from each other.

In animals meiosis occurs in *ovaries* and *testes* and yields *gametes* (eggs and sperm). In flowering plants it occurs in the stamens and ovules and produces cells that divide (mitotically) to produce gametes.

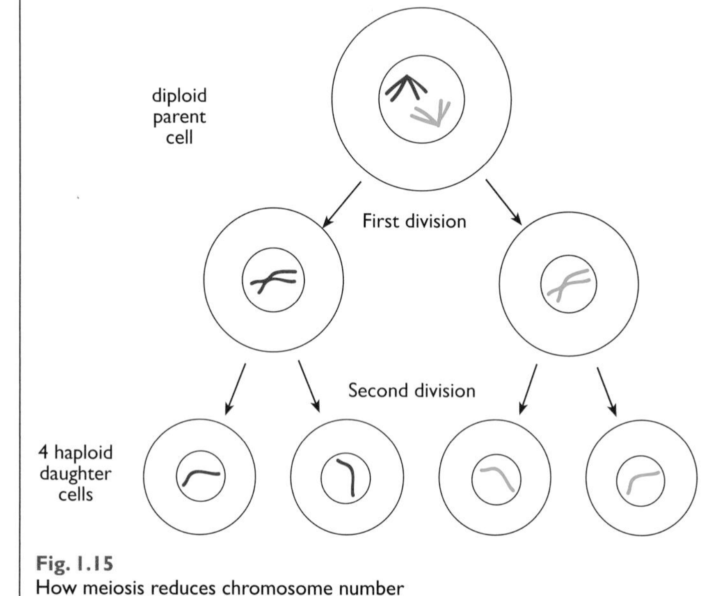

Fig. 1.15
How meiosis reduces chromosome number

Figure 1.15 shows how the chromosome number is reduced from diploid to haploid. Only one pair of chromosomes is shown; the other pairs behave similarly.

The essential point is that in the first division (meiosis I), the two members of each homologous pair are dragged to opposite poles of the spindle. This 'parting of the ways' is called **segregation**. Each daughter nucleus therefore receives only *one* of each pair. Another point to note is that in meiosis I the segregating chromosomes are *double*-stranded. The table below shows the main differences between mitosis and meiosis (the last point of difference is mentioned in the Extension).

Mitosis	Meiosis
One division	Two divisions
Daughter cells are genetically identical	Daughter cells are genetically different
Chromosome number does not change	Chromosome number is halved
Homologous chromosomes do not pair up	Homologous chromosomes pair up

In Meiosis II the chromatids of each chromosome move to opposite poles of the spindle, so the daughter nuclei now have the haploid number of *single*-stranded chromosomes.

ISBN 9780170191340

Extension: Other sources of variation

Fig. 1.15 implies that the four nuclei produced in meiosis are not all different. In fact they are, for two reasons, both of which are beyond the scope of this book. But briefly:

- In meiosis I, homologous chromosomes pair up and exchange bits of chromatid in a process called *crossing over*. As a result each member of a pair now contains bits of the other; each chromosome is no longer purely of maternal or paternal origin. As a result, the two chromatids that segregate in meiosis II are not identical (as would appear from Fig. 1.15).
- Even without crossing over, if we were to consider *more than one* pair of chromosomes, there are many different ways the chromosomes segregate. For example, a cell with two pairs of chromosomes can produce four kinds of gamete, and a cell with three pairs can produce eight kinds of gamete. A cell in a human testis could produce 2^{23} kinds of sperm. With crossing over, the number of possibilities is vastly greater than the number of electrons and protons in the universe!

Summary of key points in this chapter

- DNA is the genetic material (inherited information) and is located in the cell nucleus.
- DNA has three important features:
  1. It carries coded information for making proteins.
  2. It can be accurately *replicated* so that it can be passed from generation to generation.
  3. It can *change* slightly by mutation to enable variation among offspring.
- DNA is a two-stranded polymer, each strand consisting of a chain of *nucleotides*.
- Each nucleotide consists of a phosphate, a sugar and a base.
- The bases are of four kinds: *adenine*, *guanine*, *thymine* and *cytosine*.
- The two DNA strands are linked by *complementary* pairing between the bases, adenine with thymine and guanine with cytosine. As a result each strand contains the information for building the complementary strand.
- The 'instructions' in the DNA are the information for making proteins, and are encoded in the sequence of bases.
- The length of DNA coding for a protein is a *gene*.
- The sequence of bases in a gene represents in coded form the sequence of amino acids in a protein.
- The genetic code is a *triplet code*, meaning that each amino acid in a protein is represented by a sequence of three bases in one of the DNA strands.
- A 'mistake' in the replication of DNA is a *mutation*, and may result in the production of a protein with a different sequence of amino acids.
- Most mutations are harmful, but a small proportion are beneficial.
- Proteins are made in the cytoplasm by tiny structures called *ribosomes*.
- In making a protein, a ribosome uses a copy of a gene in the form of *messenger RNA*.
- Genes are in strands of DNA called *chromosomes*.
- The number of chromosomes is characteristic of a species.
- The gametes have one chromosome set, the *haploid* number; cells with two chromosome sets have the *diploid* number.
- In a diploid cell, chromosomes are in *homologous pairs*, one member of each pair being inherited from each parent.
- In mitotic division, the two daughter nuclei are genetically identical to each other and to the parent nucleus.
- In meiosis a diploid nucleus divides twice to form four haploid nuclei that are genetically different from each other.

ISBN 9780170191340

Test your basics

1 Copy and complete the following sentences.

a) The genetic information inherited from the previous generation is stored in a chemical called DNA, short for ___*___

b) DNA is located in threads called ___*___ which are in the ___*___ of the cell.

c) DNA consists of two chains, wrapped round each other like a twisted ladder called a ___*___ ___*___.

d) Each of the two strands in a DNA molecule consists of a chain of ___*___, each of which contains a ___*___ a ___*___, and one of four kinds of ___*___.

e) In DNA the 'rungs' of the twisted ladder are formed by ___*___ bonds between pairs of bases.

f) In the 'rungs' of the ladder, adenine pairs with ___*___ and guanine pairs with ___*___ This is called ___*___ base pairing.

g) The order of the bases along one of the chains represents, in coded form, the order of ___*___ acids in a ___*___.

h) Each ___*___ acid is represented by a sequence of ___*___ bases in a DNA strand. The genetic code is therefore a ___*___ code.

i) The sequence of bases that codes for a polypeptide is called a ___*___

j) When DNA replicates, the two strands separate and, by ___*___ base pairing, each strand is used to make a new strand.

k) Since each 'daughter' DNA molecule consists of one old and one new strand, this kind of replication is said to be ___*___.

l) Very rarely, replication is not exact and a 'mistake' is made. This gives rise to a genetic change called a ___*___.

m) DNA does not make polypeptides itself. This is the job of another nucleic acid called ___*___.

n) RNA contains the same bases as DNA, except that ___*___ replaces ___*___.

o) Making a protein consists of two stages. The first stage is called ___*___, in which a copy of the gene is made in the form of ___*___. The second stage is called ___*___; after the mRNA has been transported to the cytoplasm, tiny granules called ___*___ are used to join amino acids in the correct sequence to make the polypeptide.

p) Cells with one set of chromosomes are said to be ___*___.

q) 'Partner' chromosomes, or members of a pair, are called ___*___.

r) A nucleus can divide in two ways: by mitosis or by meiosis. Mitosis occurs as part of ___*___ and ___*___ reproduction. In a flowering plant it occurs in the tips of the ___*___ and ___*___. In adult mammals it occurs in the bone ___*___ and in the ___*___.

s) In mitosis the daughter nuclei are genetically ___*___ to the parent nucleus.

t) Meiosis occurs as part of ___*___ reproduction. In animals it occurs in the ___*___ and ___*___, and in flowering plants it occurs in the ___*___ and ___*___.

u) In meiosis there are two divisions, and the resulting four nuclei are genetically ___*___ and have ___*___ the number of chromosomes as the parent nucleus. Meiosis thus reduces the chromosome number from the ___*___ number to the ___*___ number.

2 True or false?

a) Since only one of the two DNA strands is used in making a polypeptide, the other strand has no function.

b) DNA is replicated during mitosis.

c) Mitosis only occurs in diploid cells.

d) The second division of meiosis is the same as a mitotic division.

e) A human skin cell has the same number of chromosomes as a fertilised human egg.

f) Zygotes normally have an even number of chromosomes, but gametes can have an odd or an even number.

g) Just before it divides, the nucleus of a human skin cell has 92 DNA molecules.

ISBN 9780170191340

2 Monohybrid (single character) inheritance

About the same time as DNA was isolated, an unknown monk called **Gregor Mendel** worked out some of the basic rules of inheritance. Mendel lived in Brno, in what is now the Czech Republic and he did his work by crossing different varieties of the garden pea.

The contrasting varieties he used differed sharply from one another, with no intermediates, such as tall vs short stem, yellow vs green seed, and purple vs white flower. This kind of variation is said to be **discontinuous**, and is relatively unaffected by differences in the environment. *Continuously* varying characters on the other hand are not of the 'either-or' type because there is a range of intermediates, for example skin colour in humans. Such characteristics are often affected by the environment and so are less convenient to use in genetic experiments.

Mendel was so far ahead of his time that his work was ignored for 35 years until it was independently re-discovered by three scientists. They extended Mendel's work and found that his discoveries applied to plants and animals in general.

Soon after the re-discovery of Mendel's work, Archibald Garrod, a London physician, was studying various inherited abnormal conditions in humans. Among these was *alkaptonuria*, in which the urine turns black on prolonged exposure to air (Fig. 2.1). Affected individuals also suffer arthritis and their tendons and cartilage are darkened.

Garrod couldn't do breeding experiments on humans (we prefer to choose our own mating partners), so he had to study family trees or **pedigrees**. He noticed that among the families affected by alkaptonuria, a high proportion of affected people were children of first cousin marriages. An imaginary example of such a pedigree is shown in Fig. 2.2 (A).

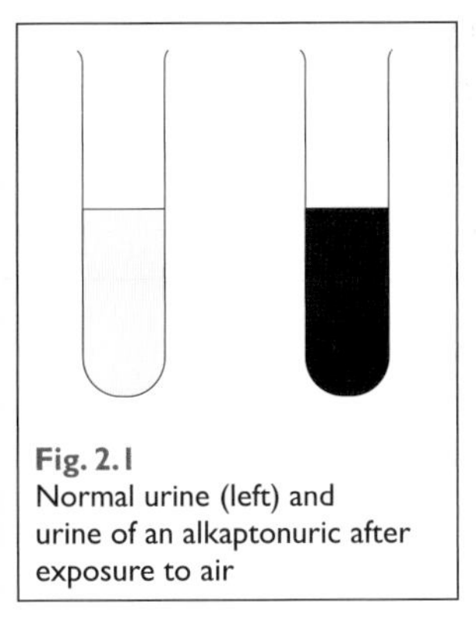

Fig. 2.1
Normal urine (left) and urine of an alkaptonuric after exposure to air

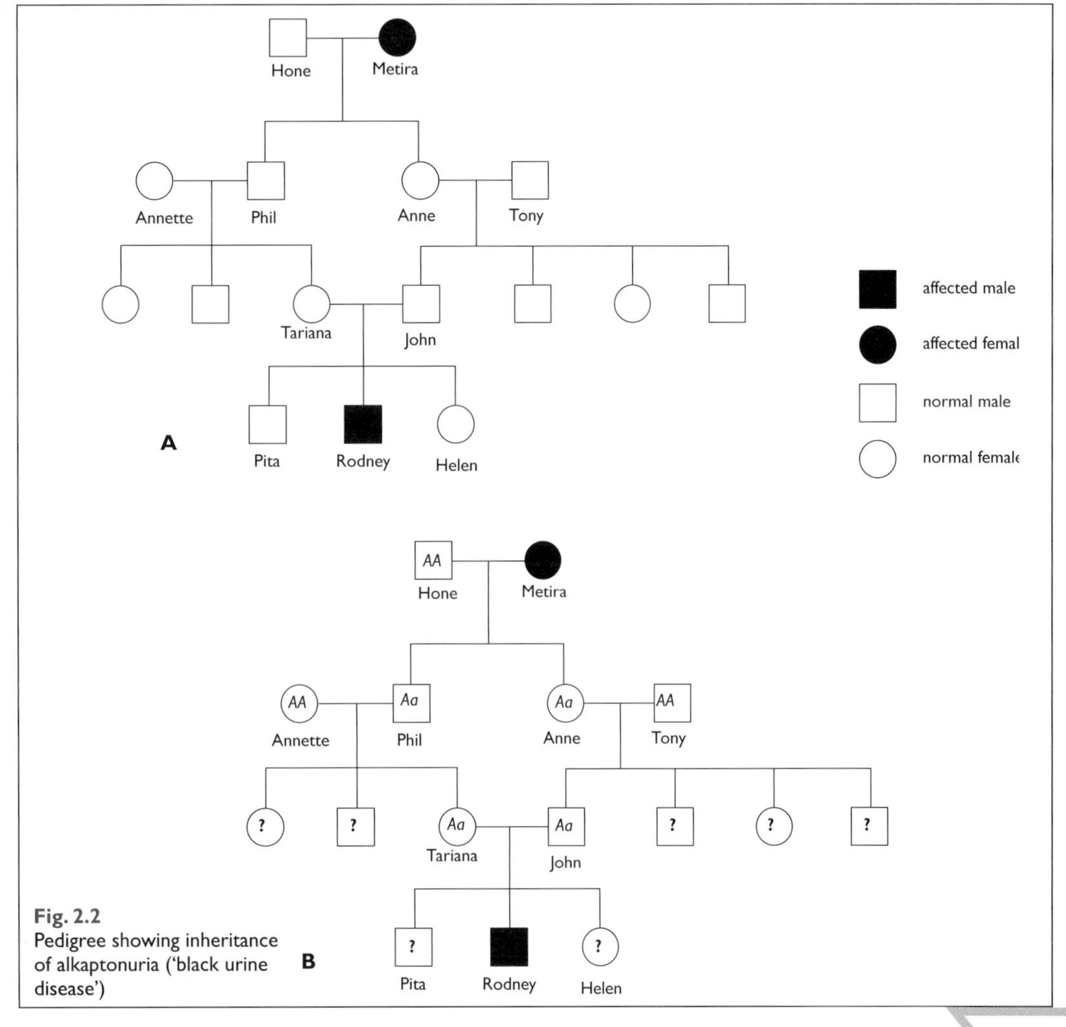

Fig. 2.2
Pedigree showing inheritance of alkaptonuria ('black urine disease')

ISBN 9780170191340

By studying the pedigree we can learn something about the rules of heredity:

- Since Rodney has the condition, he must have received the hereditary factor or **gene** from his parents, yet neither Tariana nor John showed the condition (see page 8 for a definition and explanation of the term 'gene'). The gene must have been present in the parents even though it was not expressed ('shown'). Alkaptonuria is said to be **recessive** to the normal condition, which is **dominant**. Such clearly contrasting inherited conditions are called **traits**, and the genes responsible are called **alleles**. We can thus represent the allele for the dominant (normal) trait by the symbol *A* and the recessive allele (alkaptonuria) by *a*. Symbols for alleles are normally printed in italics.
- We can represent Rodney's genetic makeup or **genotype** as *aa*, one *a* allele being inherited from each parent. Since Tariana and John were both normal, they must have had the genotype *Aa*, and so must the grandparents Phil and Anne. Figure 2.2 (B) shows the same pedigree, but with some of the genotypes shown.
- Since Rodney has two copies of the same allele, he is said to be **homozygous** (*homos* = Greek for 'same'). Tariana and John on the other hand, each have two different alleles and are **heterozygous** (*heteros* = Greek for 'other'). They are said to be *carriers*; they have the allele but do not express it.
- Alkaptonuria is very rare, affecting fewer than 1 person in 250,000. It follows that 'outsiders' (Hone, Annette and Tony), who marry into the family are most unlikely to carry the allele. These individuals must therefore have the genotype *AA*. Thus Hone and Phil must have different genotypes, even though they had the same characteristics, or **phenotype**.
- Even though a sperm is much smaller than an egg, we can deduce that both must carry genetic information, since Rodney has inherited the condition via John as well as Tariana.
- Since Rodney has the genotype *aa*, both Tariana's egg and John's sperm must have carried the *a* allele only. His parents both had the genotype *Aa*, so when they formed gametes, the *A* and *a* alleles must have **segregated**. Hence each person carries *two* alleles for relating to this condition, but each gamete only carries *one*.

Genes and chromosomes

In Chapter 1 we learned that:

- Genes are located on the chromosomes.
- Chromosomes are in *homologous pairs*.
- Gametes have only one of each pair.

As already stated, each individual has two genes for each character but each gamete carries only one of each pair. This is an exact parallel with the behaviour of the chromosomes: each person has two of each chromosome, but each gamete has only one. Chromosome number is halved in meiosis and doubled in fertilisation, as explained in Chapter 1. The explanation is that genes are situated on the chromosomes, as shown in Fig. 2.3.

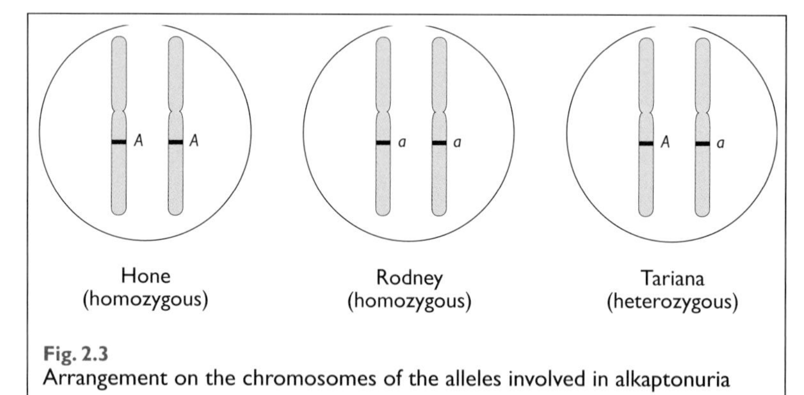

Fig. 2.3
Arrangement on the chromosomes of the alleles involved in alkaptonuria

The gene causing alkaptonuria is situated on the long arm of chromosome 3, so the arrangement of genes in a heterozygote can be represented in Fig. 2.3.

Figure 2.4 shows what happens when a heterozygous male produces sperm in meiosis. Each sperm receives only one of each chromosome pair and hence only one allele; half the sperm receive *A* and the other half receive *a*. The situation is a bit different in women, as a woman only makes one egg at a time (the other three products of meiosis die). When a woman with the genotype *Aa* produces an egg, there is a 50% chance that it will receive *A* and a 50% chance it will receive *a*. Over her lifetime, approximately half her eggs will carry *A* and the other half will carry *a*.

ISBN 9780170191340

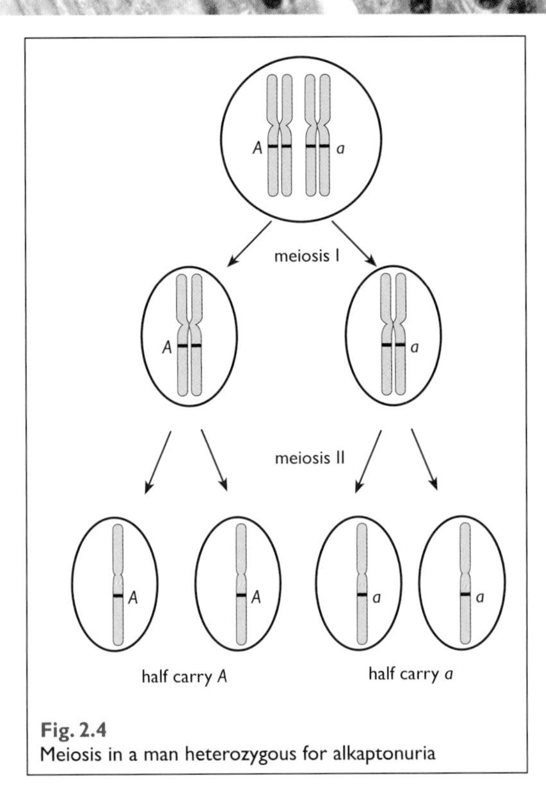

Fig. 2.4
Meiosis in a man heterozygous for alkaptonuria

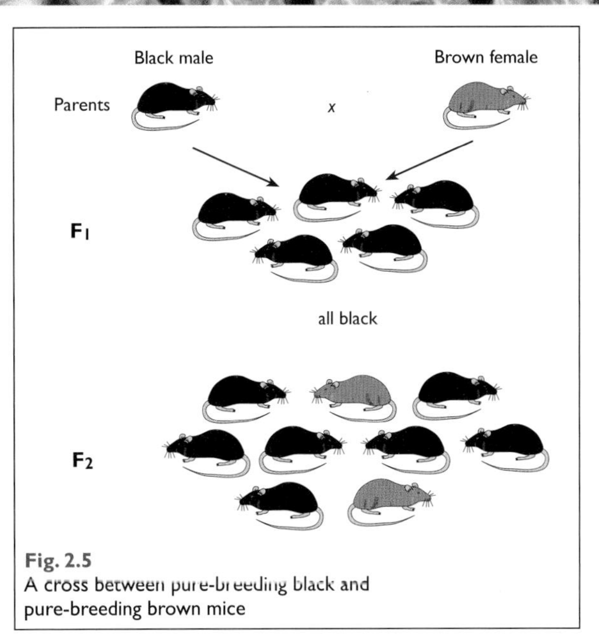

Fig. 2.5
A cross between pure-breeding black and pure-breeding brown mice

Breeding mice

Mice have larger families and shorter life cycles than humans, so breeding experiments are quicker. Figure 2.5 shows the results of a cross between a pure-breeding black male and a pure-breeding brown female.

The offspring of a cross between two differing pure-breeding parents are called the F_1 (short for 'first filial generation'), and are all *heterozygous*. When the F_1 are mated among themselves (i.e. *inbred*), an **F_2 generation** is produced. Of these, *approximately* ¾ are black and ¼ are brown. Figure 2.6 shows how we set out the explanation, using a **Punnett square**.

If properly used, a Punnett square shows two things:

1. The various different ways in which the gametes can combine.
2. Their proportions.

In this case two kinds of egg and two kinds of sperm can join in four different ways, though two of them (*B* egg + *b* sperm, and *b* egg + *B* sperm) produce the same result. Half the sperm and half the eggs carry *B* and the other half carry *b*, so the proportion of each kind of fertilisation = ½ x ½ = ¼.

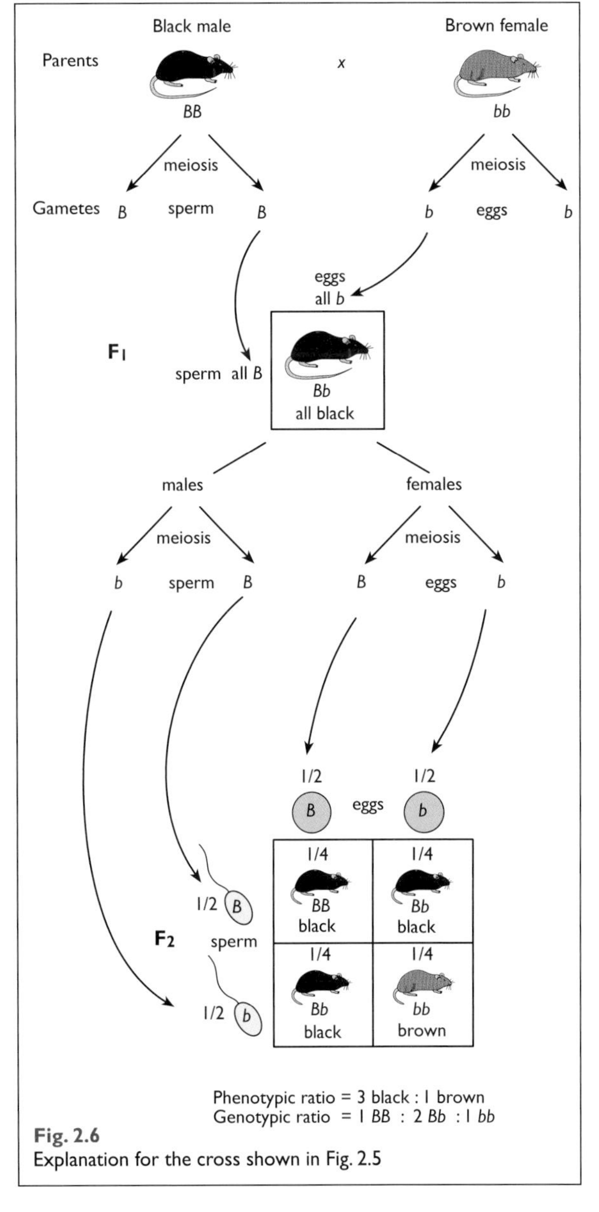

Fig. 2.6
Explanation for the cross shown in Fig. 2.5

It is incorrect to use a Punnett square with four compartments when there is only *one* kind of fertilisation event, as happens when, for example, a mouse with the genotype *BB* is mated with a mouse with the genotype *bb*.

ISBN 9780170191340

Determining an organism's genotype: The testcross

In the example above, the F_1 mice are phenotyptically indistinguishable from their pure-breeding black parent. But they are clearly different *genotypically* because when mated with their own kind they are not pure-breeding; some of their offspring are brown.

To find the genotype of an organism showing the dominant trait, a **testcross** is performed. The organism with the unknown genotype is mated with an organism with the recessive phenotype (and is therefore known to be homozygous recessive), as shown in Fig. 2.7. In this case it turns out that the mouse is heterozygous since some of the offspring are brown. If the black mouse is homozygous, all the offspring will be black.

Parents: Brown female *bb* x Black male *Bb*

Gametes: meiosis → *b*, *b* (eggs, all *b*); meiosis → *B*, *b* (sperm, 1/2 *B*, 1/2 *b*)

Offspring: *Bb* 1/2 black; *bb* 1/2 brown

phenotypic ratio = 1 black : 1 brown

Fig. 2.7
A testcross to determine the genotype of a black mouse

The essential thing about the testcross is that the parent showing the recessive trait is homozygous and thus produces gametes all carrying the recessive allele. Any variation in the offspring *must therefore be due to the parent with the unknown genotype*.

It is very important to remember that ratios in genetics are always approximate. Even if a black mouse is heterozygous, it could still produce all black offspring when testcrossed, because of the effect of chance. In the case of mice, it would therefore be better to make repeated testcrosses, ideally between a black male and a series of brown females. With larger numbers of offspring the effects of chance are reduced.

Boy or girl?

Of the 23 pairs of chromosomes in the human set, two members of one pair are different in males, and are called the X and Y chromosomes. Females have two X chromosomes. The X and Y chromosomes are called the *sex chromosomes* because they determine sex. The other 22 pairs are called **autosomes** and play no part in sex determination.

Although the X and Y chromosomes are visibly different (the Y is much smaller), they behave like a pair because in meiosis they segregate, each sperm receiving *either* an X *or* a Y chromosome. A man therefore makes two kinds of sperm in equal numbers, one kind (Y) giving rise to boys and the other (X) producing girls (Fig. 2.8).

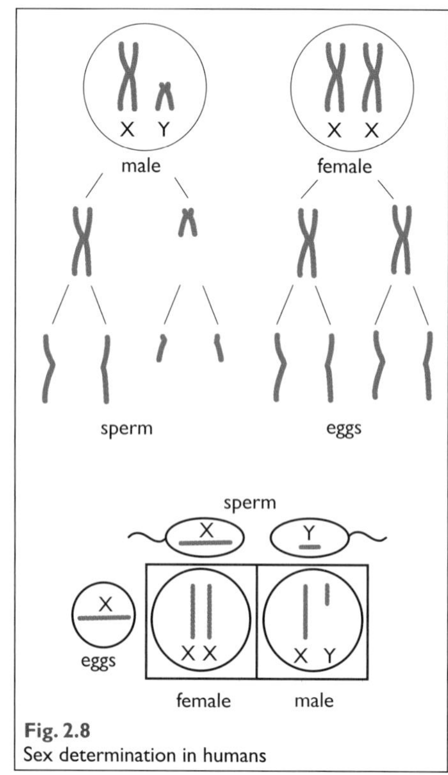

Fig. 2.8
Sex determination in humans

ISBN 9780170191340

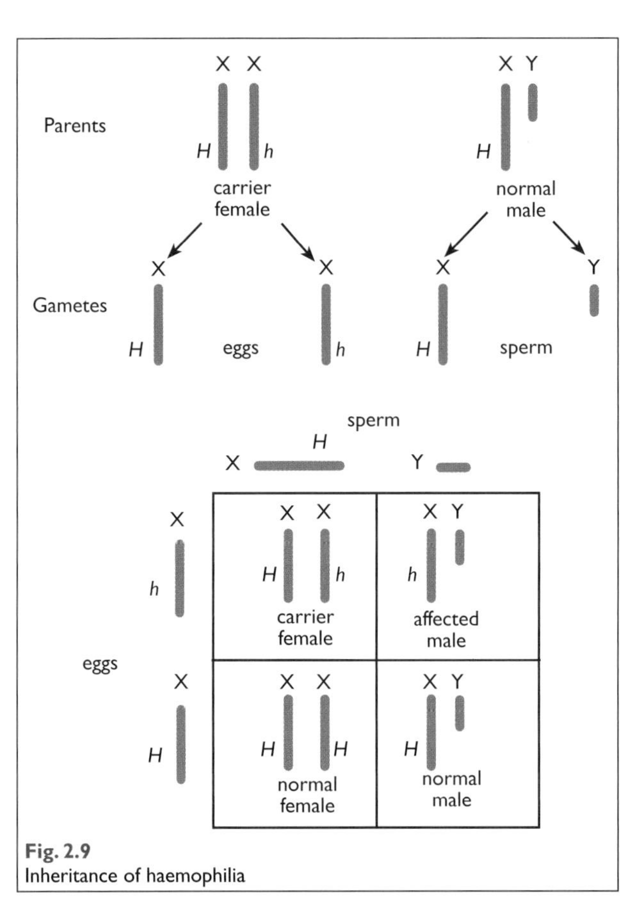

Fig. 2.9
Inheritance of haemophilia

Sex-linked inheritance

Some inherited conditions are more common in males than in females. An example is *haemophilia*, in which the blood does not clot. The genes responsible for these conditions are located on the X chromosome, of which males have only one.

The allele for haemophilia is recessive in females. Males therefore only need one copy for it to be expressed. Females need two, so they only have the condition if they inherit the allele from *both* parents. A man gives his X chromosome to his daughters and his Y chromosome to his sons. Males therefore inherit the allele from their mothers, who are usually normal, but heterozygous carriers. Figure 2.9 shows the inheritance of haemophilia.

A small number of X-linked alleles are dominant. An example is a rare kind of rickets that is not due to dietary deficiency of vitamin D. X-linked dominant conditions are more common in females, since a female can inherit the allele from either parent, but a male can only inherit it from his mother.

Extension: Probability in genetics

Most of the animals and plants used in genetics have large families, so it is reasonable to talk of ratios. Humans have small families; it would be meaningless to talk about a 3 : 1 ratio in a family of 2. It is more sensible to talk about the *probability* that a given child will have a particular characteristic. For example, approximately half of newborn babies are girls. The probability that any given child will be a girl is therefore 0.5.

When calculating probabilities, there are two rules:

The product rule. This is used when we want to know the probability that *both* of two independent events will occur. For example, we might want to know the probability that the first two children in a family will both be boys. These two events are independent of the other because the sex of the first child has no effect on the sex of the second. To find the probability that the first and the second child will both be boys, we *multiply* the two probabilities: 0.5 x 0.5 = 0.25.

The sum rule. This is used when we want to know the probability that *either* one event *or* the other will occur. In this case we *add* the two probabilities.

What is the probability that in a two-child family there will be one of each sex? Figure 2.10 shows that, as far as sexes of the children are concerned, there are four kinds of two-child family.

Two of these four kinds have one of each sex. To find what proportion of two-child families contain one of each sex, we need to know the probability of a family consisting of girl followed by boy *or* boy followed by girl. The probability of each of these is 0.5 x 0.5 = 0.25. Hence the probability that a family will consist of a boy and a girl = 0.25 + 0.25 = 0.5.

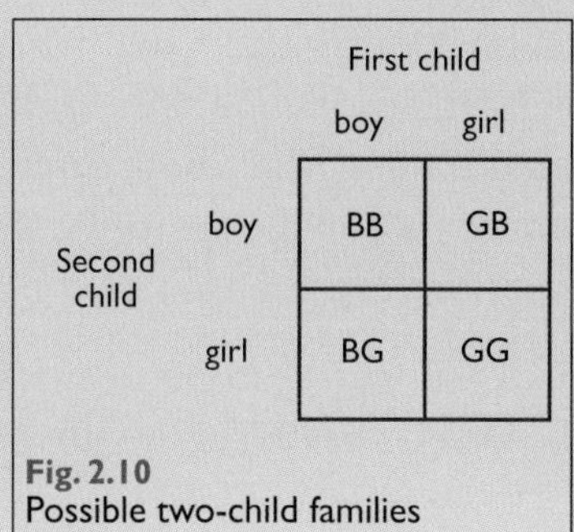

Fig. 2.10
Possible two-child families

ISBN 9780170191340

Kinds of variation

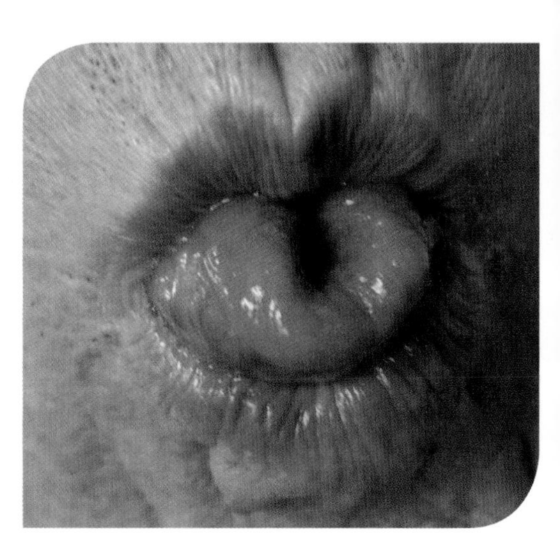

The examples described above are examples of *discontinuous* variation, in which there are no intermediates; a man is either colourblind or he is not. Other examples of discontinuous variation in humans are:

- Blood groups. Everyone belongs to one of four groups: A, B, AB or O.
- Tongue-rolling. A person either can or cannot roll the tongue.
- Red-green colourblindness, due to an X-linked allele, recessive in females.

An important feature of discontinuous variation is that it is not usually affected by differences in environmental experience.

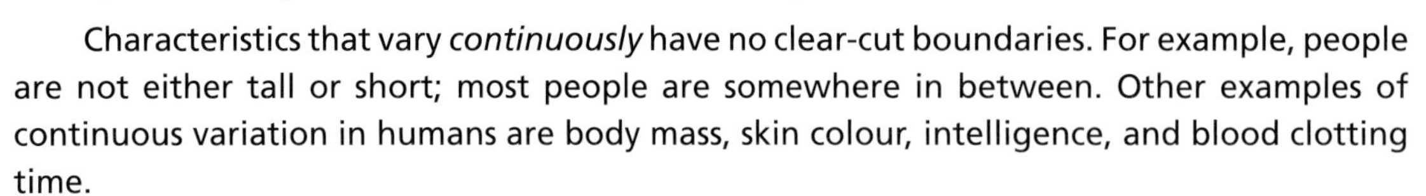

Characteristics that vary *continuously* have no clear-cut boundaries. For example, people are not either tall or short; most people are somewhere in between. Other examples of continuous variation in humans are body mass, skin colour, intelligence, and blood clotting time.

Unlike discontinuous variation, many continuously varying characters are affected by differences in environment. Your body mass, for example, may depend on your diet, and children that are not given a stimulating environment are less likely to reach their intelligence potential.

It is worth noting that many aspects of an organism's phenotype are not visible, such as blood group, reflex time, blood pressure, and so on. It is therefore not accurate to define an organism's phenotype as its 'appearance'.

Advantages and disadvantages of sexual reproduction

The 'aim' of every organism is to reproduce — in other words, to get its genes into the next generation. There are two ways of achieving this: sexual and asexual reproduction.

In sexual reproduction the result is the production of *genetic variation*. These new variants are produced in two steps: meiosis and fertilisation. The two processes are complementary; no life cycle can have meiosis without fertilisation, and vice-versa.

Suppose two organisms with the genotype *Aa* mate and have offspring. Meiosis results in *two* kinds of gamete, *A* and *a*. When these combine randomly in fertilisation, *three* genotypes are produced among the offspring (*AA*, *Aa* and *aa*).

There are potential advantages and disadvantages in variation. Successful parents have proved their *fitness* — which is their ability to survive and reproduce — otherwise they would not have become parents. If the offspring were to experience the same range of conditions as the parent, their best chance would be as genetic copies, in other words, if the parent were to reproduce *asexually*. As it is, most variants are unlikely to be as well adapted to the conditions experienced by the parents. There are two advantages of sexual reproduction:

1. Most offspring are dispersed to different environments, to which some may well be better adapted than the parents.
2. Over long periods, environments change from generation to generation, so what is the 'fittest' genotype in one generation may not be the fittest in the next.

With asexual reproduction, the situation is reversed; the offspring are genetically identical to the parent. In the same range of conditions as experienced by their parents, offspring are, however, as well adapted and so there is less waste. Species depend on sex, in the long term, to adapt to changing conditions.

ISBN 9780170191340

Summary of key points in this chapter

- The most suitable characteristics to use in breeding experiments are those that vary *discontinuously*, because the environment has little effect on such variation. The results therefore illustrate the effect of heredity rather than environment. Discontinuously varying characteristics are called *traits*.
- A unit of genetic information is called a *gene*. Most genes exist as two or more variants or *alleles*. Each individual has *two* genes controlling a given trait, but the gametes carry only *one*. This is because when animals produce gametes the alleles *segregate*. Segregation of the alleles occurs in a kind of nuclear division called *meiosis*.
- When an organism has two different alleles of a gene it is said to be *heterozygous*. When it has two of the same allele, it is said to be *homozygous*, or pure-breeding.
- When there are two allelic forms of a gene, one allele is usually *dominant* to the other, which is *recessive*.
- The information an organism inherits from its parents is its *genotype*. The resulting bodily characteristics are its *phenotype*.
- When two different, homozygous parents are crossed, the offspring are called the F_1 *generation* and are *heterozygous*.
- When individuals of an F_1 generation are mated together (inbred), the offspring are called the F_2 *generation*.
- To find out whether an organism showing the dominant trait is heterozygous or homozygous, a *testcross* is performed.
- A testcross consists of mating the organism under test with an organism showing the recessive trait, which is thus homozygous.

Test your basics

1 Copy and complete the following sentences.

a) The information that an organism inherits is its ___*___, and its physical characteristics (such as its appearance) is its ___*___.

b) Alternative forms of a gene are called ___*___.

c) Diploid organisms have ___*___ genes for each character, but each gamete carries ___*___ of each gene.

d) Organisms with two copies of the same allele are said to be ___*___, and organisms with different alleles for a given character are ___*___.

e) An allele that is expressed only in homozygotes is said to be ___*___. Alleles that are expressed in heterozygotes are said to be ___*___.

f) During meiosis, pairs of genes separate or ___*___, with the result that each resulting nucleus receives ___*___ of each pair of genes.

g) To find out whether an organism is homozygous or heterozygous, it is crossed with one showing the ___*___ phenotype. This kind of cross is called a ___*___.

h) Chromosomes that play no part in the determination of sex are called ___*___. In humans and other mammals, the chromosomes that are involved in sex determination are of two kinds, X and Y. ___*___ mammals have an X and a Y chromosome and make two kinds of gamete, half carrying X and half carrying Y. It is therefore the ___*___ sex that determines the sex of the offspring.

2 In fruit flies, grey body (*B*) is dominant to black body (*b*). A cross between a heterozygous grey-bodied male and a black-bodied female is shown below. Copy and fill in the missing letters and words in the empty boxes.

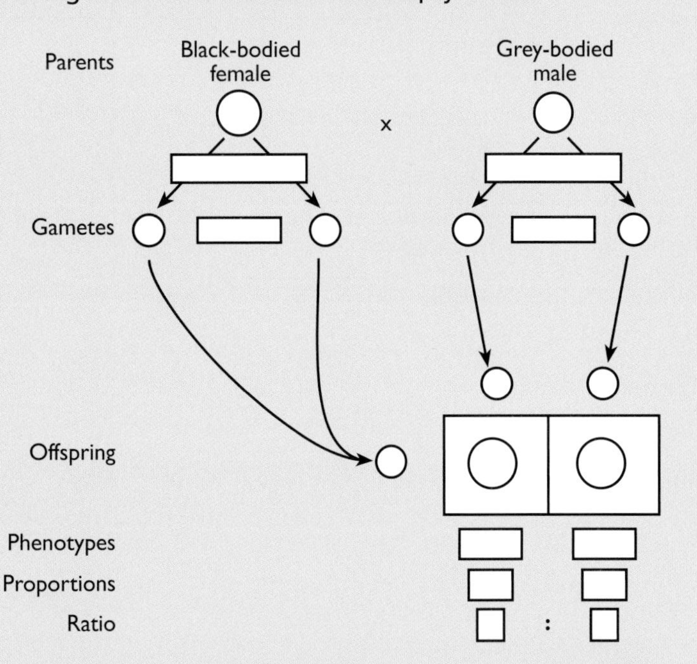

ISBN 9780170191340

Genetics: Assessment Practice

The following questions are adapted from NCEA questions. Note that the word 'Discuss' at the beginning of a question usually indicates an 'Excellence' type question.

QUESTION ONE: HAIR TEXTURE IN GUINEA PIGS

The pedigree below shows the inheritance of coat texture in guinea pigs, in which the coat may be rough or straight.

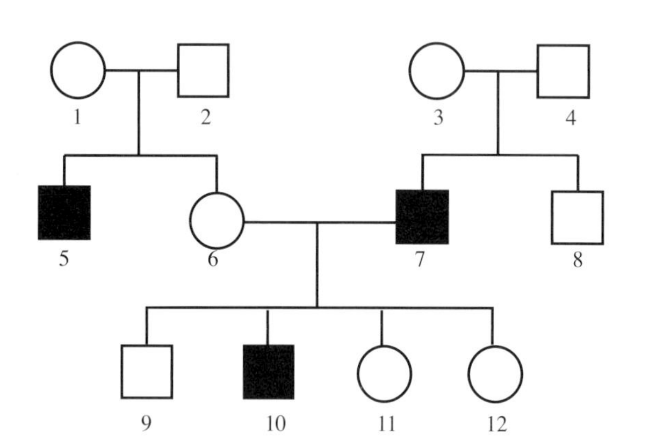

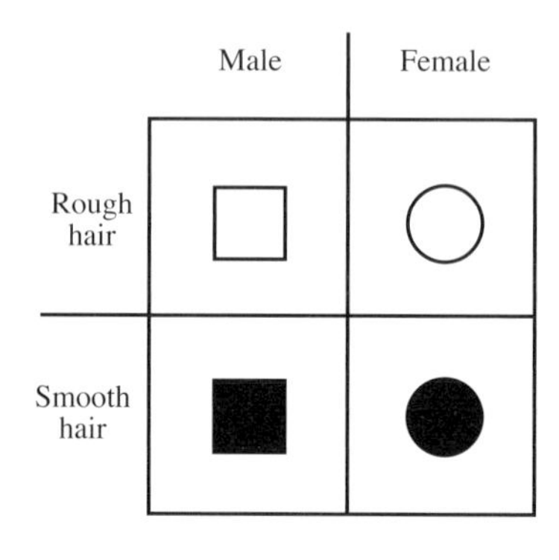

Using the information in the pedigree, determine the **genotype** of guinea pig 6 and **discuss** (i.e. *explain*) your reasoning. Use *R* for the dominant allele and *r* for the recessive allele.

In your answer:

- **Explain** how you deduce whether the rough-hair phenotype is dominant or recessive
- **Explain** how you conclude whether the genotype of guinea pig 6 is homozygous or heterozygous.

Use a Punnett square to support your answer, selecting the appropriate one from those provided below.

(Guidance: Your answer should be approximately 19 lines.)

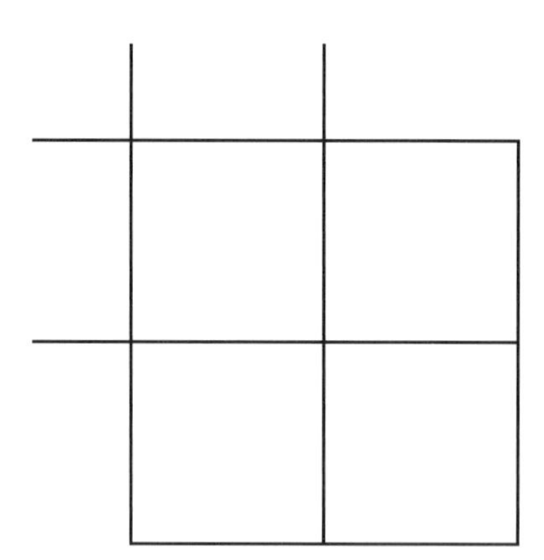

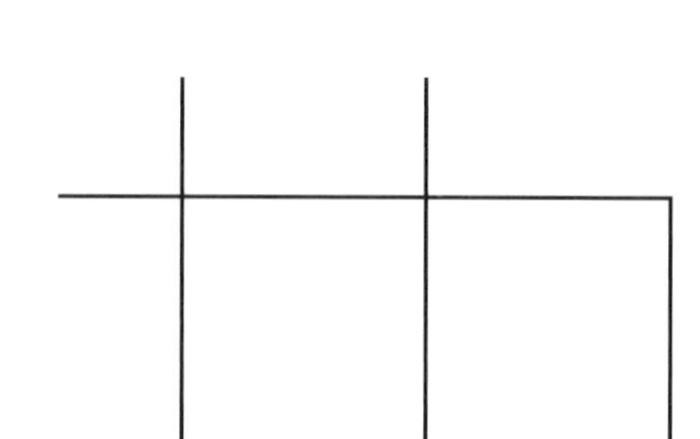

ISBN 9780170191340

QUESTION TWO: BREEDING CHRYSANTHEMUMS

A plant breeder has produced a new variety of chrysanthemum using sexual reproduction.

To produce more of this variety the breeder used asexual reproduction.

Discuss (i.e. *explain*) why the breeder used sexual methods to produce the new variety and asexual methods to produce more plants of the same variety.

In your answer you should:

- Show how meiosis and mitosis are linked to sexual and asexual reproduction.
- **Explain** how the genetic information of the parent plants is inherited in both sexual and asexual reproduction.
- **Explain** why the breeder used both sexual AND asexual reproduction to produce the new variety.

(Guidance: Your answer should be approximately 50 lines.)

QUESTION THREE

The diagram shows the sex chromosomes of two different people, Smith and Brown.

(a) What sex is (i) Smith and (ii) Brown?

(Guidance: Your answer should be approximately 2 lines.)

(b) The sex of a person is determined at the moment of fertilisation.

Explain how fertilisation determines the sex of the child.

(Guidance: Your answer should be approximately 8 lines.)

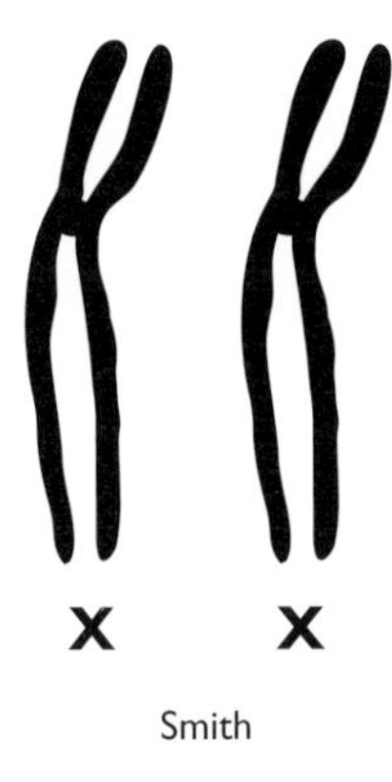

Smith

Brown

QUESTION FOUR

(a) **Describe** (i.e. *name*) the THREE components of DNA. You may illustrate your answer with a labelled diagram.

(Guidance: Your answer should be approximately 8 half-lines, + equal space for diagram.)

(b) Using one or more examples, **discuss** (i.e. *explain*) how DNA determines the characteristics of an individual person.

(Guidance: Your answer should be approximately 18 lines.)

ISBN 9780170191340

QUESTION FIVE

(a) **Describe** (i.e. *explain*) the function of **meiosis**.

(Guidance: Your answer should be approximately 2 lines.)

(b) The diagram shows the process of **mitosis**.

Explain why this process occurs. In your answer, give an example of where the process takes place.

(Guidance: Your answer should be approximately 12 lines.)

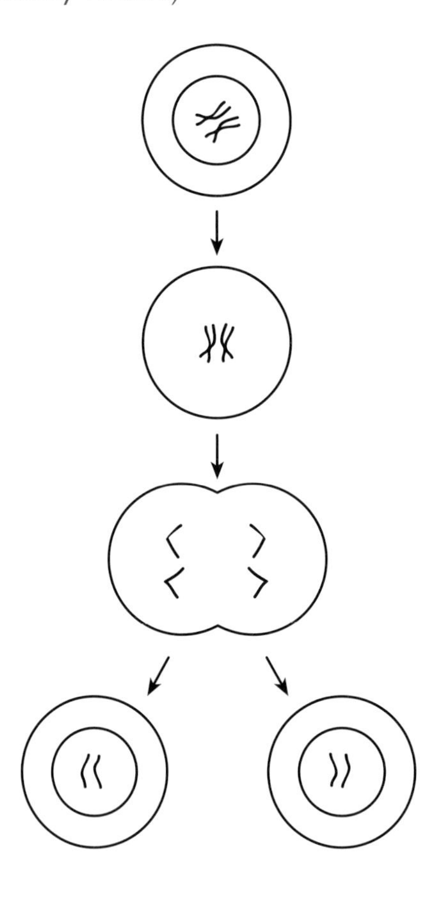

ISBN 9780170191340

QUESTION SIX

The diagram shows a short section of a DNA molecule during replication.

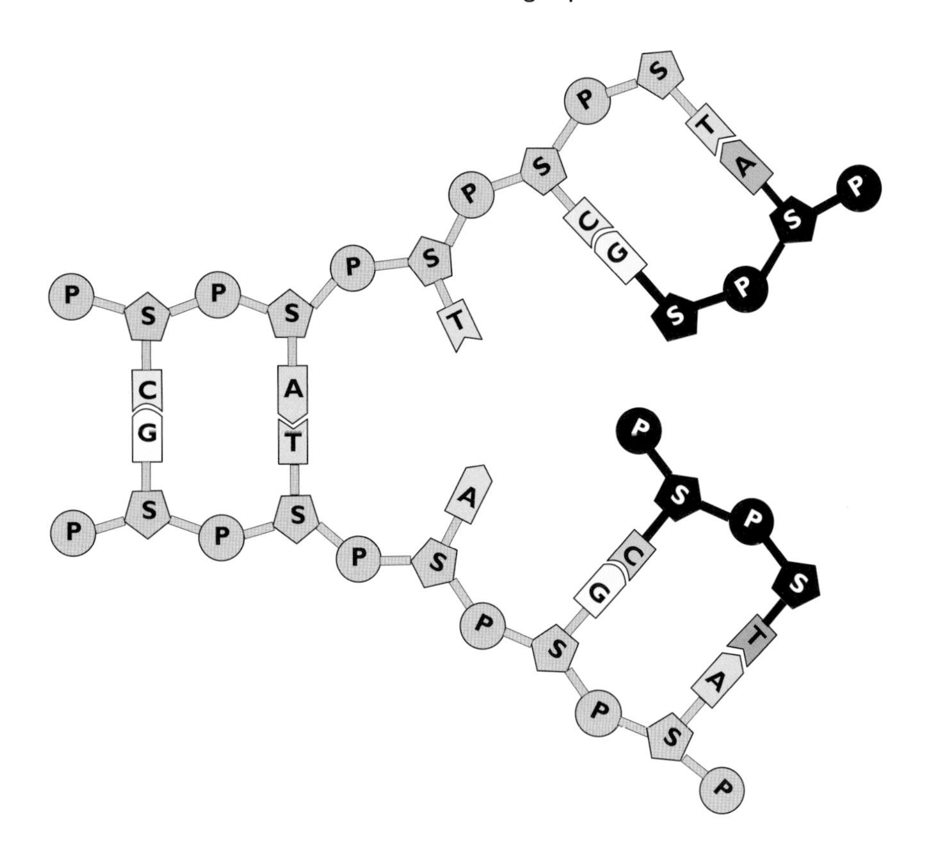

(a) **Describe** (i.e. *explain*) what a gene is.

(Guidance: Your answer should be approximately 4 lines.)

(b) **Explain** how DNA is able to be replicated accurately.

(c) **Discuss** (i.e. *explain*) why it is important that DNA replication is carried out accurately.

(Guidance: Your answer should be approximately 16 lines.)

ISBN 9780170191340

QUESTION SEVEN

In the garden pea, *tall* stem (*T*) is dominant to *short* stem (*t*), which is recessive.

(a) **Describe** (i.e. *explain*) what is meant by the terms *dominant* and *recessive*.

(Guidance: Your answer should be approximately 2 lines.)

(b) A homozygous tall pea plant was crossed with a heterozygous tall pea plant.

Copy and complete the Punnett square to show the possible genotypes of the offspring.

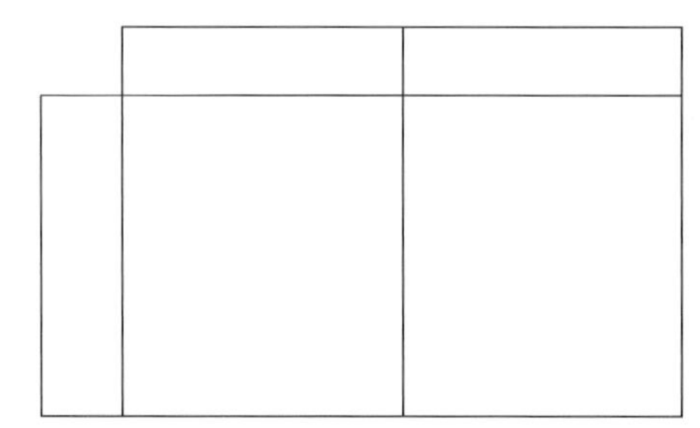

(c) Referring to your answer to part (b), **explain** the difference between *homozygous* and *heterozygous*.

(Guidance: Your answer should be approximately 8 lines.)

(d) Suppose you were given pots containing several tall pea plants.

Discuss (i.e. *explain*) how you could find out whether the plants were pure-breeding or not.

Your method should not require any specialist equipment or techniques.

(Guidance: Your answer should be approximately 12 lines.)

QUESTION EIGHT

The ability to roll the tongue, as shown in the photograph on page 22, is inherited. Bill and Nancy can both roll their tongues, but their first child cannot.

(a) Is the ability to roll the tongue a dominant or a recessive trait?

(Guidance: Your answer should be approximately 1 line.)

(b) **Explain** how you can tell from the information provided.

(Guidance: Your answer should be approximately 3 lines.)

(c) Copy and complete the Punnett square below to show the *probability* that their second child would be able to roll the tongue.

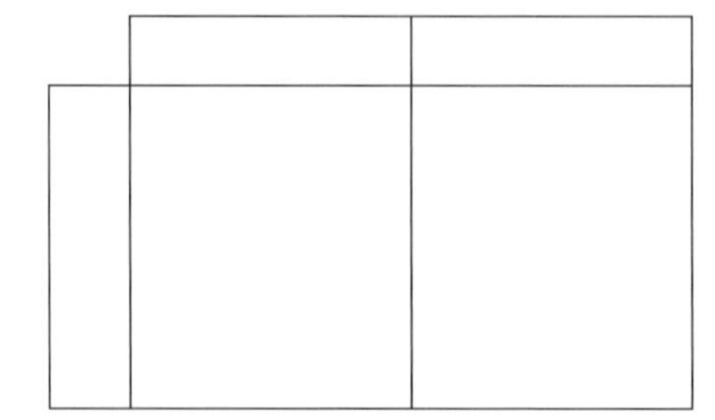

ISBN 9780170191340

Part TWO

MICROORGANISMS

3 Introduction to Microorganisms: Bacteria and other prokaryotes

WHAT ARE MICROORGANISMS?

Microorganisms, sometimes called *microbes,* are living organisms that are too small to be seen clearly with the naked eye. There are many different kinds, of which we shall be studying three:

1. **Fungi,** which include moulds, yeasts and mushrooms (though mushrooms are relatively large, most of the fungus consists of microscopic threads which extend far beyond the mushroom).
2. **Prokaryotes,** which include bacteria.
3. **Viruses.** These are not cells and have no metabolism, nor are they capable of independent growth or reproduction.

Measuring very small organisms

Tiny things need tiny units to measure them:

- The **micrometre** (**µm**) is used for things we see under the light microscope, and is a millionth (10^{-6}) of a metre, or a thousandth (10^{-3}) of a millimetre.
- The **nanometre (nm)** is used for things studied using the electron microscope, and is a billionth (10^{-9}) of a metre or a millionth of a millimetre.

The smallest thing that can be distinguished with the naked eye is about 0.1 mm, or 100 µm or 100,000 nm. Most bacteria are about a hundred times smaller than can be seen with the naked eye — about a thousand could fit across a pinhead, and viruses are even smaller.

Think of it like this. Suppose we represent a poliovirus by a blob 1 mm across. On the same scale, an average bacterium would be about 5 cm across, a human liver cell would be about 1.3 m across, and the human body would be 50 km tall!

THE DISCOVERY OF MICROORGANISMS AND THE BEGINNINGS OF MODERN SCIENCE

Until the invention of the microscope, and for some time after that, disease and decay were complete mysteries. Plague, cholera and other killer diseases were thought to be acts of God, punishing people for their sins.

The first step toward understanding these phenomena was taken by a Dutchman, Antonie van Leeuwenhoek. He used a simple microscope (one that has only one lens). Because his lenses were tiny, they had great curvature and could magnify up to 300 times. Between 1673 and 1723 he reported his observations using his microscopes. In pond water he observed tiny 'animalicules', some of which, from his descriptions, we can be fairly sure were bacteria. He found 'animalicules' wherever he looked. One of the richest sources was the scrapings from his own teeth.

One of Leeuwenhoek's microscopes

Where did these 'animalicules' come from? A popular idea was that, along with the maggots in rotting meat, they simply arose from the non-living matter in the rotting material itself. This idea of *spontaneous generation* was accepted as fact by most educated and intelligent people. There were a few, however, who did not just accept what others said without evidence. One such person was Lazzaro Spallanzani, an 18th century Italian. He designed an experiment to put spontaneous generation to the test (Fig. 3.1). Though the results of the experiment severely weakened the idea that life could develop from non-life, some people argued that although the microbes that appeared in the broth did not come from the air, contact with the air was necessary for them to develop spontaneously.

ISBN 9780170191340

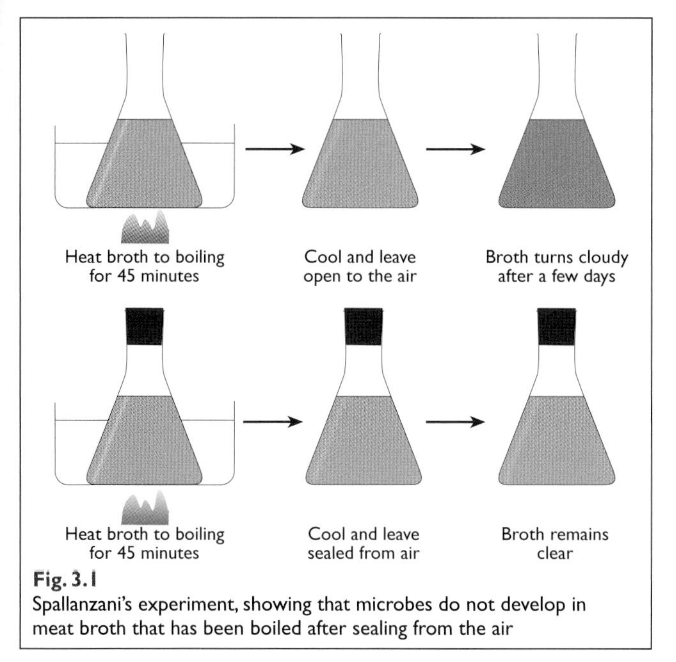
Heat broth to boiling for 45 minutes
Cool and leave open to the air
Broth turns cloudy after a few days
Heat broth to boiling for 45 minutes
Cool and leave sealed from air
Broth remains clear

Fig. 3.1
Spallanzani's experiment, showing that microbes do not develop in meat broth that has been boiled after sealing from the air

It was **Louis Pasteur** who finally proved that even bacteria have parents. He boiled some nutrient broth in a flask with a 'swan' necked flask, and allowed it to cool slowly (Fig. 3.2).

Though the broth was in contact with the air it remained clear, even after several months. When he broke the neck, it turned cloudy and smelly within days. Pasteur had thus shown that it was not the air that was important in decay, but microorganisms in the air.

In later experiments, Pasteur showed that fermentation of sugar to alcohol in wine-making was due to the presence of tiny cells, later identified as a fungus (yeast). He also showed that on those occasions when the wine became contaminated by lactic acid, it was due to a different kind of microorganism — a bacterium. As a result of this work Pasteur invented a heat treatment that killed the bacteria responsible, and a similar process, now known as **pasteurisation**, was developed to keep milk fresh.

To most people, bacteria are microscopic, disease-causing organisms. But although some bacteria are pathogenic (disease-causing), the vast majority are harmless or beneficial to humans (Chapter 7).

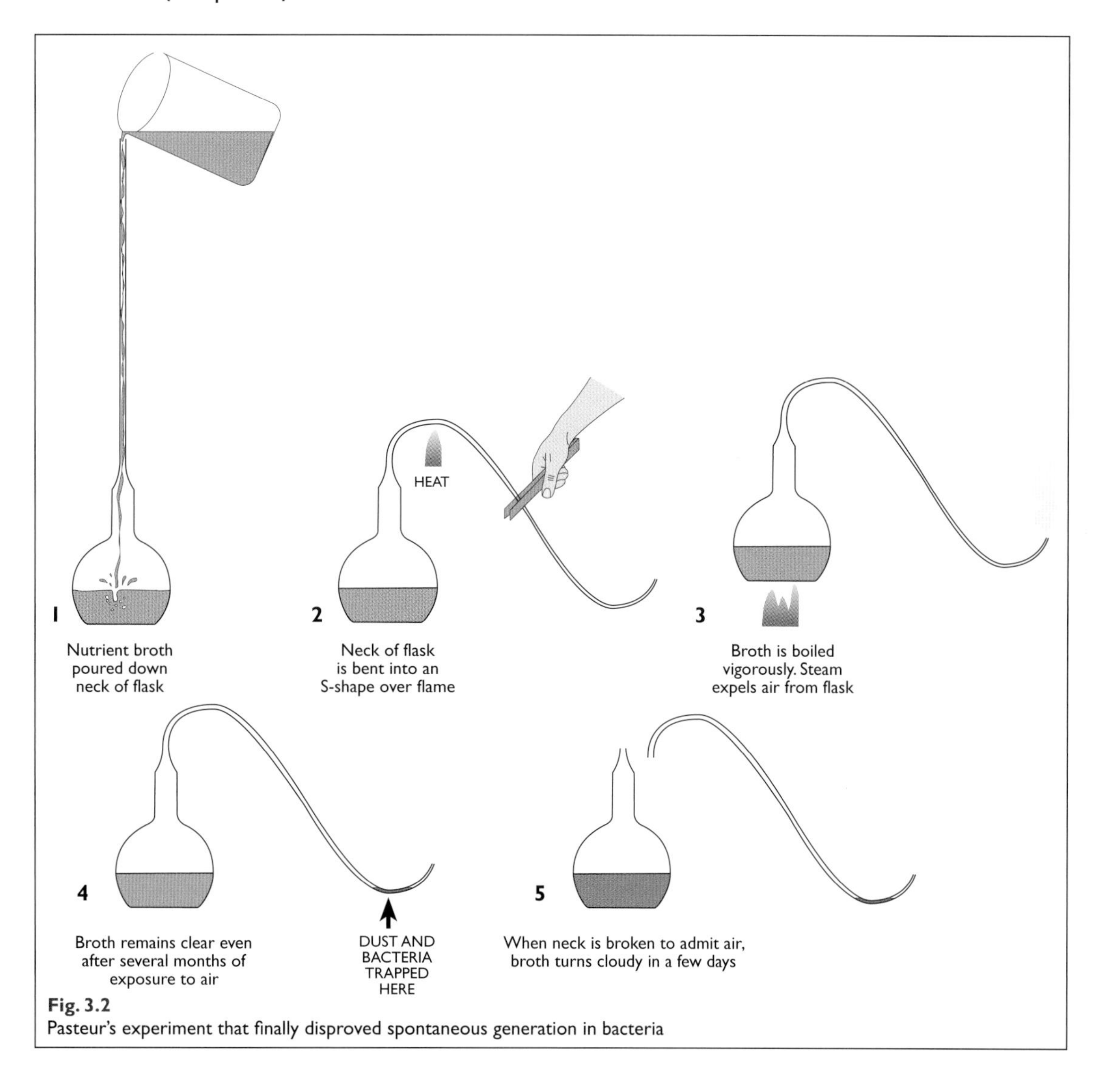
HEAT
1
Nutrient broth poured down neck of flask
2
Neck of flask is bent into an S-shape over flame
3
Broth is boiled vigorously. Steam expels air from flask
4
Broth remains clear even after several months of exposure to air
DUST AND BACTERIA TRAPPED HERE
5
When neck is broken to admit air, broth turns cloudy in a few days

Fig. 3.2
Pasteur's experiment that finally disproved spontaneous generation in bacteria

ISBN 9780170191340

Growing bacteria

Bacteria are normally grown on **agar**, a complex carbohydrate extracted from seaweed, which forms a gel (firm jelly). Agar itself cannot be used as food by most microorganisms, but nutrients are added to the agar gel for the microorganisms to use for growth. Chapter 4 describes how to grow yeast cells on agar.

Size and structure

Bacteria are single-celled organisms, and almost all are microscopic. Most are about a micrometre or so in size; the common colon bacterium *Escherichia coli* (below), is about 3 µm long and 1 µm wide.

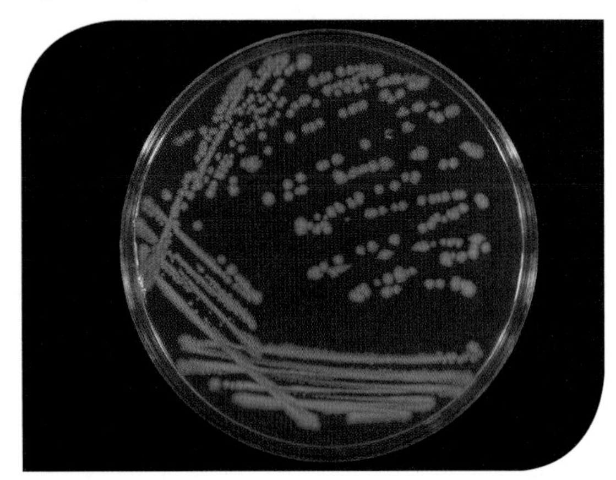

Escherichia coli colonies growing on agar. (Courtesy of Ms Anna Lau, University of Auckland; photographed by Mr Iain MacDonald)

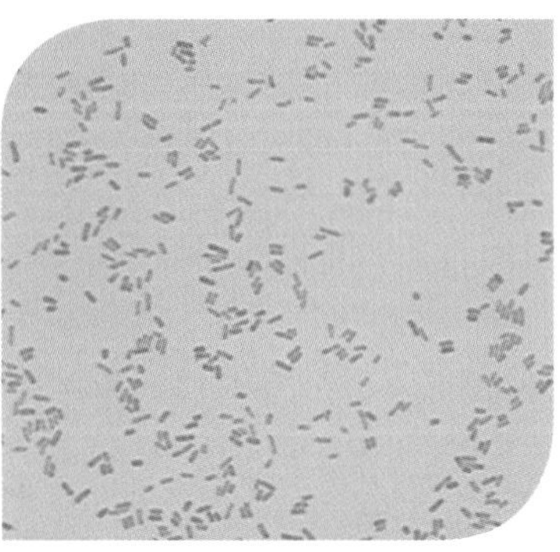

Escherichia coli bacterial cells. (Courtesy of Dr Tony Roberton, University of Auckland; photgraphed by Mrs Vidya Washington)

Bacteria can take various shapes, the most common are shown in Fig. 3.3. Bacterial cells differ in structure from those of other organisms in a number of important ways (Fig. 3.4):

- There is no nuclear envelope separating the genetic material (DNA) from the rest of the cell, so there is no clearly-defined nucleus. Organisms like this are called **prokaryotes,** in contrast to those with distinct nuclei, which are **eukaryotes**.
- Most of the DNA is in a single chromosome which forms a *closed loop*, in contrast to eukaryotes in which there are several to many chromosomes, and they are open-ended.
- Many bacteria (but not all) propel themselves by **flagella,** but these are structurally very different from the flagella and cilia of eukaryotes.
- The cell wall is chemically quite different from those of fungi and plants.

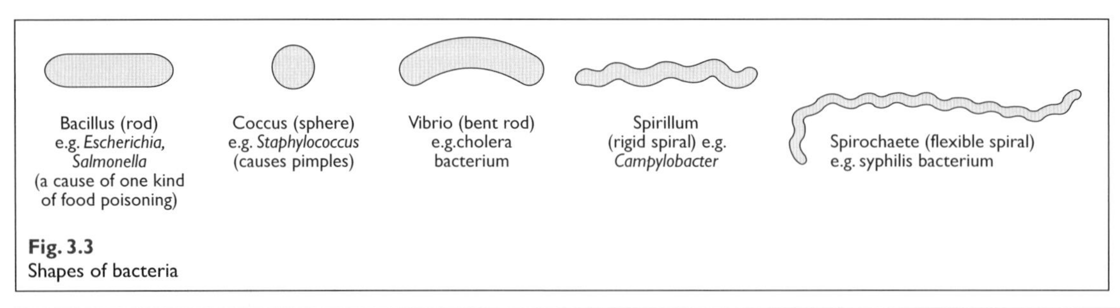

Fig. 3.3
Shapes of bacteria

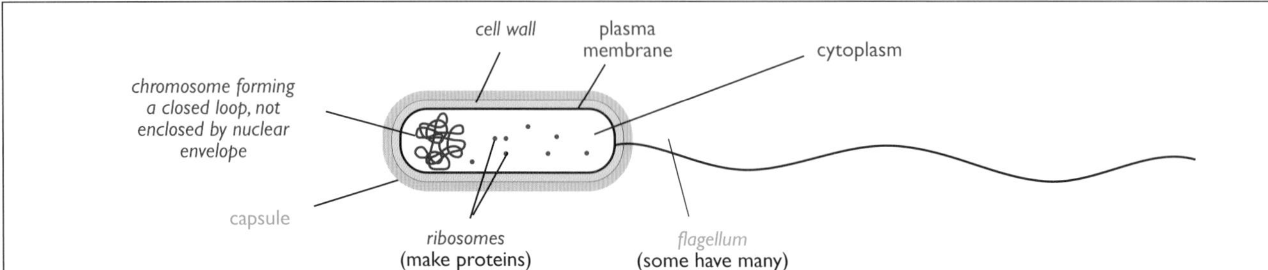

Fig. 3.4
The structure of a bacterial cell. Structures present in all prokaryotes are labelled in red type. Structures present in many prokaryotes are labelled in blue type. Structures differing from those of eukaryotes are in *italic*

ISBN 9780170191340

- The ribosomes, which make proteins, are smaller in prokaryotes than in eukaryotes.
- Eukaryotes contain additional structures not found in Prokaryotes, such as mitochondria, Golgi bodies and endoplasmic reticulum.

NUTRITION

So far as the way they feed is concerned, bacteria fall into two groups: *autotrophs* and *heterotrophs*.

Autotrophs

Bacteria that can produce their own organic matter from CO_2 and water are *autotrophic*. Some use light energy to do this by a process called **photosynthesis**; others obtain energy by oxidising inorganic substances obtained from their environment. Some of these latter bacteria play a key role in the nitrogen cycle (Chapter 7).

Heterotrophs

Heterotrophic bacteria are like fungi and animals in that they obtain their organic matter from other organisms. They fall into three groups:

1. **Saprobes** feed on dead organic matter and bring about *decay*.
2. **Parasites** feed on living organisms and may cause *disease*.
3. **Mutualists** have a close relationship with another organism in which both benefit, for example those that live in the gut of herbivores (Chapter 7).

In saprobes, digestive enzymes are secreted into the surroundings; these break down complex organic substances into simpler substances:

- proteins → amino acids
- cellulose and starch → glucose
- fats → fatty acids + glycerol

Digestion outside the cells is said to be **extracellular** (though many bacteria digest food *on* the cell surface). The simple products of digestion are absorbed into the cell and used in either of two ways:

1. Some is broken down and used for energy.
2. Some is built up into complex organic substances and used as raw materials for growth.

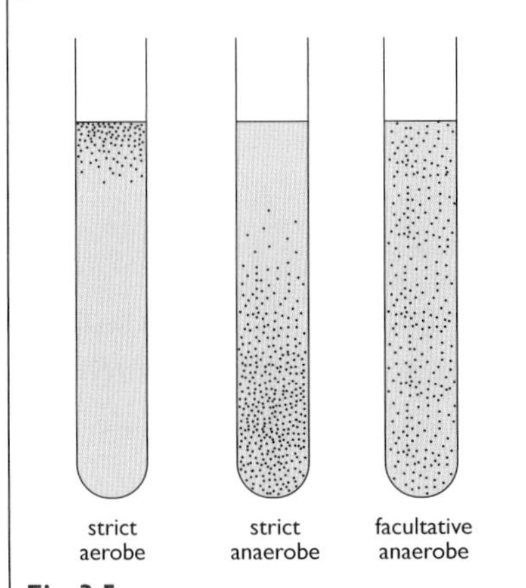

Fig. 3.5
Growth of bacteria in nutrient agar showing three different responses to oxygen

AEROBES AND ANAEROBES

In relation to the need for oxygen, bacteria fall into three groups (Fig. 3.5):

1. Some cannot grow without oxygen and are **obligate aerobes**.
2. Some cannot survive in its presence and are **obligate anaerobes**, for example the bacteria living in the guts of many herbivores such as rabbits and sheep.
3. Some are **facultatively** anaerobic; they can live with or without oxygen, for example the human colon bacterium *Escherichia coli*.

ISBN 9780170191340

HOW BACTERIA USE ENERGY

Like all organisms, bacteria need energy for cell processes such as:

- *Biosynthesis*, or making large molecules from small ones (e.g. proteins from amino acids).
- *Movement* (e.g. the use of flagella).
- **Active transport**, or moving substances across a membrane from lower to higher concentration.

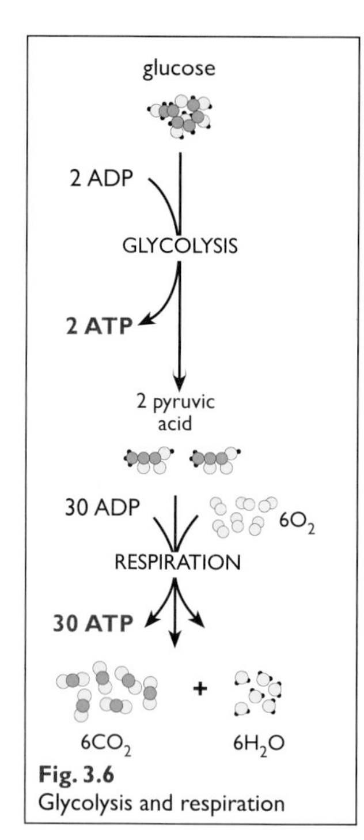

Fig. 3.6
Glycolysis and respiration

The immediate source of energy for cell processes is **ATP** (**a**denosine **tri**phosphate). This releases energy when it is converted into *ADP* (**a**denosine **di**phosphate) + phosphate:

$$ATP + H_2O \rightarrow ADP + phosphate + \textbf{energy}$$

Since ATP is continuously being 'spent' to drive cell processes, it must be continuously produced. In heterotrophic bacteria ATP can be produced in two different chemical processes (Fig. 3.6):

1. **Glycolysis**, in which glucose is converted to **pyruvic acid** + **a *little* ATP**
2. **Respiration**, in which pyruvic acid is oxidised to CO_2 and water with the production of ***lots* of ATP**.

Respiration

The 'fuels' for respiration include a variety of small molecules. One important example is pyruvic acid, made in glycolysis from glucose. In **aerobic respiration**, pyruvic acid is oxidised to CO_2 and water, using oxygen (Fig. 3.7):

pyruvic acid + oxygen → carbon dioxide + water + ***lots* of ATP**

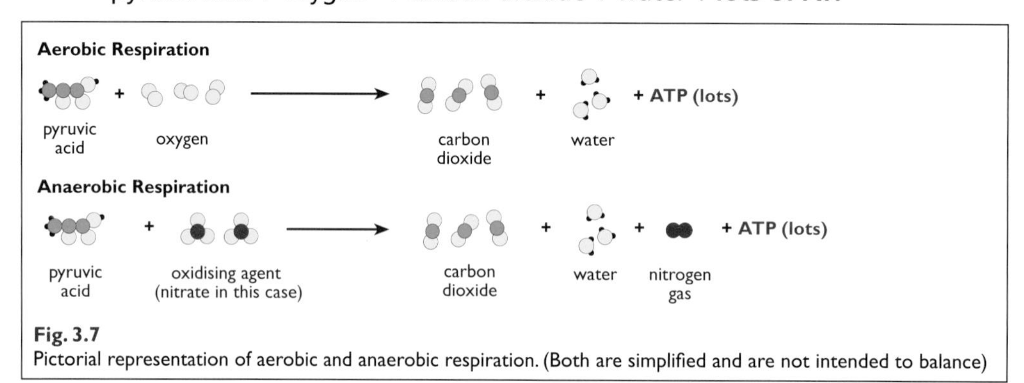

Fig. 3.7
Pictorial representation of aerobic and anaerobic respiration. (Both are simplified and are not intended to balance)

In **anaerobic respiration**, pyruvic acid is oxidised using a substance other than oxygen, such as nitrate:

pyruvic acid + nitrate → carbon dioxide + water + nitrogen gas + **lots of ATP**

This conversion of nitrate to nitrogen gas is called **denitrification** and is explained more fully in Chapter 7.

Many other examples of anaerobic respiration occur. One important example happens in the rumen of sheep and cattle, in which some of the microorganisms (actually Archaea) produce methane by anaerobic respiration. Methane is an important 'greenhouse' gas.

Fermentation

Although fermentation is often described as 'anaerobic respiration', the two processes are quite different.

In fermentation (Fig. 3.8) the *oxidation* (removal of hydrogen) in glycolysis is followed by the *reduction* of pyruvic acid by addition of hydrogen. There is thus no *net* oxidation, as occurs in both aerobic and anaerobic respiration.

ISBN 9780170191340

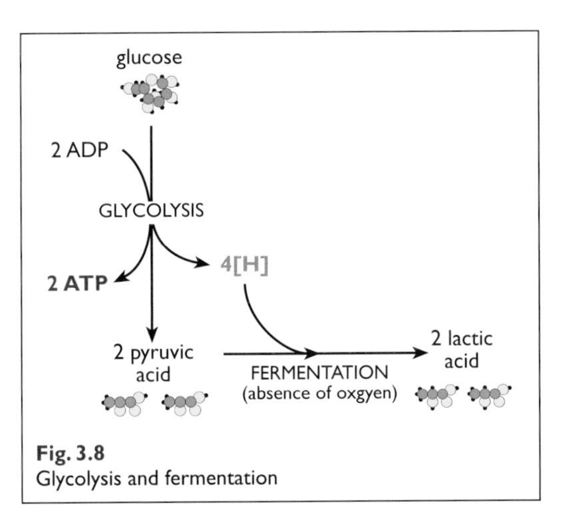

Fig. 3.8
Glycolysis and fermentation

One of the best known kinds of fermentation occurs in the souring of milk with the production of *lactic acid*:

pyruvic acid + 2[H] $\rightarrow$ lactic acid

Certain bacteria can ferment sugar to *ethanol*, as explained for yeast (not a bacterium, but a fungus) in Chapter 4.

Since fermentation releases a small amount of energy it is inefficient, so the 'fuel' is used up much more quickly. To begin with this does not matter in, say, a bottle of milk since it is plentiful, but eventually it becomes scarce, slowing growth. Also, lactic acid is slightly toxic.

There are many other end products of sugar fermentation (depending on the bacterium) besides lactic acid and ethanol, such as acetic (ethanoic) acid, propionic acid, butyric acid, formic acid, butanol.

Autotrophic bacteria

Autotrophic ('self-feeding') bacteria are of two types:

1. Photo-autotrophic bacteria use light energy to convert CO_2 to carbohydrate in **photosynthesis**.
2. Chemo-autotrophic bacteria get their energy by oxidising inorganic substances in the environment. For example, some *nitrifying bacteria* oxidise ammonia to nitrite, and others oxidise nitrite to nitrate:

ammonia + oxygen $\rightarrow$ nitrite + **ATP**

nitrite + oxygen $\rightarrow$ nitrate + **ATP**

The energy liberated is used to make carbohydrate from CO_2 and water, a process called **chemosynthesis**.

GROWTH AND REPRODUCTION

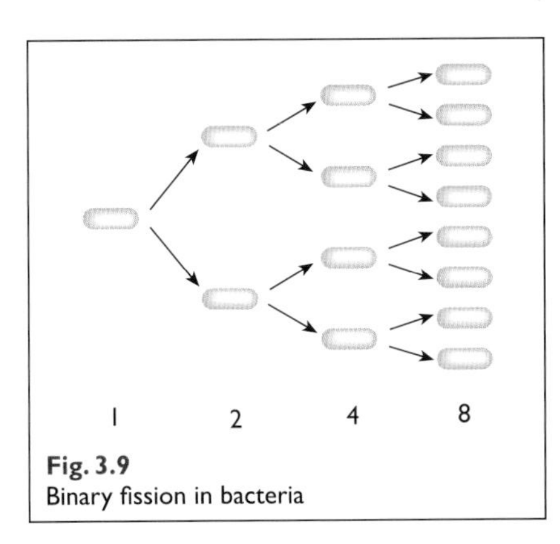

Fig. 3.9
Binary fission in bacteria

Bacteria reproduce asexually by **binary fission** (splitting in two), as shown in Fig. 3.9. The time taken for a given kind of bacterium to divide is called the **generation time**. Under ideal conditions, most bacteria have a generation time of less than an hour, but there is considerable variation, as shown in Table 3.1. This is also the *doubling time*, or the time it takes for the population to double. The doubling time depends on two kinds of factor:

1. Environmental factors such as temperature, food supply, and concentration of waste products.
2. Genetic factors. Even under identical environmental conditions, different kinds of bacteria vary in their growth rates.

Name	Note	Generation time	Temperature
Vibrio natriegens	Marine bacterium	9.8 mins	37°C
Escherichia coli	Human colon bacterium	20 mins	37°C
Mycobacterium tuberculosis	Cause of tuberculosis	12 hours	37°C
Treponema pallidum	Cause of syphilis	33 hours	37°C

Table 3.1 Generation times of some bacteria under optimal (ideal) conditions

Some bacteria can also reproduce sexually by transferring genetic material from one bacterium to another.

ISBN 9780170191340

The sigmoid ('S'-shaped) growth curve

If a single bacterium and its descendants were to divide once every hour, in less than a couple of days the total mass of bacteria would equal the mass of the earth! Obviously this does not happen, for various reasons, such as running out of food and accumulation of poisonous waste products.

To understand what happens when a population of bacteria grows, look at Table 3.2 and the graph in Fig. 3.10. They show, using imaginary numbers (real numbers would be much larger), what happens if a single rapidly growing bacterium is transferred to a fresh nutrient solution of the same kind.

Time (hr)	Number	Increase	% increase
0	1	–	–
1	2	1	100
2	4	2	100
3	8	4	100
4	16	8	100
5	32	16	100
6	64	32	100
7	128	64	100
8	256	128	100
9	456	200	78
10	756	300	65.8
11	956	200	26.5
12	1056	100	10.4
13	1106	50	4.7
14	1116	10	0.9
15	1120	4	0.03
16	1120	0	0
17	1120	0	0
18	1120	0	0

Table 3.2 Growth of an imaginary population of bacteria showing the different phases

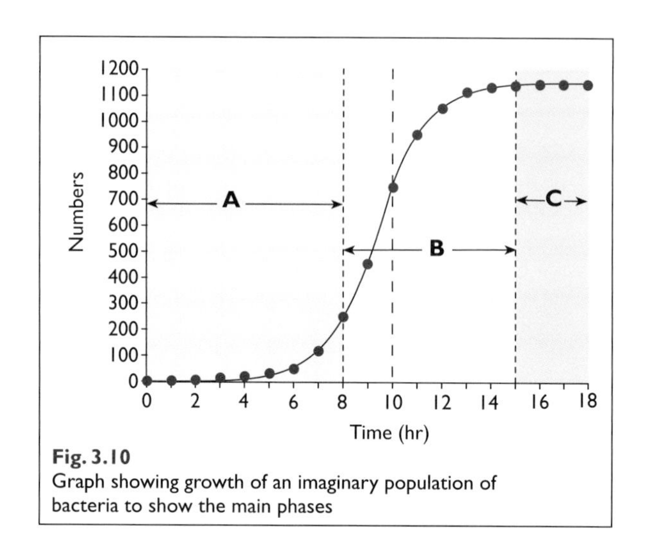

Fig. 3.10
Graph showing growth of an imaginary population of bacteria to show the main phases

The graph is 'S'-shaped (*sigmoid*), and can be divided into three distinct phases:

A. *Exponential phase*, in which the numbers increase by a constant *percentage* every given time interval. The generation time (one hour in this case) remains constant, so numbers double every hour. An increase of 5% every three weeks, or tripling every year, could also be exponential growth.

B. *Slowing phase*, in which the generation time is increasing. The graph continues to get steeper for a while because although each bacterium is taking longer to divide, this is more than offset by the increasing number of 'parents'. The *relative* growth rate (percentage increase each hour) is decreasing but the *absolute* growth rate (increase in numbers per hour) is still increasing. Eventually (after the point shown by the dotted line in the graph), the gradient begins to decrease as the absolute growth rate begins to decrease.

C. *Stationary phase*, in which growth in numbers has ceased.

When bacteria or yeast are transferred to a new medium, there may be a delay before growth resumes. This is called the **lag phase**, and is due to the bacteria having to 're-tool' their metabolism to adjust to the new conditions. This occurs, for example, when bacteria are transferred from a medium in which glucose is the energy source to one containing lactose (milk sugar).

 ISBN 9780170191340

Endospores of *Bacillus subtilis*. The bacterial cells are stained pink, the endospores are stained pale blue-green. Some bacterial cells have disintegrated, releasing free endospores. (Courtesy of Dr Tony Roberton, University of Auckland; photographed by Mrs Vidya Washington)

Dispersal and survival in harsh conditions

Many bacteria can survive drying and dispersal on air currents, as was shown by Louis Pasteur in his famous experiment. Some bacteria can form **endospores**. These are able to withstand extremely harsh conditions such as drying, or boiling at 100°C for an hour or more. Autoclaving (boiling at 121°C) kills endospores within 15 minutes.

Extension: The advantages of being small

To be able to double in size as quickly as many bacteria can (the fastest known is a bit under 10 minutes!), bacteria must be able to absorb raw materials very quickly. They can do this because, being so small, their surface is large compared with their volume.

The reason lies in the fact that surface area has two dimensions and volume has three. So if an object doubles in length and keeps the same shape, the other two dimensions must also double. Hence its volume goes up 2 x 2 x 2 = 8 times. The surface area, though, only goes up 2 x 2 = 4 times. Since the volume has increased twice as fast as the surface area, the surface/volume ratio has *halved*.

Now think of it the other way: As an object gets smaller (assuming the shape does not change), the volume shrinks much faster than the surface area, so the surface/volume ratio *increases*. A bacterium therefore has a much greater surface/volume ratio than an elephant. This is why bacteria can absorb raw materials so quickly.

Extension: The Archaea

When, a few decades ago, scientists started comparing the *molecules* different organisms are made of, there were some big surprises in store. One of the biggest was that there are really two completely different kinds of prokaryotes:

- The Eubacteria, which include most of the more familiar types.
- The Archaea, a group containing organisms that live in extreme environments such as hot springs, underwater volcanic vents at temperatures of over 100°C, the rumen of cattle, in rocks deep underground, and at temperatures below freezing.

The biggest surprise came when it was discovered that the Archaea are not only very unlike true bacteria — in certain respects they are actually more similar to the Eukaryotes (animals, plants, fungi) than they are to the bacteria.

The upshot of all this is that living organisms are now divided into three huge groups, called *domains*:

1. Bacteria
2. Eukaryotes
3. Archaea

Whereas endospores survive harsh conditions by shutting down all activity, some Archaea actually *thrive in* conditions that would kill most cells. *Pyrodictium occultum*, a prokaryote growing near hot volcanic vents in the deep ocean floor, has an *optimum* temperature of 105°C. Some archaeans can survive extreme pH values as low as pH 0, which is ten times as acidic as stomach acid! The record holder for harsh conditions is *Picrophilus torridus*, an archaean that thrives in extreme temperatures and pH in hot springs in Japan (optimum temperature of 60°C *and* pH of 0.7).

ISBN 9780170191340

Summary of key points in this chapter

- Bacteria are *prokaryotes*; they lack a clearly-defined nucleus.
- Bacteria may be *autotrophic* or *heterotrophic*.
- Like fungi, heterotrophic bacteria may be *saprobes*, *parasites* or *mutualists*.
- Like fungi, saprobic bacteria digest their food *extracellularly* or outside the cell.
- Bacteria may be *aerobic* (able to grow in the presence of oxygen) or *anaerobic* (able to grow in the absence of oxygen).
- Some bacteria obtain energy by *fermentation*; others by *respiration*.
- When inoculated into a fresh culture medium, the graph of numbers against time is typically S-shaped (*sigmoid*), and is divided into three distinct phases.
- Some bacteria can resist boiling for short periods because of their ability to produce highly resistant *endospores*.

Test your basics

1 Copy and complete the following sentences.

a) Prokaryotes are organisms whose cells lack a clearly defined ___*___.

b) The chromosomes of bacteria are in the form of a ___*___.

c) Organisms that cannot make organic compounds from CO_2 are ___*___.

d) An ___*___ is an organism that can grow in the absence of oxygen.

e) A thousandth of a millimetre is a ___*___. A millionth of a millimetre is a ___*___.

f) Many bacteria can move by means of whip-like ___*___.

g) Growth in which numbers increase by a constant percentage in a given time is said to be ___*___.

h) The chemical processes in a cell are its ___*___.

i) Bacteria and fungi are grown on jelly called ___*___, extracted from seaweed.

j) Organisms that feed on dead matter are called ___*___.

k) In all cells, the *immediate* source of energy is a chemical called ___*___ ___*___ (ATP).

l) The first stage in which energy is obtained from glucose is called ___*___, and yields a little ATP.

m) Many organisms can make ATP from organic molecules without oxidising them. This is called ___*___.

n) Bacteria reproduce by dividing into two, a process called ___*___ ___*___.

o) When a nutrient solution is inoculated with bacteria, a graph of the increase in numbers is 'S'-shaped or ___*___.

p) Equipment used in microbiology is sterilised using a special kind of pressure cooker called an ___*___.

q) Some bacteria can form ___*___, which are highly resistant to heat.

r) Because bacteria are very small, they have a large ___*___ compared with their ___*___, so they can absorb raw materials very rapidly.

s) Organisms that live in a close relationship with other organisms, both of which obtain benefit, are called ___*___.

t) An organism that lives at the expense of another organism without killing it is called a ___*___.

u) Digestion that occurs outside a cell is said to be ___*___.

2 True or false?

a) Bacteria normally divide every 20 minutes.

b) All bacteria are heterotrophic.

c) Boiling at 100°C kills all bacteria.

d) The cell walls of prokaryotes are chemically quite different from those of fungi and plants.

e) Bacteria lack mitochondria.

ISBN 9780170191340

4 Fungi

Fungi (Latin: *fungus* = mushroom) include mushrooms, moulds, and yeasts. Like the Animal and Plant Kingdoms, they are **Eukaryotes** — so-called because their cells have a distinct nucleus. Eukaryotes form one of three huge groups of organisms called *domains*. The other two domains are Bacteria and Archaea.

Fungi have the following distinguishing features:

- They lack chlorophyll and so cannot carry out photosynthesis (in the green mould *Penicillium* the pigment is not chlorophyll). As a result they are **heterotrophic**, meaning that they cannot produce their own organic matter from CO_2 and water, but feed off organic matter made by other organisms.
- Their cell walls are made of **chitin**, the same material that the skeletons of insects are made from.
- Most fungi have bodies consisting of microscopic threads or **hyphae** (singular: *hypha*). The whole network of hyphae forms the body of the fungus, or **mycelium**.

Fungi can be both useful and harmful to humans:

- Some are **saprobic** (Greek: *sapros* = rotten), feeding on dead organic matter and bringing about their decay, helping to release the simple inorganic materials that are used by plants.
- Others are **parasitic**, feeding on other living organisms. Some of the most serious diseases of crop plants are caused by fungi, and some fungi are parasitic on animals e.g. the cause of athlete's foot.
- Others enter into **mutualistic** relationships with other organisms, in which *both* benefit. The roots of most flowering plants, for example, contain fungi that help them absorb minerals. This relationship is called **mycorrhiza**.
- Yeast is used in making wine, beer and bread, and some of the most useful *antibiotics* are made by fungi.

Rhizopus, a common 'pin' mould

Rhizopus (and its close relative *Mucor*) is a mould commonly found on old bread and decaying fruit (Fig. 4.2).

The mycelium forms a dense branching network in the food material or *substrate*. In most of the mycelium the hyphae have no cross walls, so many nuclei share the same cytoplasm.

Fig. 4.1
Mould growing on an orange

Nutrition

Because *Rhizopus* feeds on dead matter, it is a *saprobe*. The hyphae branch throughout the substrate and secrete digestive enzymes into it (Fig. 4.3):

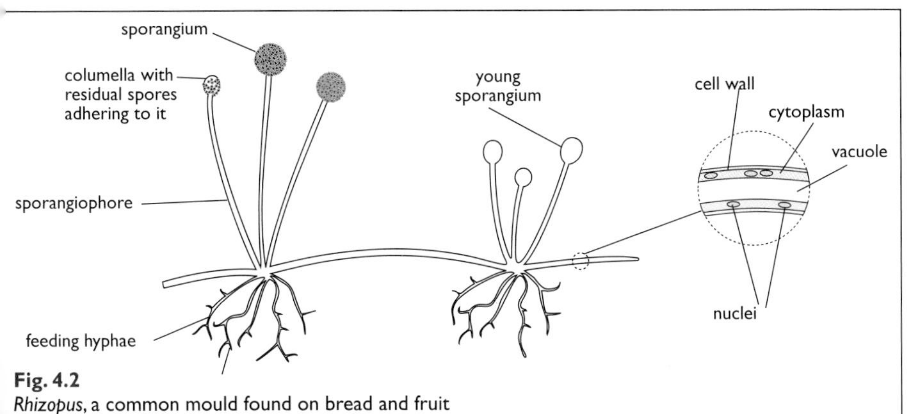

Fig. 4.2
Rhizopus, a common mould found on bread and fruit

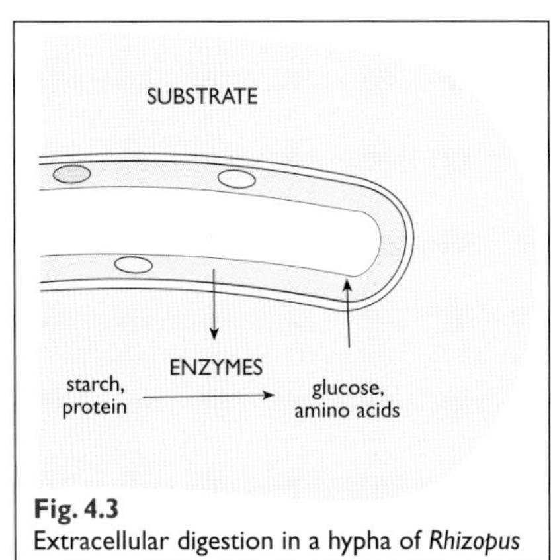

Fig. 4.3
Extracellular digestion in a hypha of *Rhizopus*

ISBN 9780170191340

- Carbohydrases convert complex carbohydrates (e.g. starch) into simple sugars (e.g. glucose).
- Proteases convert proteins into amino acids.
- Lipases convert fats into fatty acids and glycerol.

Because it occurs outside the hyphae, digestion is said to be *extracellular* ('outside the cell'). The products of extracellular digestion are absorbed by *active transport*, which requires energy, obtained from respiration.

Growth

If a spore lands on a suitable substrate, it germinates as shown in Fig. 4.4. By growing at the tips of the hyphae, the fungus can rapidly extend over a wide area. Depending on the temperature, a fungus can spread at over a millimetre per hour — far faster than bacteria can spread through a substrate.

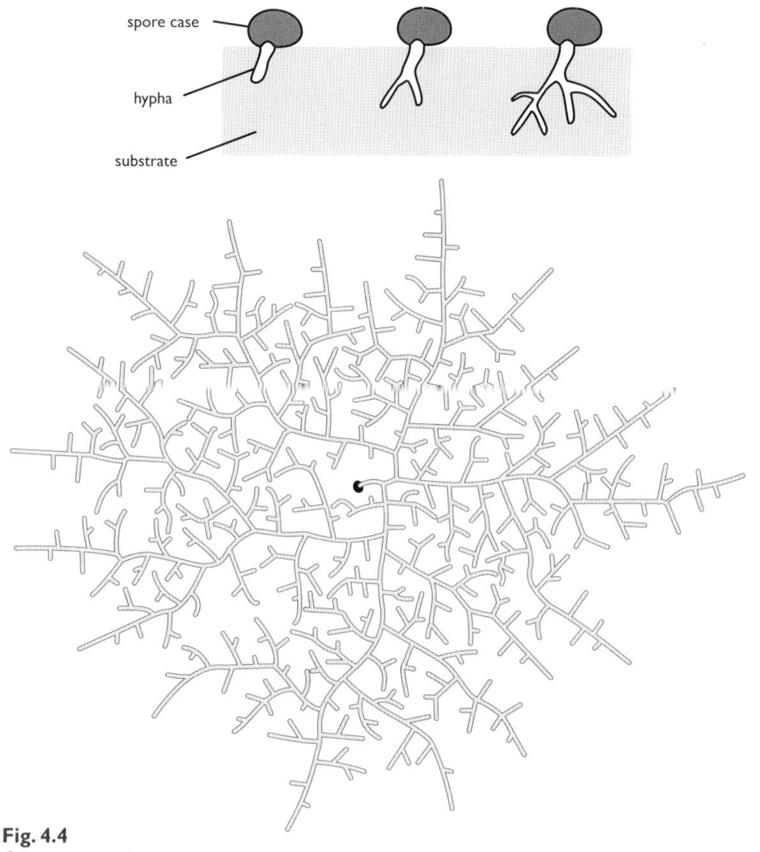

Fig. 4.4
Germination of a spore of *Rhizopus* (above) and early growth of the mycelium (below)

Many fungi produce **antibiotics**, which are chemicals that reduce competition from bacteria by inhibiting (preventing or slowing down) the growth of bacteria. Many antibiotics have been discovered and are used in treating infections. Most are now synthesised in the laboratory.

Reproduction

Reproduction is usually by the production of **spores** and is mainly asexual (without fertilisation), but can also be sexual. A few days after a spore germinates, certain hyphae called **sporangiophores** grow upward (against gravity) and develop swellings called **sporangia** at their tips. Each sporangium becomes separated from the rest of the mycelium by a dome-shaped wall called the *columella*. In each sporangium there develop up to 50,000 tiny spores, each with many nuclei. As the sporangium ripens, its wall turns black and disintegrates, releasing the spores. These may be carried by wind or stick to the bodies of crawling insects.

Because fungal spores are so small they have an extremely large surface compared with their volume. This means the air resistance (which depends on surface area) is large compared with the gravitational pull (depends on the mass and thus volume).

It is important for the fungus to produce as many spores as possible as quickly as possible for two reasons:

1. The chances of a given spore landing on a suitable food supply are very small.
2. Each potential food source is soon colonised by fungi and/or bacteria. The more spores a fungus can produce the better its chances of getting there before competitors.

Each mycelium can produce millions of spores within a few days; clearly only a tiny fraction find a suitable substrate, so the mortality (death rate) is extremely high.

Penicillium — a mould that changed medicine

Penicillium was the fungus that led to the discovery of the first antibiotic. Alexander Fleming, a Scottish bacteriologist, found that one of his bacterial cultures had become contaminated by *Penicillium*, a green mould. He also noticed that there was a clear zone round the fungus, with no bacteria growing. He concluded that the bacteria were prevented from growing by a substance diffusing out from the fungus. Later, two other scientists identified the substance and named it *penicillin*.

ISBN 9780170191340

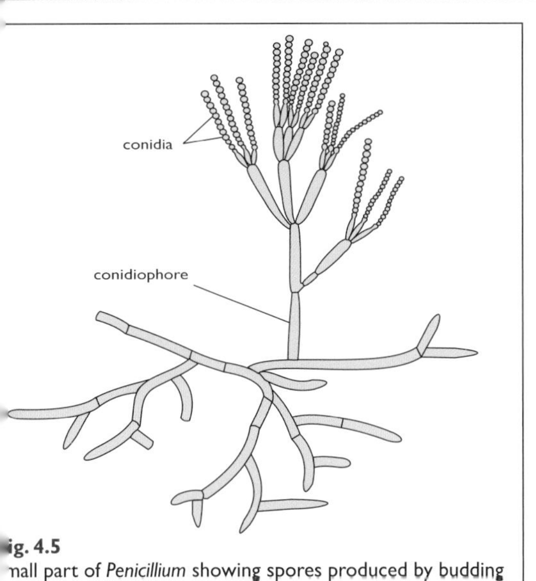

ig. 4.5
nall part of *Penicillium* showing spores produced by budding

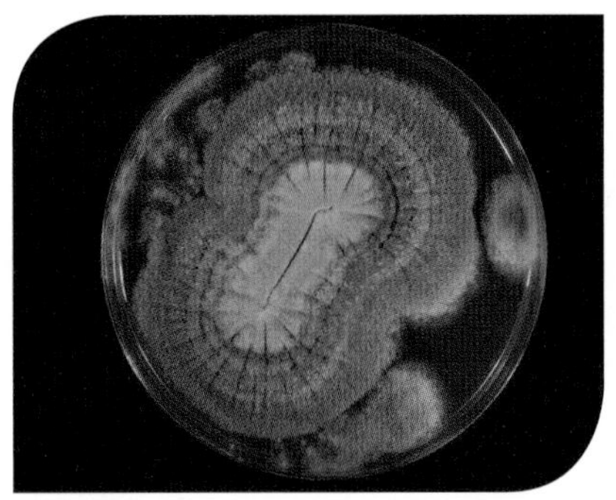

Culture of *Penicillium* fungus growing on agar (Courtesy of Mrs Yimin Dong, University of Auckland; photographed by Mr Iain MacDonald)

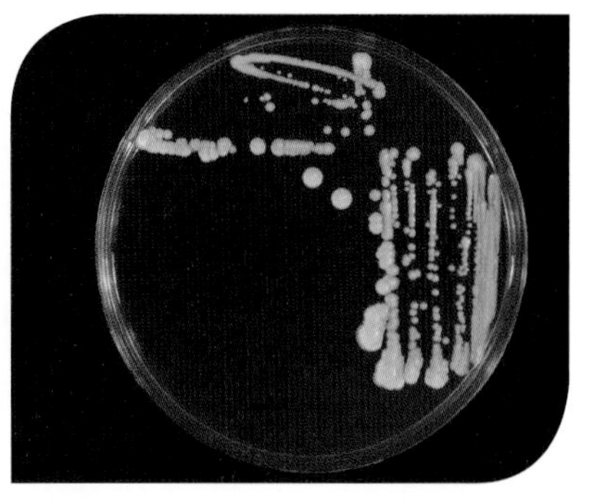

Baker's yeast (*Saccharomyces cerevisiae*) growing on agar (Courtesy of Mrs Yimin Dong, University of Auckland; photographed by Mr Iain Macdonald)

Penicillium is a common mould often found on rotten fruit and bread (Fig. 4.5). Unlike *Rhizopus,* the hyphae are subdivided by cell walls. The spores (called **conidia**) are not produced in sporangia, but are budded off from vertically-growing hyphae called **conidiophores**. This method of spore production is actually much more common than the method seen in *Rhizopus*.

Yeast

Yeast is a microscopic fungus that has been used for thousands of years to make wine and bread. Unlike other fungi, yeasts (there are many species) do not consist of hyphae but are *unicellular* (single-celled) and reproduce by *budding* (Fig. 4.6).

How yeast obtains energy

Yeast obtains energy from glucose in two stages (Fig. 4.7):

1. In the first stage (*glycolysis*), each glucose molecule is converted into two molecules of pyruvic acid with the release of a little energy in the form of ATP:

   glucose $\rightarrow$ 2 pyruvic acid + 4[H] + a little ATP

2. What happens next depends on whether oxygen is present or not. In the presence of oxygen, **respiration** occurs: pyruvic acid is converted to CO_2 and water:

   pyruvic acid + oxygen $\rightarrow$ carbon dioxide + water + lots of ATP

   In the absence of oxygen, **fermentation** occurs, in which pyruvic acid is converted to ethanol and carbon dioxide:

   pyruvic acid + 2[H] $\rightarrow$ ethanol + carbon dioxide

   In the conversion of pyruvic acid to ethanol there is no release of useful energy, so in the absence of oxygen the sole source of energy release is glycolysis.

new bud
'parent' cell

g. 4.6
ast cells budding

glucose
GLYCOLYSIS
2 ATP
little energy
4[H]
2 pyruvic acid
RESPIRATION
FERMENTATION
O_2
O_2 + $6H_2O$ + lots of energy
2 ethanol + $2CO_2$
30 ATP

g. 4.7
ow yeast obtains energy from glucose

Fermentation is sometimes called 'anaerobic respiration'. *This is quite incorrect*; the only organisms that can carry out anaerobic respiration are certain bacteria (see Chapter 3).

Since the ethanol can be burnt, much of the original energy in the sugar is still present in the ethanol. The ethanol is thus 'partially-burnt fuel'. If oxygen becomes available, ethanol can be converted in respiration to CO_2, water and *lots* of energy.

Because glycolysis produces much less ATP than respiration, glucose must be used up more quickly if ATP is to be produced at the same rate. This is no disadvantage if sugar is abundant, for example in fruit that has fallen off a tree.

ISBN 9780170191340

GROWING YEAST ON AGAR

Yeast and many other microorganisms can easily be grown on **agar**, a firm jelly to which nutrients such as glucose are added. Pure agar is a powder. This is dissolved in hot water, the flask covered with cooking foil and sterilised in a steam **autoclave**. This is simply a kind of *pressure cooker*, in which water boils at higher than 100°C. Whereas boiling at 100°C will kill most microorganisms, autoclaving at 121°C for 15 minutes or more kills all of them.

While still hot, the agar is poured into a sterile **Petri dish** (named after its inventor) and the lid quickly replaced. After it has set the resulting *agar plate* is placed upside down in a refrigerator until ready for use. There may be some condensation on the lid, but since the plate is upside down it cannot drip onto the agar.

To grow any kind of microorganism on the nutrients added to the agar, you need to exclude unwanted types, so all equipment must be *sterile* i.e. free of contamination by other microorganisms. All the equipment you use will have been previously sterilised in an autoclave. Before beginning work, all surfaces should be mopped with disinfectant solution.

You will need the following equipment (Fig. 4.8):

- Yeast suspension in sterile water in sterile beaker, covered with sterile cooking foil.
- Sterile Petri dish containing nutrient agar.
- Wire loop.
- Bunsen burner and matches.

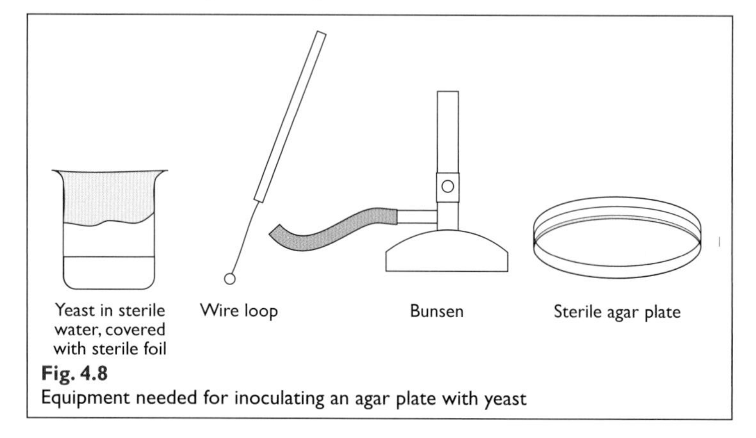

Fig. 4.8
Equipment needed for inoculating an agar plate with yeast

Inoculation

The process of introducing a microorganism to an agar medium is called *inoculation*. Hold the wire loop in the flame until it glows, and then dip in the yeast suspension. Quickly raise the lid of the Petri dish and *gently* wipe the loop over the agar in zig-zag fashion as shown in Fig. 4.9. Then re-sterilise the loop, cool, and re-streak through the inoculated portion. Repeat once more. Quickly replace the lid and seal with two pieces of sellotape. Label the base with your name, the date and type of microorganism, and place in an incubator set to 25°C. If an incubator is not available, room temperature will do, but you will have to wait longer to see the results (Fig. 4.10).

After 24 hours at 25°C or about two days at room temperature, you should see hundreds of white specks. Each speck is a **colony** of millions of yeast cells, derived from a single yeast cell.

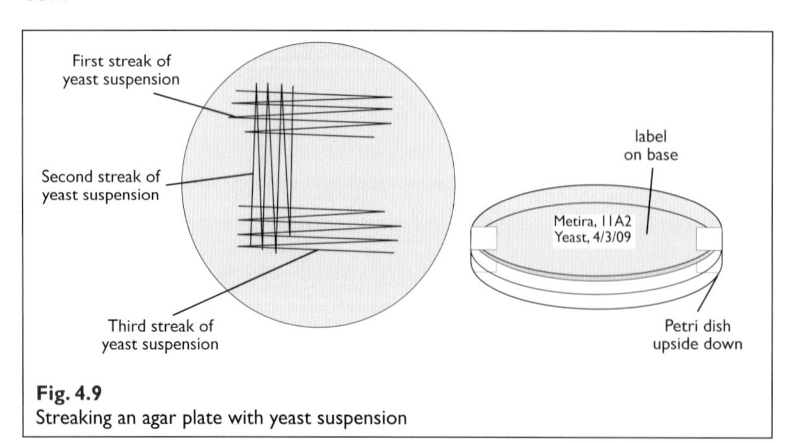

Fig. 4.9
Streaking an agar plate with yeast suspension

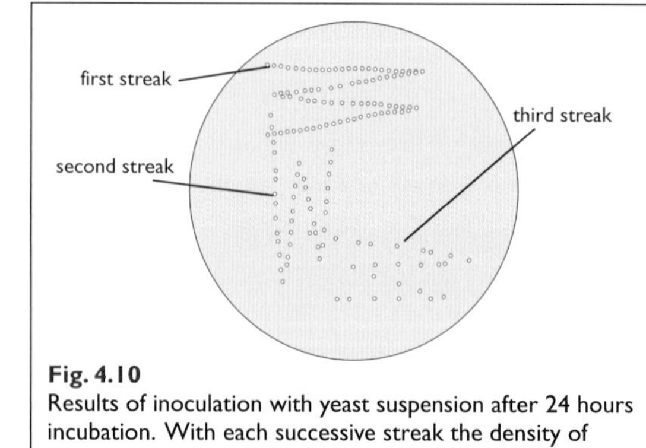

Fig. 4.10
Results of inoculation with yeast suspension after 24 hours incubation. With each successive streak the density of colonies decreases

ISBN 9780170191340

Mushrooms

An edible mushroom is only the reproductive part of the fungus; most of the mycelium extends over many square metres of soil, feeding on dead matter. The spores are produced sexually by meiosis, on the surfaces of the 'gills' beneath the cap.

Some mushrooms form mutualistic relationships with plants, in which both organisms benefit. The fungus is able to extract minerals from the soil more efficiently than the plant, and fungus obtains carbohydrate from the plant. Some mushrooms are parasitic on trees.

Summary of key points in this chapter

- Microorganisms include fungi, bacteria and viruses.
- Special units are used to measure them, such as *micrometres* and (in the case of viruses) *nanometres*.
- Fungi are *eukaryotes*, which are organisms whose cells have clearly-defined nuclei.
- All fungi are *heterotrophic*; meaning that they cannot produce organic compounds from carbon dioxide.
- Fungi are either *saprobic* (feed on dead matter), *parasitic* (feed on living matter), or *mutualistic* (live in intimate relationship with another organism, both obtaining benefit).
- Most fungi have bodies consisting of fine threads or *hyphae*, the whole network being called a *mycelium*.
- Unlike plant cells, fungal hyphae have walls of *chitin*.
- Saprobic fungi such as *Rhizopus* digest their food by secreting enzymes into it and absorbing the digested products. This is called extracellular ('outside the cell') digestion.
- Fungi reproduce by producing vast numbers of microscopic *spores*, in a variety of different ways.
- Some fungi (e.g. *Mucor, Rhizopus*) produce spores in sac-like *sporangia* at the tips of vertically-growing hyphae called *sporangiophores*.
- Many other fungi produce spores called *conidia*, by budding from vertically growing hyphae called *conidiophores*.
- *Fermentation* is a process which enables yeast and many other organisms to obtain energy from food without net oxidation. The process is *anaerobic* ('without oxygen').
- Fermentation by yeast produces carbon dioxide and ethanol, and is used in making bread and alcoholic drinks.
- Yeast and other fungi (also bacteria) are cultivated on *nutrient agar* jelly in *Petri dishes*.
- When growing microorganisms, equipment must be sterilised using an *autoclave*.

Test your basics

1 Copy and complete the following sentences.

a) Most fungi consist of microscopic branching threads called ___*___, collectively called a ___*___.

b) The cell walls of fungi differ from those of plants in that they consist of ___*___.

c) All fungi are ___*___, because they cannot make their own organic compounds from carbon dioxide.

d) Many fungi live in close relationship with a flowering plant in a mutualistic relationship called ___*___.

e) Fungi like *Mucor* and *Rhizopus* reproduce asexually by producing microscopic, single-celled structures called ___*___, inside sac-like structures called ___*___ at the top of vertically growing hyphae called ___*___.

f) Fungi like *Penicillium* produce spores called ___*___ which are budded off vertically growing hyphae called ___*___.

g) Yeast can obtain energy from glucose in an anaerobic process called ___*___, in which the glucose is converted into ___*___. This is then converted to ___*___ + ___*___.

h) Yeast and other microorganisms can be grown on agar jelly in a ___*___ dish.

i) The process of introducing microorganisms into a nutrient medium is called ___*___.

j) Enzymes that digest carbohydrates are called ___*___.

ISBN 9780170191340

5 Viruses: Living or non-living?

Viruses are not cells, and in some respects they are not like living organisms at all. For example, they can be crystallised. They have no chemical processes of their own, but use a host cell to provide the raw materials and energy for reproduction. This is why they can only reproduce inside a host cell. The only thing they provide in their reproduction is the genetic material ('instructions').

In colonising a host cell, a virus is acting as a parasite and in many cases viruses cause disease. Viral infections affect all groups of organisms: animals, plants, fungi, and even bacteria (Fig. 5.1). Some of the most deadly human diseases are caused by viruses.

A virus consists of a strand of nucleic acid surrounded by a protein coat or **capsid**. In *enveloped viruses* there is an additional, outer lipid membrane that had been part of the plasma membrane of the host cell in which it had been produced (Fig. 5.2).

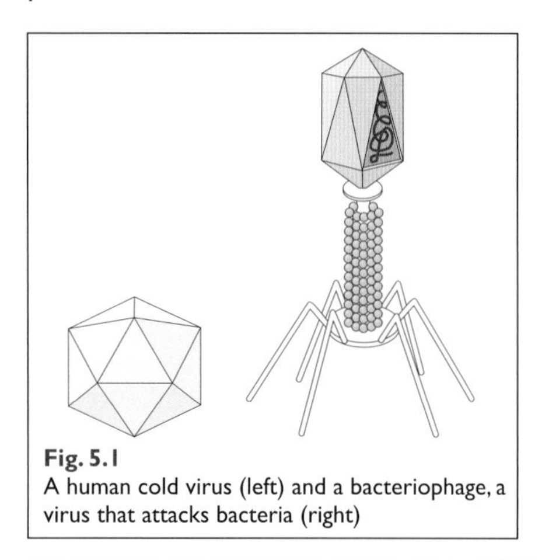

Fig. 5.1
A human cold virus (left) and a bacteriophage, a virus that attacks bacteria (right)

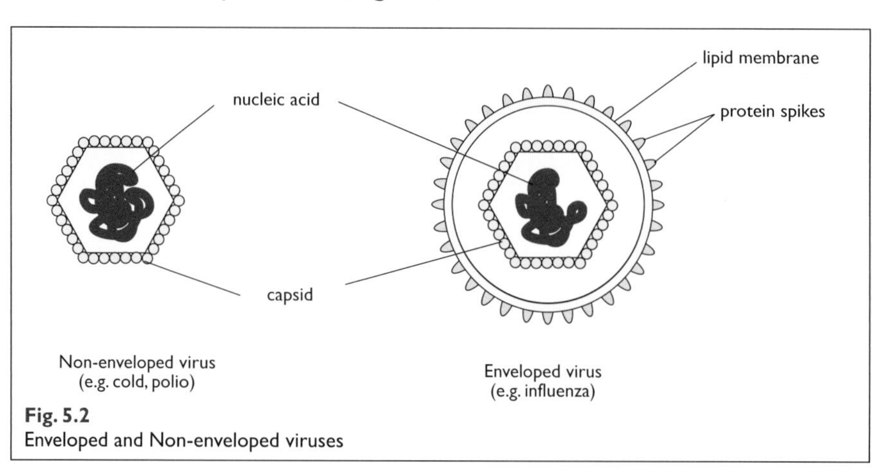

Fig. 5.2
Enveloped and Non-enveloped viruses

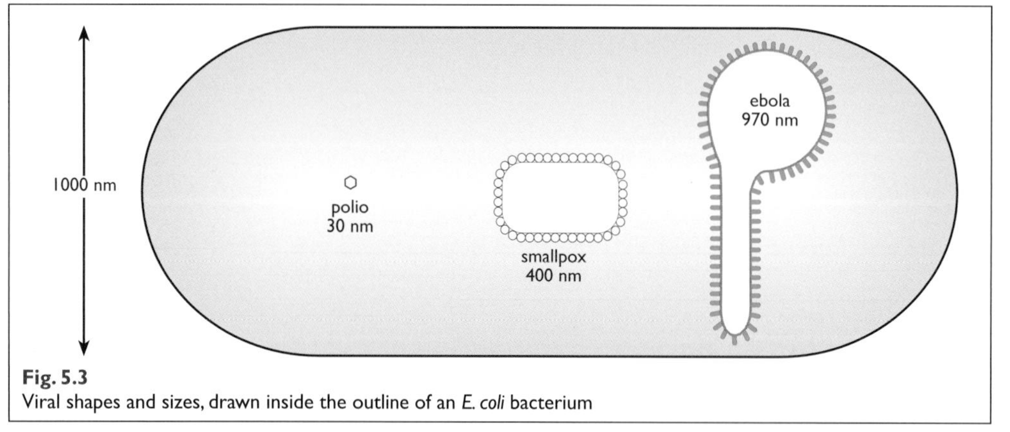

Fig. 5.3
Viral shapes and sizes, drawn inside the outline of an *E. coli* bacterium

A virus particle is called a **virion** and most can only be seen under the electron microscope. The smallest are about 20 nm (about 1/50th the size of the bacterium causing pimples), and the largest are about 400 nm (e.g. smallpox virus) and are *just* visible under the light microscope (Fig. 5.3).

The genetic material of the virus is a **nucleic acid**, which can be **DNA** (e.g. polio virus) or its close relative, RNA (e.g. HIV).

REPRODUCTION

Viruses can only reproduce inside a host cell, which provides the necessary raw materials and energy. The virus only provides the genetic 'instructions' for making more virus particles. This is why viruses cannot be grown on an agar medium; they must have living host cells.

ISBN 9780170191340

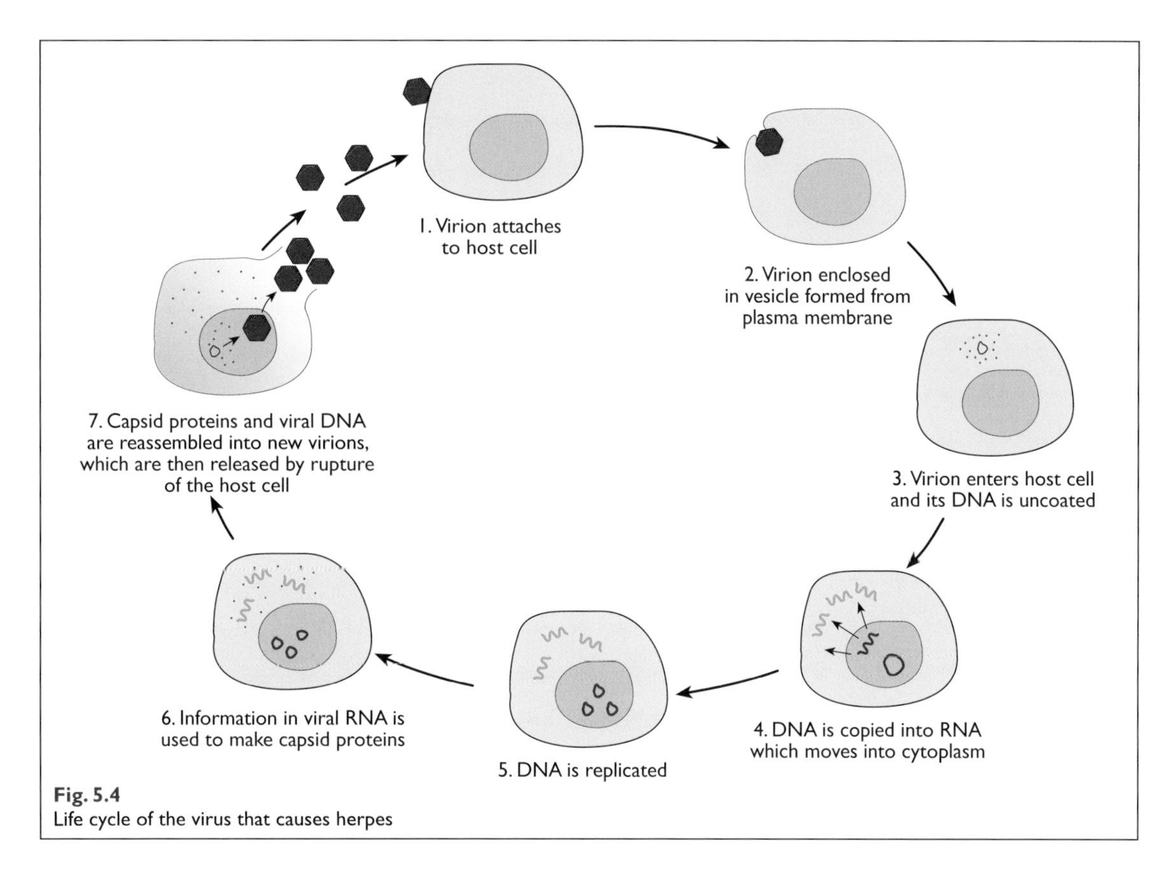
1. Virion attaches to host cell
2. Virion enclosed in vesicle formed from plasma membrane
3. Virion enters host cell and its DNA is uncoated
4. DNA is copied into RNA which moves into cytoplasm
5. DNA is replicated
6. Information in viral RNA is used to make capsid proteins
7. Capsid proteins and viral DNA are reassembled into new virions, which are then released by rupture of the host cell

Fig. 5.4
Life cycle of the virus that causes herpes

Like all parasites, viruses can only exploit particular types of host. The polio virus, for example, only infects humans and chimpanzees. Moreover, only certain types of cell are affected. HIV, for example, only affects a certain kind of cell in the immune system.

The basis for this specificity lies in the proteins in the outer layer of the virus. These proteins are able to link up with specific proteins in the plasma membrane of the host cell. The combination between viral and host cell proteins can be compared to a key fitting in a lock, and explains why each virus can only enter certain kinds of host cell.

Surface proteins on the virus are the 'labels' by which the immune system recognises a virus as foreign. 'Foreign' substances that trigger defensive responses in the immune system are called **antigens**. Antigens stimulate the body to produce defence proteins called **antibodies**, which combine with the antigens (Chapter 6).

The surface proteins of the virus are coded for by the virus's genetic material. This can **mutate**, resulting in a slightly different protein. Influenza viruses mutate frequently, which is why you can have influenza more than once; you are being infected by a different strain of the virus that your immune system does not recognise.

The mechanism of entry is different in animal cells and in bacteria. In bacterial viruses (called **bacteriophages**), the virus injects just its nucleic acid, the protein coat remaining outside. In animal viruses, the attached virus is taken in by **endocytosis** (similar to the way white blood cells engulf ('eat') bacteria). The plasma membrane gets to work, enclosing the virus in a vesicle that enters the cell (Fig. 5.4). In plant viruses, the virus enters the host cell via tiny strands of cytoplasm that connect adjacent cells.

Once inside the host cell, the viral coat is digested, exposing the genetic material. This is then replicated (copied) and new viral proteins are made using the viral genes. New viral particles are assembled, and in most animal viruses the host cell bursts open to release the viral particles. In enveloped viruses, the new viral particles become coated with plasma membrane of the host cell just before release and leave the host cell by **exocytosis**, the reverse of endocytosis.

ISBN 9780170191340

Summary of key points in this chapter

- Viruses are not cells, and are more like complex chemicals than living things.
- They are much smaller than bacteria; most are between 20 nm and 400 nm.
- A virus consists of DNA or RNA surrounded by a protein coat.
- In *enveloped viruses* there is an additional lipid layer derived from the plasma membrane of the host cell.
- Viruses can only reproduce inside a host cell, which provides raw materials, energy and enzymes. The virus provides only the information (genetic material) in the form of DNA or RNA.

Test your basics

1 Copy and complete the following sentences.

a) A virus particle consist of two chemicals: a ___*___ acid (RNA or DNA), surrounded by a ___*___ coat called a ___*___.

b) Viruses that attack bacteria are called ___*___.

c) Viruses enter animal cells by ___*___ and enveloped viruses leave by ___*___.

d) The protein part of animal virus is the part that may be recognised by the immune system as 'foreign' and thus acts as an ___*___.

2 Give an example of:

a) A human disease caused by a virus.

b) A DNA virus.

c) An RNA virus.

3 True or False? (Write T or F)

a) Viruses are the smallest known cells.

b) Viruses have no metabolism.

c) Viruses can be crystallised.

d) Most viruses can only be seen under the electron microscope.

ISBN 9780170191340

6 Microorganisms that can be harmful to humans

Though microorganisms play an essential role in ecosystems, they can also be harmful to humans in a number of ways:

- They may be **pathogenic**, causing disease.
- They may cause **food spoilage**.
- Though **decay** is an essential process in nature, it can be harmful in some situations. For example, human activity can result in excess organic matter in rivers and lakes. When this decays the resulting loss of dissolved oxygen can cause the death of fish, and is explained in Chapter 7.

MICROORGANISMS AND DISEASE

Many bacteria and fungi live as **parasites**, feeding off a **host** organism, usually without killing it. The most successful parasites do little harm; it is not in the parasite's interest to kill its source of food. Less well-adapted parasites are **pathogenic** because they harm the host, causing disease.

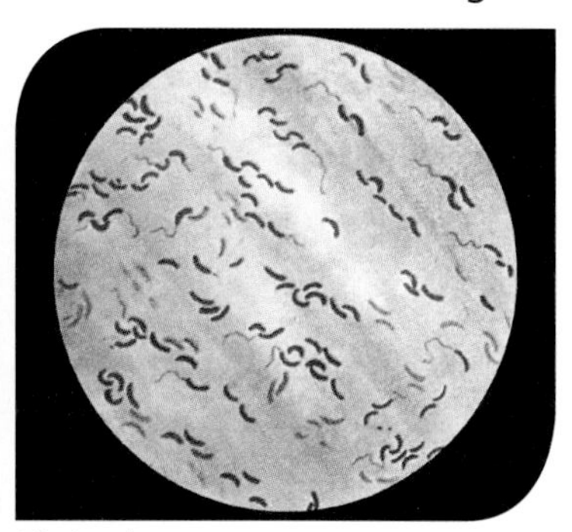

Cholera bacteria

How pathogens are spread

Like all parasites, pathogens have to move from one host to another. Knowledge of the method by which diseases are *transmitted* is a crucial part of prevention.

Before a disease can be combated, scientists need to know how it is spread. This is the job of **epidemiologists**, who study the pattern of diseases occuring within populations.

The science of epidemiology began in London in 1854 during an outbreak of cholera. This is a disease in which diarrhoea is so severe that death often results from loss of fluid. A doctor called John Snow found that most of the victims took their water from the same well. This well drew its water from the river Thames at a point below where raw sewage was discharged. People who drew their water from a well using water further upstream seldom caught the disease. Snow concluded that in some way cholera was spread by drinking water contaminated by raw sewage. He did not know that it was tiny microbes in the water that were responsible; it was another 20 years before the link between microbes and disease was proven.

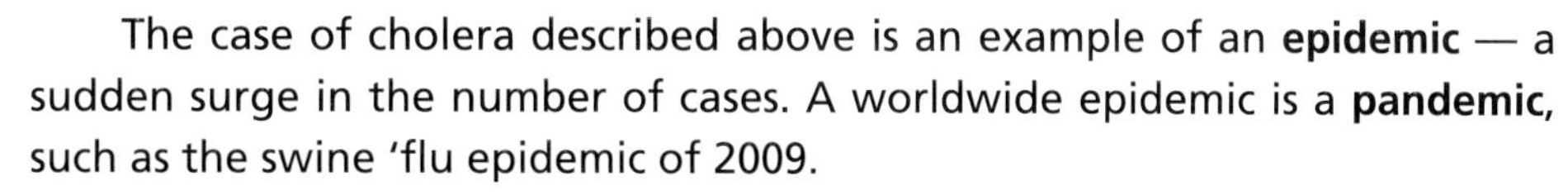

The case of cholera described above is an example of an **epidemic** — a sudden surge in the number of cases. A worldwide epidemic is a **pandemic**, such as the swine 'flu epidemic of 2009.

This mid 19th century sketch is called 'Death's Dispensary'. In London at that time drinking water came from the contaminated river Thames.

The role of carriers

Typhoid fever is a disease spread by infected water and food. Some people recover from the disease but continue to infect others and are called **carriers**. A famous case was 'Typhoid Mary' Mallon, in New York. Around the turn of the last century there were 28 cases of typhoid fever in homes where Mary had worked as a cook. When she was forcibly confined to hospital it was found that her faeces were heavily infected with typhoid bacilli, even though she showed no symptoms. Evidently she was not washing her hands after using the toilet. She was released after agreeing not to work as a cook again, but she promptly changed her name and disappeared. There were further outbreaks and she was eventually tracked down and spent the rest of her life in prison. A post-mortem showed that her gallbladder was infested with the bacterium *Salmonella typhi*.

ISBN 9780170191340

Typhoid is an example of a **notifiable disease**, in which doctors must, by law, inform the authorities because the right to privacy is considered less important than the safety of others.

Safe drinking water

Once it was realised that diseases could be spread by infected drinking water, two public health measures began to be taken:

1. Rather than discharging sewage into rivers, it began to be treated to make it less harmful (see pages 56–58).
2. Drinking water was treated with chlorine, which kills bacteria

Safe food

Campylobacter jejuni is a bacterium that naturally infects farm animals, especially chickens. In humans it causes severe gastro-enteritis (stomach cramps, vomiting and diarrhoea). In 2006 it was found that the incidence of *Campylobacter* infection in New Zealand was the highest in the world, with over 400 cases per 100,000 of the population (including over 800 people in hospital and at least one death). Since then stringent safety measures in poultry production have resulted in a 50% reduction in cases. These measures include freezing of meat, which kills most of the bacteria, and thorough cooking, which kills all of them.

Contact transmission

Contact transmission can occur in three ways:

1. By direct contact, such as touching, kissing or by sexual intercourse. Examples are the common cold, influenza, hepatitis A, HIV/AIDS, syphilis and other venereal diseases.
2. By droplets sprayed. Examples include influenza, measles, whooping cough, pneumonia and tuberculosis that are spread by microscopic droplets sprayed at high speed (over 100 km per hour!) during coughing and sneezing.
3. By indirect contact with objects such as towels, bed linen, handkerchiefs, drinking cups and clinical thermometers.

How the body defends itself

Body defences are of two general kinds: the outer defences and the immune response.

The outer defences

These take various forms, illustrated in Fig. 6.1:

- The outer layer of the skin (**epidermis**) is tough and fairly dry, though bloodsucking insects that carry pathogens may penetrate it.
- The stomach secretes *hydrochloric acid*, which kills most bacteria.
- The lining of the breathing passages secretes **mucus** that traps bacteria and is slowly swept to the back of the throat by the beating of tiny *cilia*. It is then swallowed.
- Tears contain the enzyme **lysozyme**, which digests the cell walls of many bacteria.

Defence by mutualistic bacteria

Living on and in the human body are vast numbers of bacteria. Most are harmless, and some are beneficial because they prevent certain pathogenic microbes from getting established. These bacteria are thus in a **mutualistic** relationship with the body they inhabit.

This balance can be upset when a person takes a course of antibiotics, with the result that many of the normal gut bacteria are killed. Thrush, for example is a condition caused by fungi infecting the back of the throat. Normally these fungi are kept in check by bacteria. If these bacteria are killed, the fungus grows unchecked, producing a very sore throat.

Other mutualistic bacteria inhabit the large intestine. If these are removed following treatment with antibiotics, pathogenic bacteria may increase in numbers. Pathogens are

ISBN 9780170191340

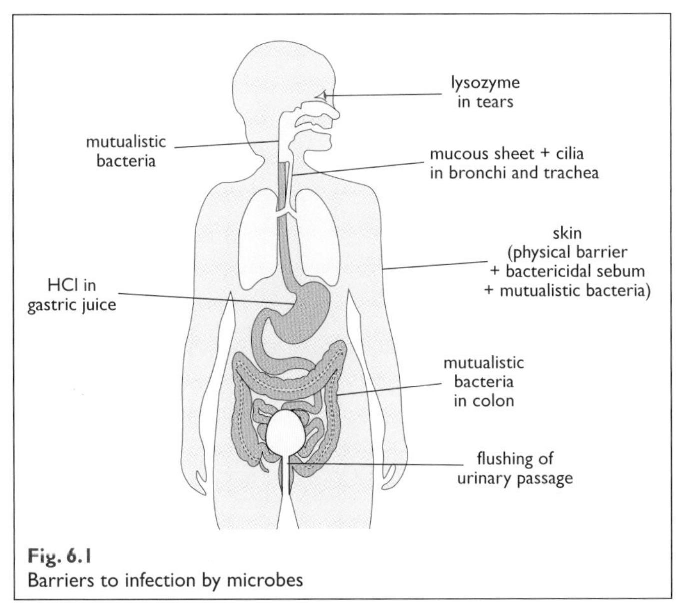

Fig. 6.1
Barriers to infection by microbes

often normally present, but are suppressed by competition with normal bacteria.

Incubation period

For a while after infection the host shows no symptoms. This is the **incubation period**, which may be as short as a day or two, or may be several weeks or months as in hepatitis B. During the incubation period the numbers of the invading microorganisms are increasing. The rate of increase is usually much slower in the body than it is *in vitro* (i.e. when grown in the laboratory). For example, tuberculosis bacteria have a doubling time of 12 hours *in vitro* but in the body it is several days.

The immune response

Microbes that get through the outer defences are attacked by the immune system. Certain cells 'recognise' specific substances on the surface of the invader as 'foreign'. These foreign substances are called **antigens**. They stimulate certain cells in the immune system to make defence proteins called **antibodies**. Each antibody combines with a specific antigen on the surface of the invader. These 'labelled' microbes are then vigorously attacked and eaten by white blood cells called **phagocytes**. Other antibodies combine with and neutralise toxins made by the bacteria. It is a race against time between the immune system and the pathogen's ability to multiply. If the body loses, disease is the likely result.

How bacterial pathogens harm the body

One of the most important ways in which bacteria can harm the body is by the production of poisonous substances called **toxins**. Some toxins are among the most deadly poisons known. Diphtheria toxin for example inhibits protein synthesis (stops the cells making proteins). A single molecule of the toxin will kill a cell. The bacterium that causes cholera (*Vibrio cholerae*) produces a powerful toxin that causes massive loss of salts and water from the small intestine. Death by dehydration may occur unless the patient is given salt solution. This gives the body time to produce an antibody called an **antitoxin**, which neutralises the poisonous substance.

Viruses do not produce toxins since they use the host cell's metabolism.

How pathogens leave the body

Having reached a host, a pathogen must reproduce as quickly as possible before the host either recovers or dies. Many pathogens produce effects on the body that help them escape from the host; gut pathogens often produce diarrhoea, and those infecting the breathing passages usually cause the host to cough and sneeze. Venereal diseases are spread by sexual intercourse.

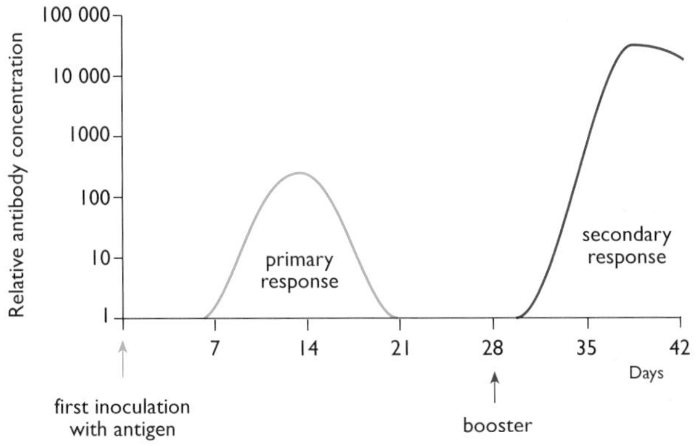

Fig. 6.2
Primary and secondary responses in vaccination. Note the logarithmic scale.

Help from medicine: Vaccination

The ability of the body to produce antibodies is exploited in the production of *vaccines*. A vaccine is an attenuated (weakened) extract of the bacterium or virus which can still cause an immune response. Some weeks after the initial vaccination a 'booster' dose is given. In response to this the output of antibodies is much greater and more rapid (Fig. 6.2).

ISBN 9780170191340

Some vaccines consist of a toxin that has been slightly altered chemically, and is called a **toxoid**. The toxoid is harmless but is sufficiently similar to the toxin to stimulate the body to produce antibodies that are effective against the disease. An example is the toxoid used in vaccinations against diphtheria.

In the case of diseases that spread directly from person-to-person, it is not necessary for *every* individual in a community to be immunised. If more than a certain proportion is immune, the disease tends to peter out because the pathogen usually comes into contact with individuals who are immune (Fig. 6.3). For polio this figure is about 70%, but for highly infectious diseases such as measles it is nearer 90–95%.

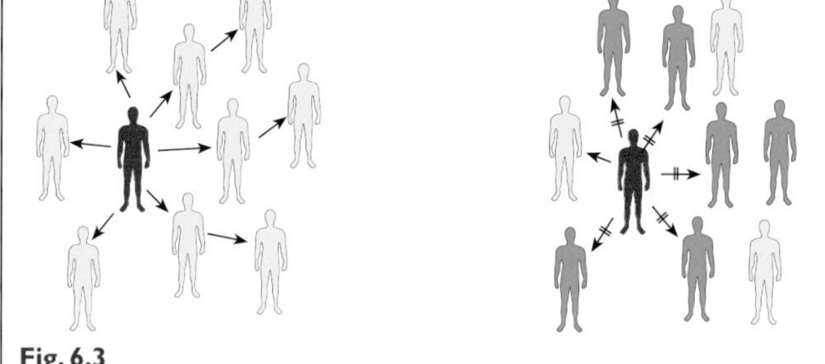

Fig. 6.3
'Herd' Immunity. Immune individuals shown in green, susceptible ones in yellow

Antibiotics

In 1928 Alexander Fleming noticed that one of his cultures of the bacterium *Staphylococcus aureus* had become contaminated by a green mould, *Penicillium chrysogenum*. There was a clear area around the fungus with no bacterial growth. He suggested that the fungus was producing a substance that was inhibiting the growth of the bacteria. Later, two scientists working in Britain, Howard Florey and Ernest Chain, identified the substance and called it **penicillin**, and by 1944 it was soon in mass production.

Penicillin was one of the first **antibiotics** to be used in medicine. These are substances produced by microorganisms that kill or prevent the growth of bacteria. Many others have been discovered since. Antibiotoics depend on the fact that the metabolism of bacteria differs in a number of ways from that of eukaryotes. It is therefore possible to interfere with the metabolism of bacteria without affecting the patient.

Viral diseases are not affected by antibiotics, since viruses use the host metabolism to reproduce, but a number of anti-viral molecules have been developed.

Antibiotic resistance

At first, antibiotics were thought to be 'the answer' to bacterial disease; millions of lives were saved. Then in 1947 it was found that some bacteria were becoming *resistant* to certain antibiotics. Since then, every antibiotic that has been introduced has been followed by a bacterium that is resistant to it.

Resistance to antibiotics is an excellent example of evolution. Like all organisms, bacteria vary as a result of *mutation*. When a person begins a course of antibiotics, some bacteria are less sensitive than others and take longer to be affected. If the patient does not complete the course, the resistant bacteria may survive and escape to infect other people (Fig. 6.4).

To minimise the risk of antibiotic resistance developing, two rules should be followed:

1. Doctors should not prescribe antibiotics for trivial infections.
2. Patients should always complete the course of antibiotics, even if they feel better.

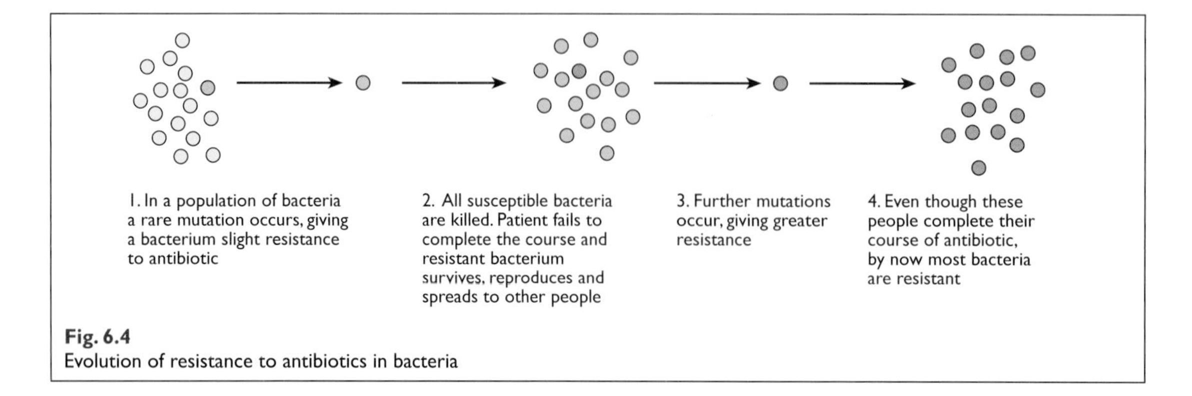

1. In a population of bacteria a rare mutation occurs, giving a bacterium slight resistance to antibiotic
2. All susceptible bacteria are killed. Patient fails to complete the course and resistant bacterium survives, reproduces and spreads to other people
3. Further mutations occur, giving greater resistance
4. Even though these people complete their course of antibiotic, by now most bacteria are resistant

Fig. 6.4
Evolution of resistance to antibiotics in bacteria

ISBN 9780170191340

Fungal diseases of humans

Certain fungi can cause infections of the skin, for example:

- Athlete's foot, caused by either of two fungi, leading to the skin between the toes becoming inflamed.
- Thrush, caused by *Candida albicans*, a yeast infecting the mouth and throat. *Candida* is normally present in a high proportion of people, but other microbes compete strongly with it and normally keep it in check.

Fungal diseases of plants

One of the most serious and economically costly fungal diseases in New Zealand is 'bunch rot' of grapes, caused by the fungus *Botrytis cinerea* (Fig. 6.5). This fungus not only attacks grapes but other fruit crops, causing stem-end rot in kiwifruit, grey mould in strawberries and dry eye rot in apples.

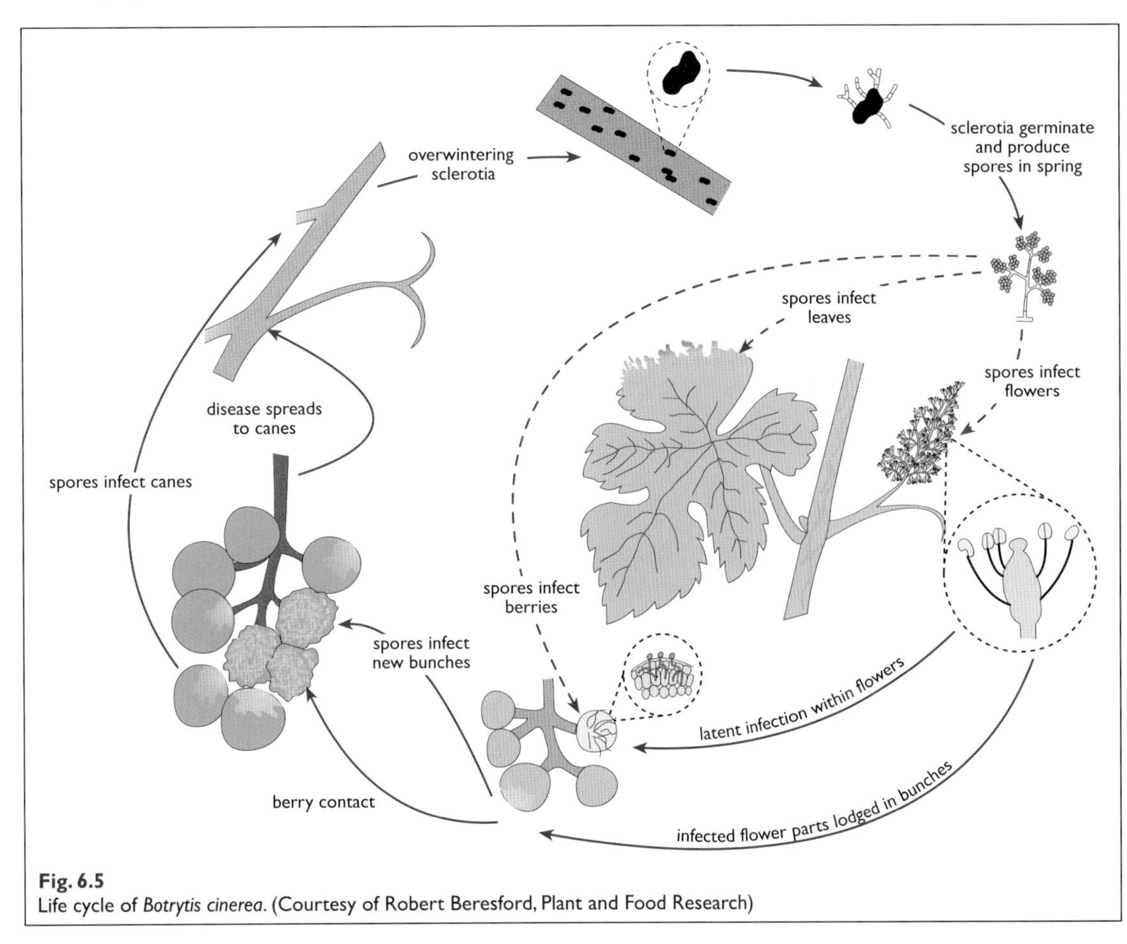

Fig. 6.5
Life cycle of *Botrytis cinerea*. (Courtesy of Robert Beresford, Plant and Food Research)

Botrytis bunch rot. (Courtesy of Robert Beresford, Plant and Food Research)

The fungus spends the winter as mycelium feeding on decaying plant matter. It also forms compact masses of hyphae called *sclerotia* that germinate in the spring to produce spores on branching, tree-like *conidiophores*. The spores or *conidia* are detached and carried by wind. Botrytis most easily enters a new host through injured tissue. The spores develop hyphae which invade the tissue, causing the cells to disintegrate. The fungus goes on to produce more tree-like conidiophores, and even more spores that can invade more host tissues. This process continues until colder weather sets in and sclerotia are formed.

Although damage to grapes occurs late in the season, if there is heavy enough rainfall during ripening, botrytis invades the plant tissues much earlier by infecting flowers and young berries. It then remains latent ('hidden') until ripening occurs, when it begins to rot the fruit.

Botrytis can be controlled by fungicides, though at considerable cost;

ISBN 9780170191340

a single application all of New Zealand's vineyards can cost $3–5 million, and standard treatment uses up to five applications per season in regions with wet conditions.

Plant viruses

Some of the most serious diseases of agricultural crops are caused by viruses. Aphids (greenfly) are a major **vector** of plant viruses.

Grapevine leafroll-associated virus

This RNA virus causes a serious disease in grapevines in most parts of New Zealand, as well as in other countries. The effect of the virus is to interfere with the movement of sugar in the phloem, followed by a rolling up of the leaves and reduced chlorophyll content. At the same time as the chlorophyll content is reduced, anthocyanin pigments accumulate, giving the leaves a red colour. The decreased photosynthetic ability reduces growth and up to 60% loss in yield, and the berries have a lower sugar content. The virus is transmitted by mealybugs (insects related to aphids), which are said to act as **vectors**.

Effect of leafroll virus on grapevine. (Courtesy of Nick Sage)

FOOD SPOILAGE

Besides the fact that food can be contaminated by pathogenic bacteria, microbes can cause it to *decay* ('go bad'). Microbes that do this are **decomposers**, and reduce the value of food in two ways:

1. Food that has begun to decay is unpalatable (though not necessarily harmful).
2. Some microbes produce toxic (poisonous) chemicals when they grow. *Food poisoning* results from eating food containing such bacteria. Even if the food is cooked (thus killing the bacteria) the toxins may be unaffected, and symptoms (vomiting and diarrhoea) occur within a few hours. The two main offenders are the bacteria *Staphylococcus aureus* and *Clostridium botulinum*. The latter produces botulinum toxin, one of the most poisonous substances known, causing numerous deaths each year; 1 mg is enough to kill 20,000 people!

FOOD PRESERVATION

Food may be protected from decay in two basically different ways: *microbicidal* and *microbiostatic*.

Microbicidal methods

Microbicidal methods of food preservation involve *killing* the microorganisms.

Canning

In this method the food is heated to kill microbes and then sealed in a tin can to prevent airborne microorganisms getting in. It is highly successful, but there have been occasional failures leading to food poisoning.

Pasteurisation

By *briefly* heat-treating foods such as milk, yoghurt or beer with heat, the 'shelf life' can be considerably increased. This process is called *pasteurisation* (after its discoverer), and kills most (but not all) the microorganisms in the food. In the pasteurisation of milk an important additional aim is to kill bacteria causing tuberculosis. *Low temperature* pasteurisation involves treating the milk at 62.8°C for 30 minutes. An alternative is *ultra high temperature* (UHT) processing, in which the milk is maintained at 141°C for two seconds.

ISBN 9780170191340

Irradiation

Treating food with radiation such a gamma rays from a radioactive source kills *all* microbes, though this is not used in New Zealand. The method works well, but many people believe (incorrectly) that the treated food gives off radiation.

Though not suitable for food because it penetrates very poorly, ultraviolet radiation (UV) is deadly to bacteria and very effective in sterilising drinking water and surfaces on which food is prepared. UV is carcinogenic (causes cancer), so it must be switched off when humans are present.

Microbiostatic methods

These methods keep food in conditions that slow or prevent growth of microorganisms.

Refrigeration

Refrigeration depends on the fact that metabolism (chemical reactions inside cells) is extremely sensitive to temperature. This is because enzyme activity is sensitive to temperature. For every 10°C decrease in temperature, the rate of metabolism (and therefore growth) is approximately halved. This means that at 5°C, bacteria grow half as fast as they do at 15°C and a quarter as fast as they do at 25°C.

You might think that keeping food in a fridge at 5°C would result in only a quarter as many bacteria as there would be had the food been kept in a warm room at 25°C. This is not so, as the following example shows.

Assuming that bacteria in food divide every hour at 25°C, at 15°C the generation time would be two hours, and at 5°C it would be four hours. Suppose a single bacterium is placed in each of two pieces of meat and kept at two temperatures: in a fridge at 5°C and in a warm room at 25°C. Table 6.1 shows the dramatic difference in numbers of bacteria after 12 hours.

Time (hours)	Numbers of bacteria	
	5°C	25°C
0	1	1
1		2
2		4
3		8
4	2	16
5		32
6		64
7		128
8	4	256
9		512
10		1024
11		2096
12	8	4192

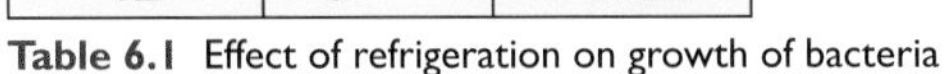

Table 6.1 Effect of refrigeration on growth of bacteria

It is therefore *very* important that if you warm food from the refrigerator, it should be eaten immediately rather that returned to the fridge.

It is worth noting that the example above refers to *mesophilic* bacteria (which include most pathogens). These are bacteria that grow at moderate temperatures. *Psychrophilic* bacteria grow best at low temperatures, and these can even spoil food kept in the fridge.

Drying and salting

Microorganisms need a certain amount of moisture to grow, so if food is kept dry it does not go bad. If meat is kept in concentrated salt solution, bacteria cannot grow because water tends to pass from the bacteria to the salt solution by **osmosis**.

ISBN 9780170191340

Pickling

Food kept in vinegar can be preserved for long periods. This is because vinegar is ethanoic ('acetic') acid, and most microbes do not grow in very acidic solutions. Pickled onions are an example.

Summary of key points in this chapter

- Most microorganisms are either harmless or beneficial to humans, but some cause disease or render food inedible.
- The most successful parasites do little harm to the host; those that cause disease are less well-adapted.
- The cause of a disease can often be identified by studying its pattern of occurrence (*epidemiology*).
- Diseases that are spread by infected water or food can easily be prevented.
- The surfaces (both outer and inner) of the body are protected by a variety of physical and chemical barriers.
- A pathogen that gains entry is recognised as 'foreign' by the immune system, which responds by producing *antibodies*, coupled with activation of phagocytes that eat bacteria.
- A person can be protected from a disease by *vaccination*, in which the person is given a weakened dose of the antigen.
- Antibiotics have been very successful in treating infection, but many bacteria have become resistant.
- Food can be preserved either by killing bacteria and then sealing the food (*microbicidal methods*), or by preventing or slowing the growth of bacteria (*microbiostatic methods*).

Test your basics

1 Copy and complete the following sentences.

a) Organisms that cause disease are called ___*___.

b) An organism on which a parasite feeds is called a ___*___.

c) The study of the pattern of occurrence of diseases is called ___*___.

d) A disease in which there is no right to privacy (i.e. must be reported to the authorities) is a ___*___ disease.

e) Tears and some other body fluids contain an enzyme called ___*___ that digests the outer walls of many bacteria.

f) Many pathogenic bacteria harm the body by producing poisonous chemicals called ___*___.

g) The body responds to the presence of invading organisms by producing special proteins called ___*___. The substances that trigger the formation of these defence proteins are called ___*___.

h) A toxin that has been chemically altered to make it harmless, but without altering its ability to produce an immune response, is called a ___*___.

i) Penicillin was the first ___*___ to be discovered.

j) Refrigeration is a ___*___ method of food preservation because it slows the growth of microorganisms rather than killing them.

k) Canning is a ___*___ method of food preservation because it kills microorganisms.

2 Give an example of:

a) A bacterium that causes human disease.

b) A fungus that causes disease in humans.

c) A fungus that causes disease in plants.

d) A plant virus.

ISBN 9780170191340

7 Microorganisms that are useful to humans

Humans benefit from the activities of microorganisms in two ways:

1. Though decay can be a nuisance and even a major problem (Chapter 6), decomposers supply plants with the simple inorganic substances they need for growth. Decay is therefore essential for human survival.
2. Microorganisms are also used in the production of a number of important foods and drinks, such as bread, cheese, yoghurt, wine and beer.

DECOMPOSITION

Decomposers are *fungi* and *bacteria*. They have three features in common:

1. They feed on dead matter.
2. They digest their food outside the cells, or extracellularly. To achieve this, a decomposer needs a large surface area compared with its volume. Bacteria have a large surface/volume ratio because they are very small. Fungi have a large surface because they have finely divided bodies in the form of thin filaments or hyphae, or are very small as in the case of yeasts.
3. They must be able to reach new food sources. In this respect they are in competition with other decomposers — the successful ones are those that get there first. Most bacteria and fungal spores are dispersed randomly by wind or water, so the majority fail to reach a suitable food source. The 'winners' therefore tend to be those that produce the greatest number of offspring.

The role of microbes in the circulation of nitrogen

The element nitrogen is an essential constituent of living matter, for example:

- It is a constituent of all *amino acids* and thus all *proteins*.
- It is a constituent of the *nucleic acids DNA* (and in some viruses, *RNA*).

Though nitrogen gas makes up about 80% of the atmosphere, plants cannot use it. Instead, plants absorb nitrogen in the form of **nitrates**. Nitrates are thus being continuously removed from the soil. The return of nitrates to the soil is carried out by microorganisms. The circulation of nitrogen in nature is called the *nitrogen cycle* and shown in Fig. 7.1.

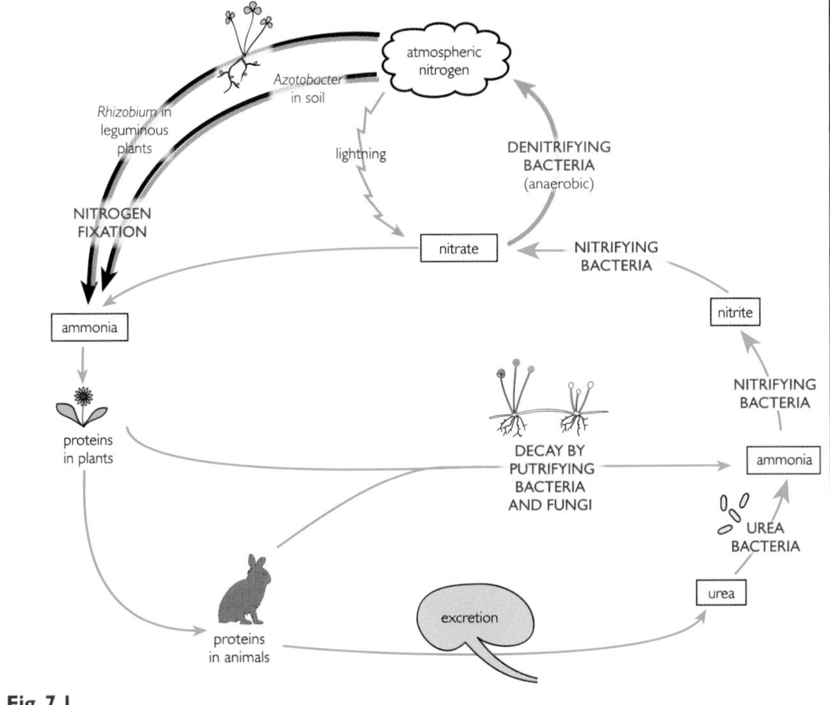

Fig. 7.1
The nitrogen cycle

The terrestrial (land) nitrogen cycle actually consists of two sub-cycles, each of which involves microorganisms:

1. A *rapid* cycle confined to the soil.
2. A much *slower* cycle involving the atmosphere.

ISBN 9780170191340

Each cycle can be divided into two further parts: a stage in which nitrates are removed from the soil, and a stage in which they are returned from the soil.

In the rapid cycle, plants absorb nitrate and use it to make amino acids and proteins. To do this they also need organic compounds made in photosynthesis. When the plant dies, decomposers digest the plant proteins and use the amino acids in one of two ways:

1. Some are used to make decomposer proteins.
2. Most are broken down and used in respiration to provide energy. Before an amino acid can be used for energy it must be **deaminated** by removal of the nitrogen-containing part as **ammonia**, which diffuses into the soil.

The ammonia is then oxidised to **nitrate** (via nitrite) by **nitrifying bacteria**. This oxidation releases energy, which the bacteria use to make organic compounds from CO_2 and water in the process of *chemosynthesis*.

Animals also deaminate amino acids, but in mammals the ammonia is converted to urea, which is then excreted. The urea is used as an energy source by *urea bacteria*, which convert it to ammonia.

The slower part of the cycle involves atmospheric nitrogen. In **nitrogen fixation**, nitrogen gas is converted to a form that plants can use. This occurs in three ways:

1. Leguminous plants (plants of the pea, bean, gorse and clover family) have *root nodules* containing *Rhizobium* bacteria that can *fix* nitrogen, converting it to ammonia. In return the bacteria obtain carbohydrate from the legume. This kind of relationship is called *mutualism* because both organisms benefit. Each can do something the other cannot do: the bacteria converts nitrogen gas to ammonia, and the legume converts CO_2 to organic matter in photosynthesis. In combination, the two organisms can thus make amino acids and proteins from nitrogen and CO_2.
2. Some nitrogen-fixing bacteria (e.g. *Azotobacter*) live freely in the soil and convert nitrogen to ammonia using dead organic matter as an energy source.
3. By lightning. The extremely high temperatures created by a lightning flash cause nitrogen to combine with oxygen, forming (after several reactions) dilute nitric acid, which is a form of nitrate.

The other part of the slow cycle is **denitrification**, in which bacteria convert nitrates to nitrogen gas. It occurs in soils that have become depleted of oxygen, such as occurs in soil that has been flooded. These bacteria obtain their energy using *anaerobic respiration*, using nitrate instead of oxygen to oxidise organic matter:

organic matter + nitrate $\rightarrow$ carbon dioxide + nitrogen (gas) + water + energy

Because nitrogen is present in enormous amounts in the atmosphere, it takes millions of years for nitrogen to 'travel' around this part of the cycle.

Sewage treatment

Until the mid 19th century (and much later in many parts of the world) sewage used to be 'dumped', often piped into rivers or into the sea. This can lead to several bad results:

- Epidemics of water-born diseases such as cholera.
- Even if it contains no pathogenic bacteria, raw sewage is rich in organic matter and thus food for saprobic bacteria, which use up oxygen in their respiration. When sewage is piped into rivers, the water becomes anoxic (depleted of oxygen), which kills fish and other aquatic animals. The smell is horrible.

ISBN 9780170191340

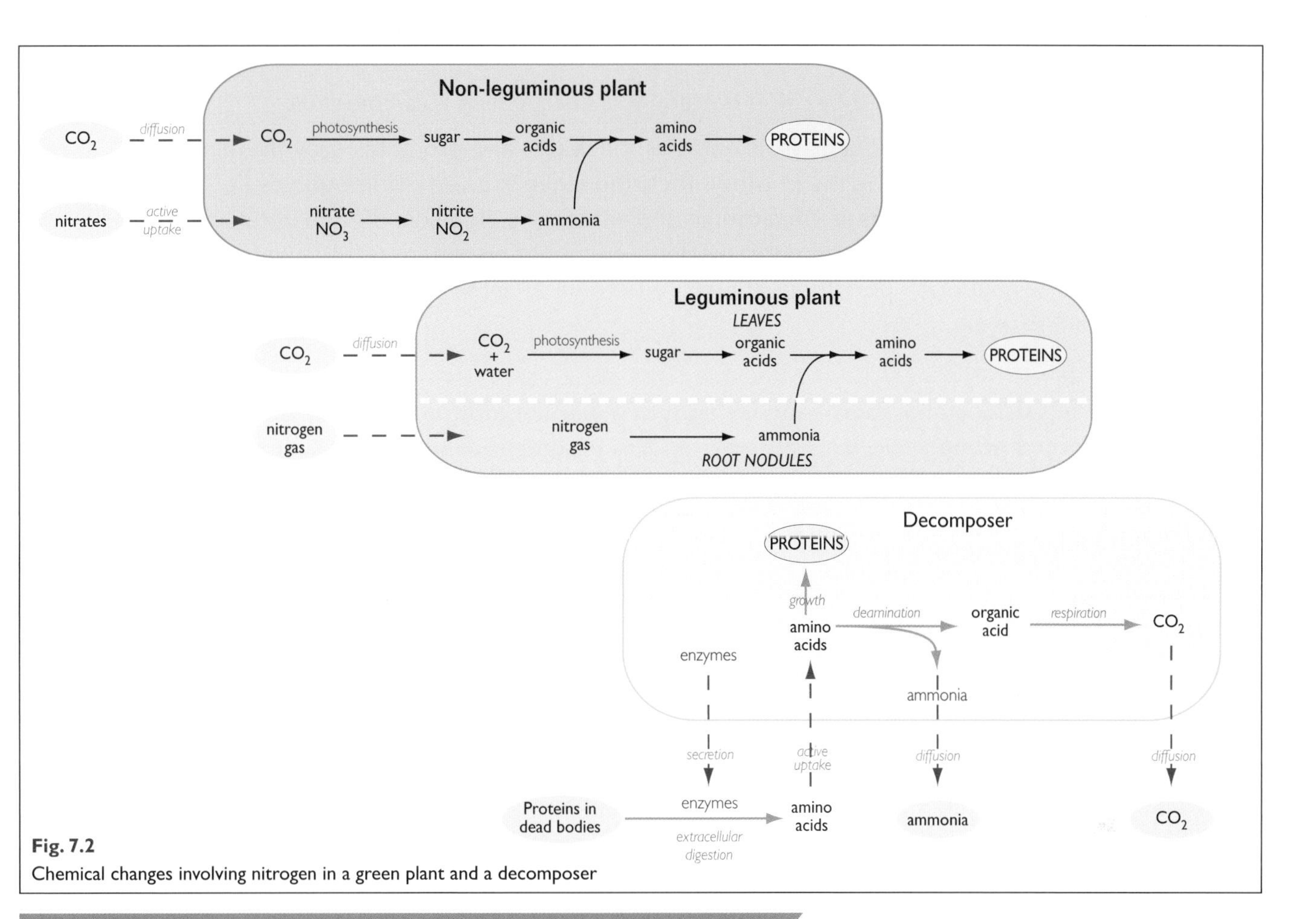

Fig. 7.2
Chemical changes involving nitrogen in a green plant and a decomposer

Extension: A closer look at the nitrogen cycle

Figure 7.2 shows some of the key processes of the nitrogen cycle in more detail. Notice the following:

- **The conversion of nitrate to ammonia in plants is the reverse of nitrification in the soil. Just as nitrification is an energetically 'downhill' process (releasing energy), the conversion of nitrate to ammonia requires an input of energy (from photosynthesis).**
- **The production by plants of amino acids from ammonia and organic acids is the reverse of deamination in decomposers and animals.**

Nowadays, sewage is treated to get rid of the organic matter and to make it safe. In a modern sewage treatment plant the sewage passes through three main stages. The first stage, or *primary treatment*, is purely mechanical and involves separating solids from liquids.

Secondary treatment

The aim of secondary treatment is to get rid of the organic matter by using bacteria to convert it into CO_2 and nitrogen gas. It is a complex process, but follows several essential steps:

1. Amino acids from proteins are deaminated by saprobic bacteria to form organic acids and ammonia.
2. As occurs naturally in well-aerated soils, the ammonia is oxidised by *nitrifying* bacteria to nitrate.
3. Under anaerobic conditions, *denitrifying* bacteria convert nitrate to nitrogen gas.

ISBN 9780170191340

Stage 2 is necessary because nitrate can be harmful in two ways:

1. In drinking water it is converted by colon bacteria to nitrite, which is *carcinogenic*.
2. If discharged into lakes, rivers or bays, it enriches the water and promotes growth of algae. The oxygen produced by the photosynthesising algae is given off in bubbles (it is only slightly soluble in water). In the autumn and winter, the algae die off and decay. The oxygen needed to oxidise the organic matter is no longer present (it left the water in the summer as bubbles). Conditions therefore become anaerobic, causing fish and other animal life to die.

Tertiary treatment

The final stage consists of sterilisation using UV light of a wavelength that is deadly to life because it is absorbed strongly by DNA, damaging it. This treatment eliminates 99% of the bacteria and viruses.

FOOD PRODUCTION

For thousands of years, humans have used microorganism in the production of bread, cheese and wine.

Bread

There are countless different kinds of bread, yet they all are made from flour, water, sugar and yeast. Water containing suspended yeast is added to flour and sugar, and kneaded to make a *dough*. This is left for an hour, during which time yeast feeds on the sugar, producing bubbles of CO_2 which are trapped in the dough, causing it to rise. Ethanol is also produced, but this is driven off during subsequent baking.

Cheese

There are over 2000 different kinds of cheese, but the manufacture of all of them involves bacteria that ferment milk sugar (lactose) to lactic acid. The first step is the precipitation of the milk protein as *curds*, using the enzyme *rennin* and a bacterium such as *Lactobacillus* or *Lactococcus*. The curd undergoes a prolonged ripening process, during which other microbes may be inoculated, such as the green mould *Penicillium* if one wants a blue vein cheese.

Wine

Most wines are made from fermented grape juice and the microorganism used is yeast. This converts the sugar in the fruit juice into ethanol; the CO_2 is allowed to bubble away. In sparkling wines, extra sugar is added and the CO_2 is retained in sealed vessels, building up pressure.

ISBN 9780170191340

Summary of key points in this chapter

- Bacteria and fungi play an essential part in the circulation of nitrogen and other nutrient elements.
- The *nitrogen cycle* consists of two interlocking cycles; a rapid cycle and a much slower one.
- Some steps in the nitrogen cycle can be carried out by fungi or bacteria, but some can only be performed by bacteria.
- Bacteria and fungi play an essential part in the treatment of sewage.
- Bacteria and fungi are used extensively in the production of certain foods, such as cheese, bread and wine.

Test your basics

Copy and complete the following sentences.

a) The element nitrogen is a key constituent of two kinds of complex biological molecules: ___*___ and ___*___ acids (e.g. DNA).

b) Plants absorb their nitrogen mainly in the form of ___*___, which is used to make ___*___ acids that are built up into proteins.

c) Proteins in dead organisms are digested by ___*___ to produce ___*___ acids. These are either used by the ___*___ to make their own proteins, or are ___*___ to release ammonia.

d) Ammonia in the soil is oxidised by ___*___ bacteria, first to ___*___ and then to ___*___.

e) Some bacteria live in a ______ relationship with leguminous plants and are able to ___*___ nitrogen, converting it into ___*___. This is then used with ___*___ provided by the legume to make ___*___ acids and proteins.

f) Some bacteria can oxidise food using nitrate instead of oxygen, converting it into ___*___ gas. This process is called ___*___ and results in a loss of soil fertility. It occurs especially in ___*___ conditions.

g) In bread making, ___*___ is added to the dough, together with some sugar. This is ___*___ by the ___*___, producing bubbles of ___*___ gas, causing the dough to rise.

h) In the manufacture of wine, the useful product of fermentation is ___*___.

i) In cheese making, bacteria are used to ferment ___*___ to ___*___.

ISBN 9780170191340

Microorganisms: Assessment Practice

QUESTION ONE: USING MICROBES TO MAKE FOOD

Fermentation is a process that enables many organisms to obtain useful energy from organic matter without oxidising it. It is therefore anaerobic. This process is widely used in the manufacture of certain foods.

Name ONE kind of microorganism that obtains energy using fermentation and **discuss** (i.e. *explain*) how this microorganism is used in the manufacture of a **named** food.

In your answer:

- **Name** the **microorganism** and the **food** it is used to manufacture.
- **Explain** how the **activity** of the microorganism helps in the manufacture of the named food.
- **Explain** the conditions necessary for the microorganism to carry out fermentation.
- **Link** the process of **fermentation** to the **survival** of the named microorganism.

(Guidance: Your answer should be approximately 45 lines.)

QUESTION TWO: GROWTH IN BACTERIAL POPULATIONS

A bottle of milk was opened and left open for a week in the kitchen. The graph shows how the number of bacteria changed during that period.

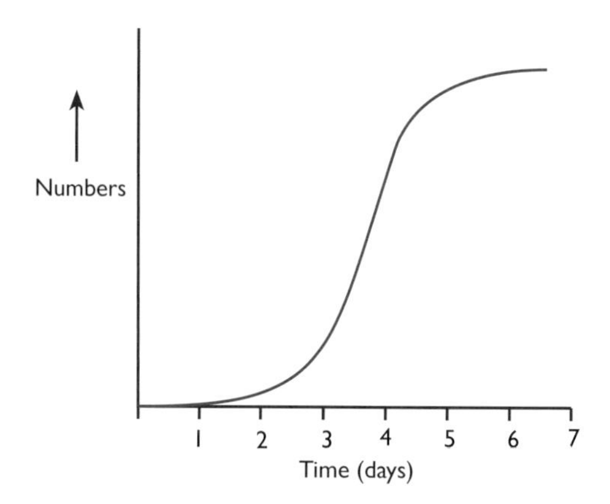

Discuss (i.e. *explain*) why the graph has this particular shape.

In your answer:

- **Explain** how bacteria reproduce.
- **Explain** how the conditions in the milk bottle lead to the growth curve shown in the graph.
- **Explain** how the rate of reproduction of bacteria can be slowed so as to reduce the rate at which the milk becomes unfit for consumption.

(Guidance: Your answer should be approximately 33 lines.)

ISBN 9780170191340

QUESTION THREE: ANTIBIOTICS

Bill felt unwell and went to the doctor, who said it was probably influenza, a viral disease for which there was no treatment. The doctor told Bill to go to bed until he felt better.

Discuss (i.e. *explain*) the reasons why the doctor did not prescribe an antibiotic.

In your answer:

- **Describe** (i.e. *explain*) what an antibiotic is.
- **Explain** how antibiotics **work**.
- **Explain** the possible **consequences** of the misuse of antibiotics.

(Guidance: Your answer should be approximately 15 lines.)

QUESTION FOUR: USEFUL FUNGI

(a) *Clematis vitalba* ('Old Man's Beard') is an introduced vine that is threatening native forests by smothering trees and denying them light. A number of parasites have been introduced in attempts to control this pest, including a fungus *Phoma clematidina*. Infected leaves show black areas and tend to drop off prematurely.

(i) **Describe** (i.e. *explain*) what is meant by the term 'parasite'.

(Guidance: Your answer should be approximately 3 lines.)

(ii) **Explain** how the fungus may help to reduce the effect of Old Man's Beard on the growth of native trees.

(Guidance: Your answer should be approximately 10 lines.)

(b) Many fungi play an important part in decomposition and the cycling of nutrients.

Discuss (i.e. *explain*) how fungi act as decomposers and how the process of decomposition helps in the cycling of nutrients.

(Guidance: Your answer should be approximately 24 lines.)

ISBN 9780170191340

Part Three

PLANT PROCESSES

8 How plants grow

PLANT CHARACTERISTICS

Plants are very different from animals. Their most fundamental feature is that they make their own organic compounds from CO_2 and water, using sunlight energy, in **photosynthesis**, which can be summarised:

$$\text{light} + \text{carbon dioxide} + \text{water} \xrightarrow{\text{chlorophyll}} \text{sugar} + \text{oxygen}$$

Because plants make their own organic compounds from inorganic carbon in the form of CO_2, they are said to be **autotrophic** (meaning 'self-feeding').

Besides carbon dioxide and water, plants also absorb minerals — simple inorganic mineral ions (charged particles) from the soil. Like carbon dioxide, these are present at very low concentrations; the atmospheric CO_2 concentration is 385 parts per million, and some minerals are present in the soil at concentrations measured in parts per *billion*.

Closely linked with their autotrophic nutrition, plants are built in a very different way from animals:

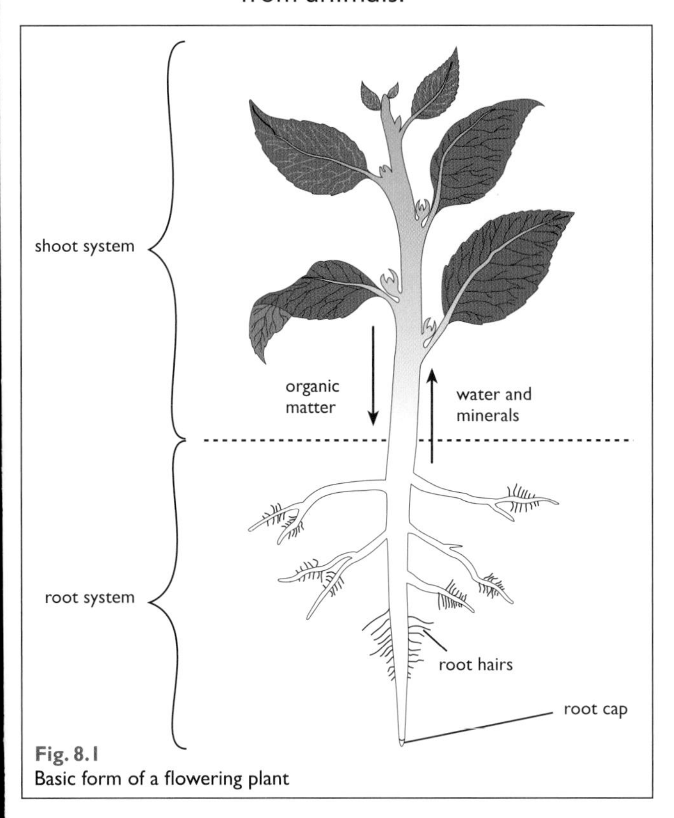

Fig. 8.1
Basic form of a flowering plant

- To absorb raw materials in such dilute form and to absorb energy in the form of sunlight, plants must expose a very *large surface area* to the environment. They achieve this by having a *finely divided body* (Fig. 8.1).
- They are fixed to one spot and move slowly by growth, whereas animals can move rapidly by muscles and most can move from place to place.
- A plant grows throughout its life, whereas animals generally cease growth as adults.
- Plant cells are surrounded by a cell wall consisting largely of **cellulose**.
- Plant cells typically have large internal spaces called **vacuoles**.

The body of a land plant is typically divided into two parts:

1. An above-ground **shoot system**, which absorbs light and carbon dioxide.
2. A below-ground **root system**, which absorbs water and minerals and provides anchorage.

The root and shoot systems are *interdependent*; the shoot imports water and minerals from the root system, and the root system imports organic matter from the shoot system. This necessitates a two-way transport system:

- **Xylem** tissue transports water and minerals from root to shoot.
- **Phloem** tissue transports organic matter from shoot to root.

Plants also differ from animals in several ways at the level of their cells:

- Because plants are surrounded by their raw materials (CO_2, water and minerals), they do not need to be able to move about.

ISBN 9780170191340

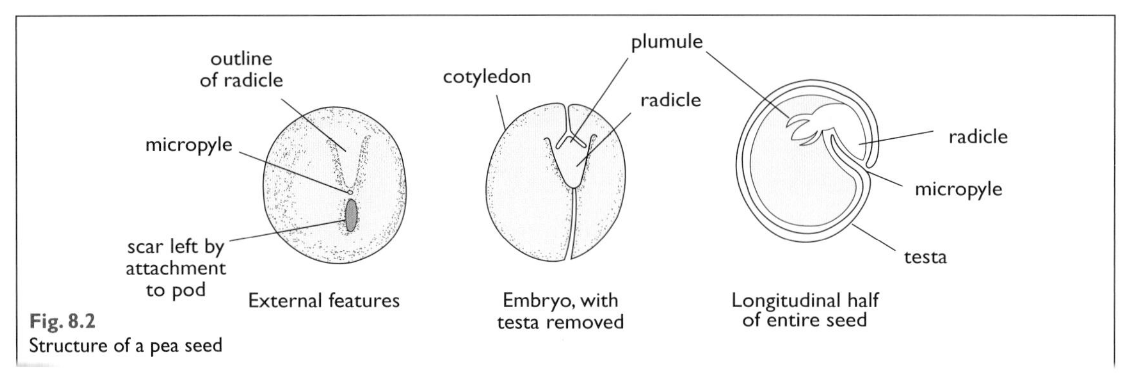

Fig. 8.2
Structure of a pea seed

Growth in a flowering plant

A flowering plant begins life as an **embryo**, protected within a **seed** (Fig. 8.2). The seed is dispersed from the parent and, if it reaches a suitable habitat it *germinates*, growing into a young plant or seedling.

Before the young plant can become a parent, it must first go through a period of *vegetative* (non-reproductive) growth. This provides the energy reserves needed for later *reproductive* growth — the production of *flowers*, *fruits* and *seeds*.

germination → vegetative growth → reproductive growth

The embryo of a plant such as a garden pea is surrounded by a seed coat or **testa** and consists of a tiny shoot or **plumule**, a tiny root or **radicle**, and two embryonic leaves or **cotyledons**. These do not look like 'normal' leaves because they are distended with stored starch and protein. (The garden pea is a *dicotyledon*, because the embryo has two cotyledons; in seeds of *monocotyledons* there is only one cotyledon - see Box 1.)

Box 8.1 Monocots and dicots

Flowering plants fall into two major groups, monocotyledons (e.g. native flax) and dicotyledons (e.g. kowhai). The differences between the two groups are summarised below.

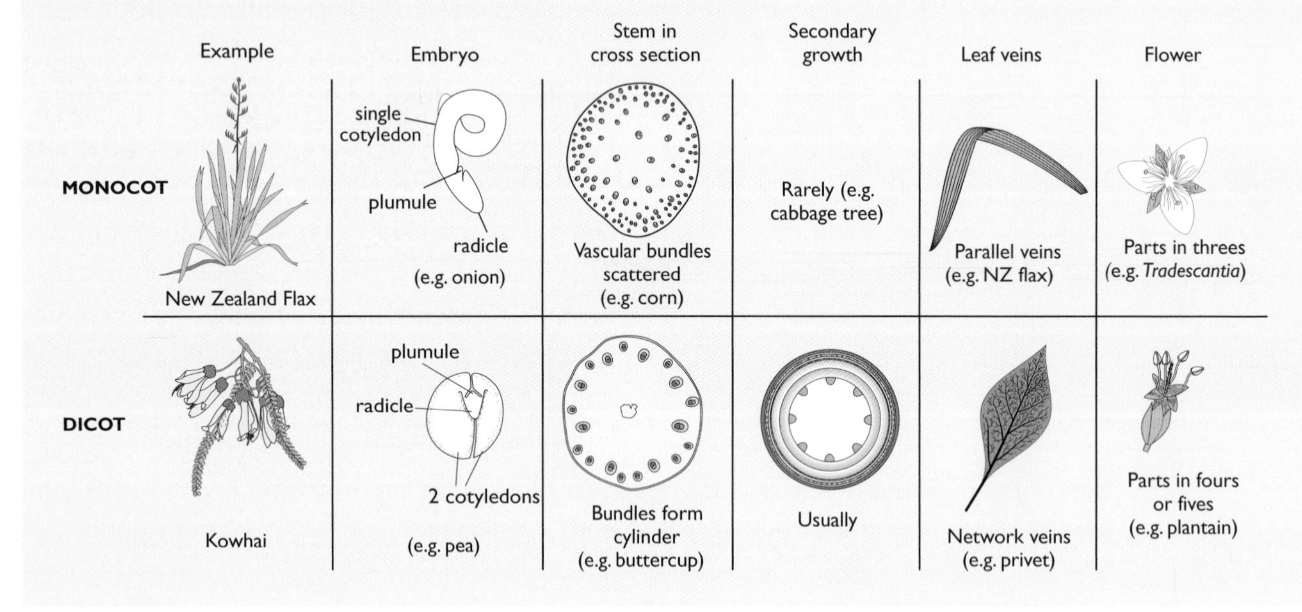

Example
Embryo
Stem in cross section
Secondary growth
Leaf veins
Flower
MONOCOT
New Zealand Flax
single cotyledon
plumule
radicle
(e.g. onion)
Vascular bundles scattered
(e.g. corn)
Rarely (e.g. cabbage tree)
Parallel veins
(e.g. NZ flax)
Parts in threes
(e.g. Tradescantia)
DICOT
Kowhai
plumule
radicle
2 cotyledons
(e.g. pea)
Bundles form cylinder
(e.g. buttercup)
Usually
Network veins
(e.g. privet)
Parts in fours or fives
(e.g. plantain)

Germination

When the seed leaves the parent plant it has a very low water content (about 10%) and shows none of the activities of life (such as respiration). If it lands in a favourable environment it *germinates*, in the following stages (Fig. 8.3):

1. During the first 24 hours or so it absorbs water causing an increase in *live mass*.
2. Enzymes begin to digest the starch and protein stored in the cotyledons.
3. The products of digestion (sugar and amino acids) are *translocated* in the phloem from the cotyledons to the radicle and plumule.

ISBN 9780170191340

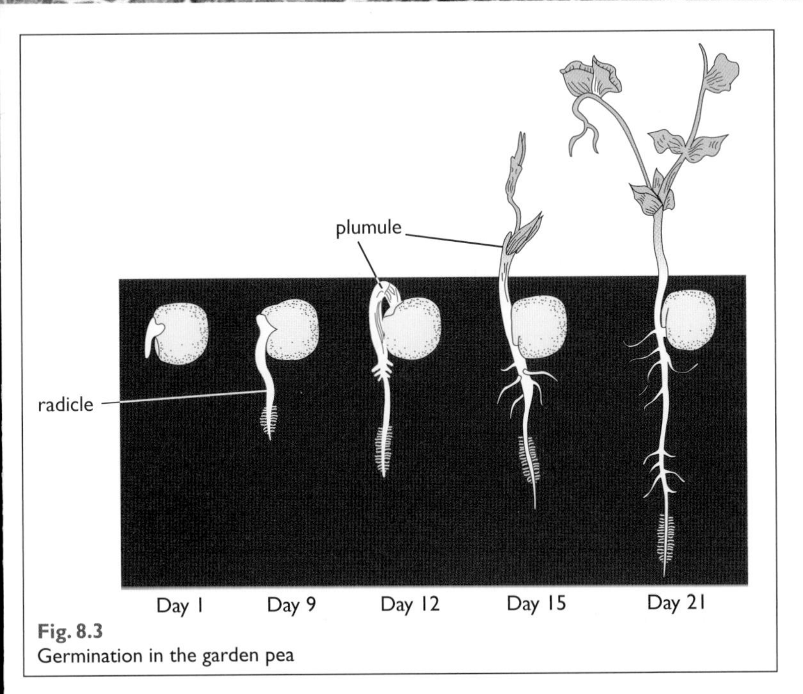

Fig. 8.3
Germination in the garden pea

4. The amino acids and some of the sugar are used as raw materials for growth (e.g. synthesis of proteins and cellulose).
5. Most of the carbohydrate is used in respiration to supply energy for growth.
6. The radicle grows out, rupturing the testa. It grows *downward*, a response called **positive gravitropism**. This enables it to grow towards a more reliable water supply and gain better anchorage.
7. The plumule grows **negatively gravitropically** (*upward*, away from gravity). This is the only way it can grow towards light (which it cannot 'see', since it is in darkness).
8. As the plumule grows upward it is hook-shaped, protecting its tiny leaves.
9. Once the plumule reaches the surface, light stimulates a number of changes:
   - It straightens out.
   - The leaves expand, develop chlorophyll and turn green
   - Rate of elongation of the plumule slows.
10. Once in the light the plumule grows towards it, a response called **positive phototropism**.
11. Until the plumule reaches the light, the seedling has been living on its stored energy reserves and the total quantity of organic matter (its *dry mass*) has been decreasing. Once photosynthesis gets under way, carbohydrate is produced faster than it is used in respiration, so the dry mass of the plant begins to rise (Fig. 8.4).

If a seedling is allowed to germinate in continuous darkness, the plumule continues to grow in length and the leaves remain small and pale yellow. A plant in this condition is said to be **etiolated**. Something like this happens if a seed germinates too deeply in the ground; the plant puts all its resources into elongation of the plumule, for unless it reaches the light before its energy reserves run out, it will die.

Figure 8.5 illustrates the conditions necessary for germination. In most seeds, light is not required but it is needed for photosynthesis once the plumule reaches the surface. This is why very small seeds can only germinate successfully near the surface; they only have sufficient energy reserves to grow upward in darkness for a short distance. The small seeds of some plants (e.g. foxglove) require light to germinate, ensuring they only germinate *very* close to the surface, where some light penetrates.

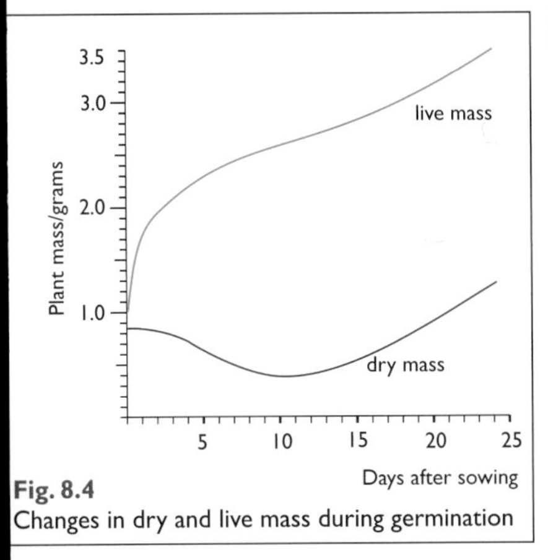

Fig. 8.4
Changes in dry and live mass during germination

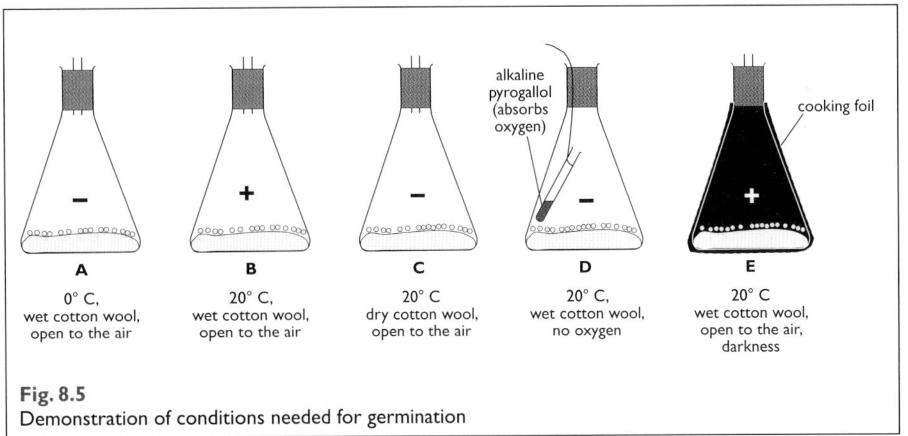

Fig. 8.5
Demonstration of conditions needed for germination

ISBN 9780170191340

To summarise, germination requires the following environmental conditions:

- Water as a solvent for biochemical reactions, and to transport materials from the cotyledons to the growing points.
- Oxygen, for respiration to provide energy.
- A suitable temperature; enzymes work slowly if the temperature is too low, and are inactivated if the temperature is too high.

The tropic responses seen in germination continue to be important throughout the life of the plant. For example, phototropism enables a plant to respond to changes in the direction of incident light that may result from shadowing by neighbouring plants.

Primary and secondary growth

The growth that occurs in a root tip and shoot tip result in increase in *length*. This is called **primary growth** and gives rise to tissues that carry out all the fundamental processes: *protective*, *conducting*, *photosynthetic*, *supporting* and *storage*.

In most dicotyledons primary growth is followed by **secondary growth**, which results in growth in thickness and the addition of extra conducting tissues (secondary xylem and secondary phloem). In trees and shrubs, secondary protective tissues that make up the *bark* are produced.

Box 8.2 Some useful definitions

An **organelle** is a structure within a cell that carries out a particular function. Examples are *mitochondria*, which carry out respiration, and *plastids*, which are of various kinds. *Chloroplasts* carry out photosynthesis, *amyloplasts* store starch, and brightly coloured *chromoplasts* are often used in advertising in flowers and some fruits.

A **tissue** is a group of cells that are specialised to carry out a particular function. In many cases the cells are organised in such a way that they 'work as a team', carrying out a function that none of the individual cells can. An example is the *epidermis* of a leaf.

An **organ** is a group of tissues that work together, carrying out a function that none of the individual tissues can. An example is a *leaf*.

Only some of the tissues in a leaf actually carry out photosynthesis — the other tissues are necessary for the photosynthesising tissue to function.

What do cells *do* during growth?

Growth is not just a matter of getting bigger; it involves three kinds of cell activity:

1. Division — increase in cell *number*.
2. Enlargement — increase in cell *size*.
3. Differentiation — increase in cell *diversity*.

Cell division

In growth, cell division involves *mitosis*, in which the nucleus divides to produce genetically identical daughter nuclei. In flowering plants, cell division occurs in particular regions called **meristems**. In a germinating seed these are at the tips of the radicle and plumule. The cells in a meristem all look more or less alike.

Cell enlargement

When a cell in a meristem has ceased to divide, it begins to enlarge (Fig. 8.6). It does so by taking in water by *osmosis*. As water enters it forms spaces called **vacuoles**, which eventually join up to form a single large vacuole. The enlargement of the cell stretches the wall, but since new cellulose is deposited, its thickness is maintained.

ISBN 9780170191340

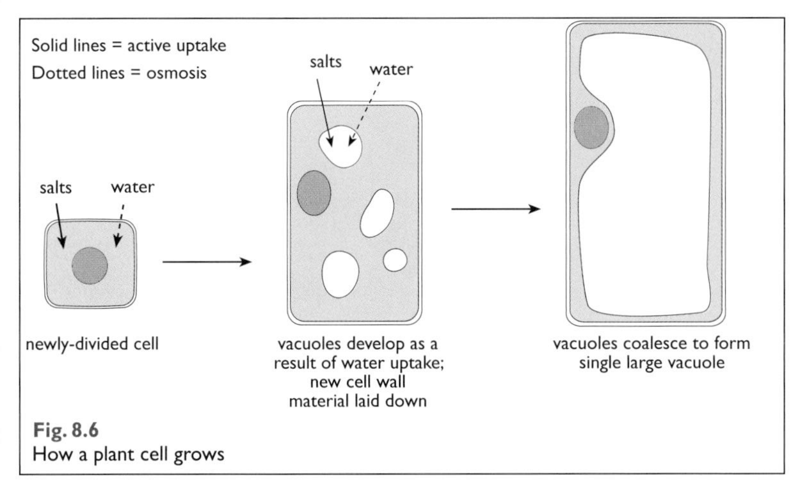

Fig. 8.6
How a plant cell grows

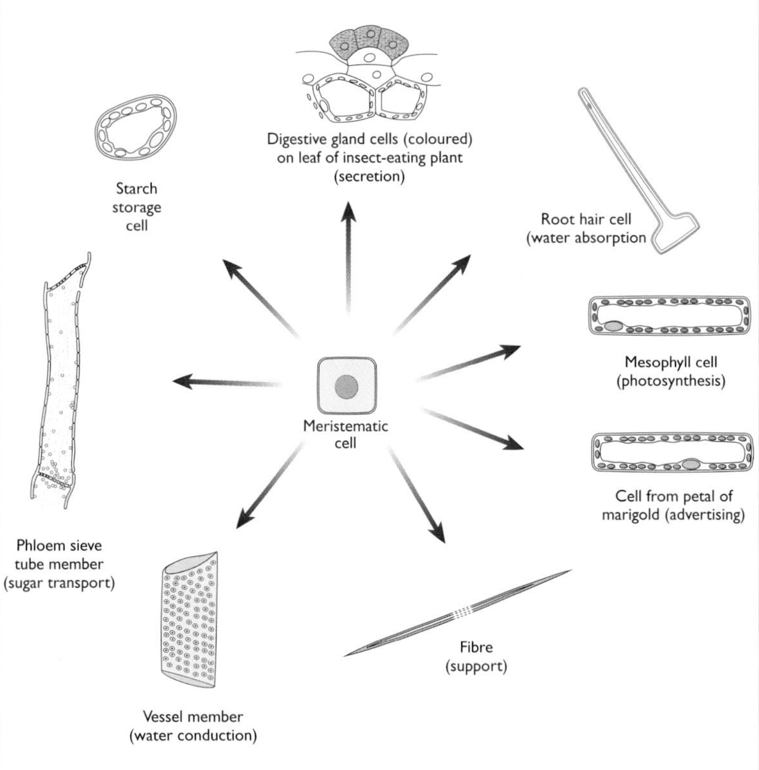

Fig. 8.7
A single meristematic cell has the potential to differentiate into a range of cell types

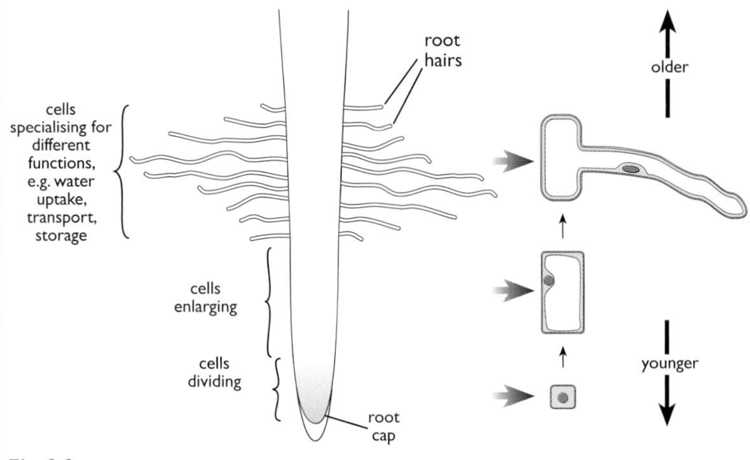

Fig. 8.8
Longitudinal (lengthwise) section through a young root to show the zones of growth

The uptake of water would cause the water concentration inside the cell to rise and eventually osmosis would cease, were it not for the *active uptake* of mineral salts. Because of this, a difference in water potential (see Chapter 9) is maintained, so water uptake continues.

Cell differentiation

Plant cells take on different functions by:

- Becoming different sizes and shapes.
- Producing different enzymes in their cytoplasm.
- Developing different organelles (Box 8.2). Meristematic cells contain *proplastids*, which can develop into *chloroplasts* in photosynthetic cells, *amyloplasts* ('starch grains') in starch storage cells, or *chromoplast*s in the roots of carrots, in the skins of tomato fruits, or in the petals of yellow and orange flowers.

The different kinds of plastids can sometimes change from one type to another. If potatoes are left in the light, they become green as starch-storing amyloplasts develop into chloroplasts, and when a tomato fruit ripens, chloroplasts develop into chromoplasts.

Although all the living cells in a plant are *genetically* identical, they become *phenotypically* different because each cell uses only some of its genes, which are different for each cell type. For example an epidermal cell in a leaf uses the genes for producing a waterproof cuticle, and a root hair cell uses the genes needed for producing a root hair. 'Root hair' genes are present in a leaf cell, but are not used. Figure 8.7 shows some examples of different kinds of plant cell.

Cell division, enlargement and differentiation in plants tend to occur in different regions (though with some overlap). Figure 8.8 shows the situation in a root tip.

As in a root, the youngest part of a shoot is nearest the tip and older parts are further back. The shoot tip is more complicated than a root tip because it produces leaves, the youngest of which arch over and protect the apical meristem from desiccation.

ISBN 9780170191340

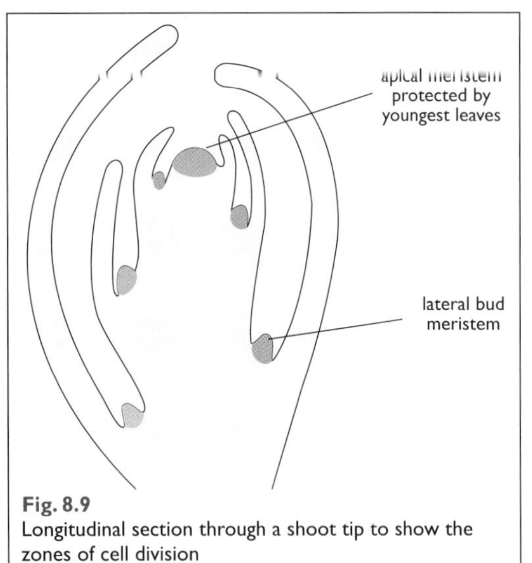

Fig. 8.9
Longitudinal section through a shoot tip to show the zones of cell division

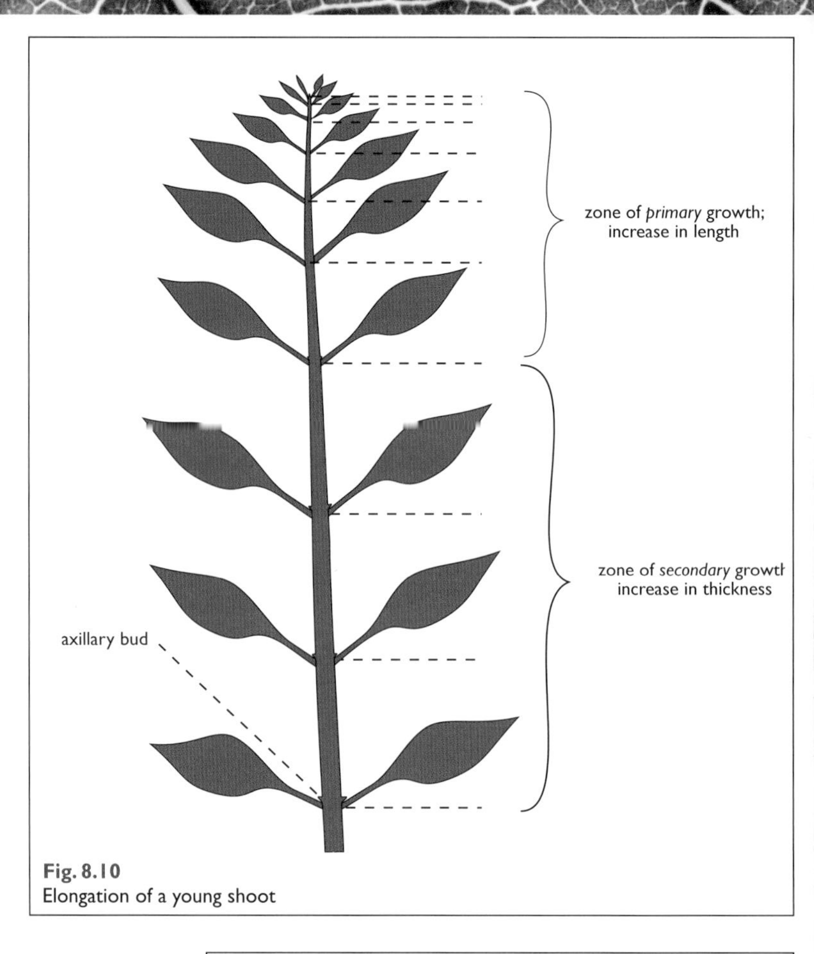

Fig. 8.10
Elongation of a young shoot

Growth in stems

Growth in length can be seen in a leafy shoot because the leaves act as markers (Fig. 8.10).

The point of attachment of a leaf to a stem is a *node*, and the region of stem between two nodes is an *internode*. At the base of a leaf is an *axillary bud*, so-called because the angle between the stem and the base of a leaf is the *axil* (Latin: *axilla* = armpit). In the upper (younger) part of the shoot, the internodes are growing further apart as the stem elongates. In the lower part, primary growth has ceased and the stem is getting thicker in secondary growth. The tissue layout in a primary (young) stem is shown in Fig. 8.11.

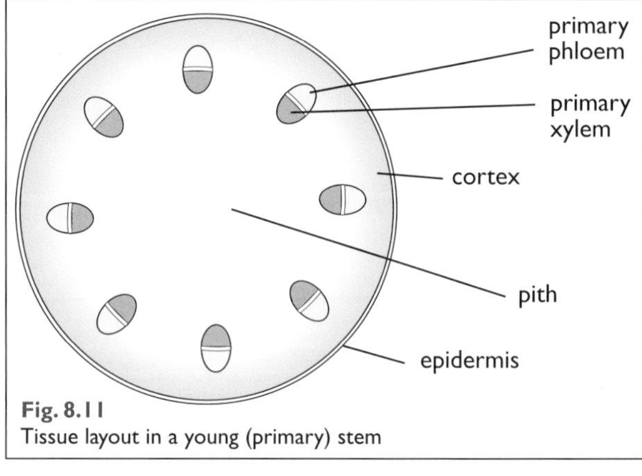

Fig. 8.11
Tissue layout in a young (primary) stem

Secondary growth in stems

As a shoot grows in length, the increasing number of leaves means there is need for extra conducting and supporting tissue. In most dicotyledons and a few monocotyledons these are supplied by **secondary growth** that results in growth in *thickness* (Fig. 8.12).

The meristem responsible for secondary growth is the **cambium**. The cells of the cambium divide to produce **secondary xylem** (wood) on the inside and **secondary phloem** on the outside (Fig. 8.12 and Fig. 8.13).

Each time a cambial cell divides, one of the two daughter cells remains meristematic as a cambial cell, and the other begins to differentiate into either a xylem cell (towards the inside) or a phloem cell (towards the outside). The result is that a ring of secondary phloem is produced outside the cambium and a ring of secondary xylem is formed on the inside.

In trees and shrubs the cambium continues to produce secondary xylem and secondary phloem year after year (Fig. 8.14). Whereas the secondary xylem accumulates year after year, the secondary phloem is stretched and crushed, so only the current year's phloem is functional.

ISBN 9780170191340

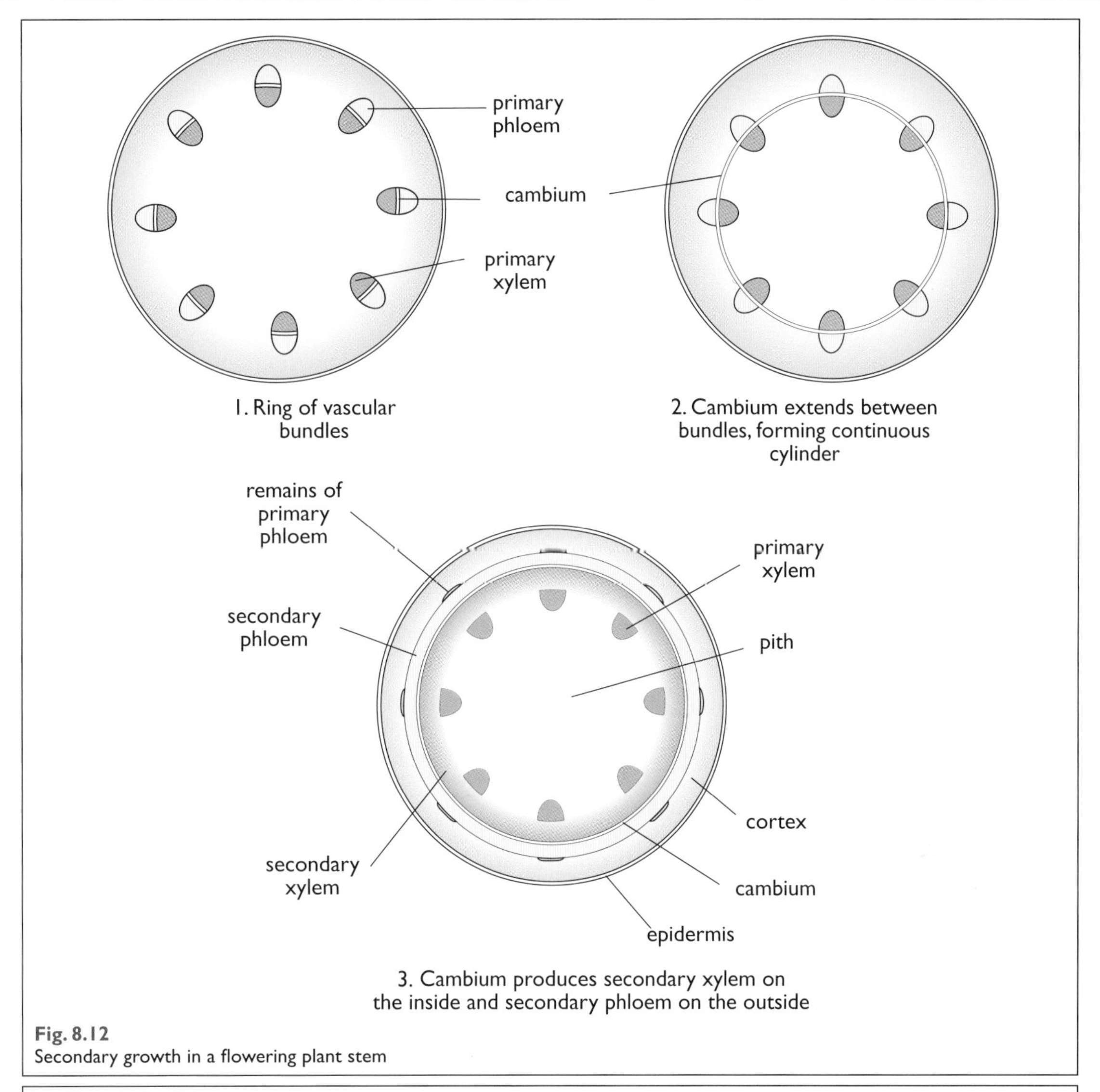

Fig. 8.12
Secondary growth in a flowering plant stem

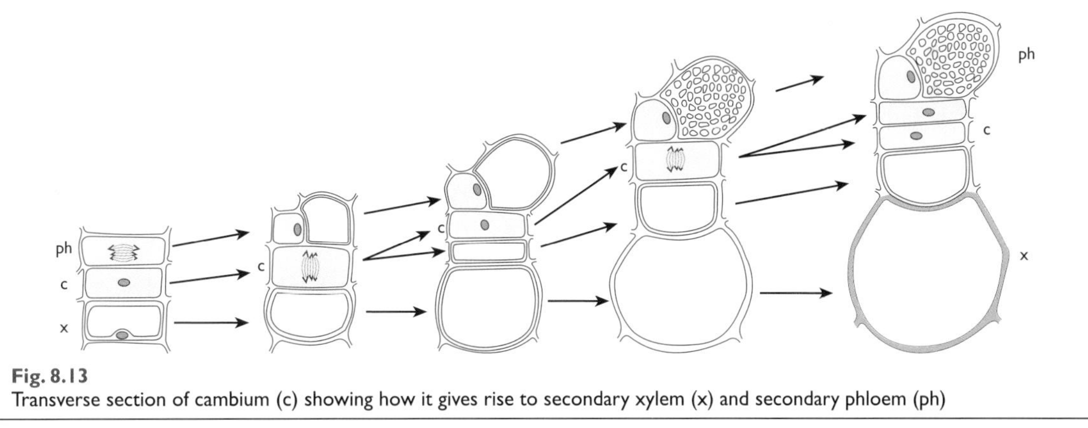

Fig. 8.13
Transverse section of cambium (c) showing how it gives rise to secondary xylem (x) and secondary phloem (ph)

Trees and shrubs are *woody* plants because secondary growth in the stems continues for more than one year. In herbaceous plants there is never more than one year's secondary growth. In herbaceous plants that live from year to year, the aerial stems die back at the end of each year.

In woody plants, secondary growth stretches the epidermis and cortex and they die in the first or second year. The role of protection is taken over by **bark**. This consists largely of **cork**, which is produced by the **cork cambium**. Though cork is dead and therefore cannot grow, more cork is continually being produced by the cork cambium.

Although cork is waterproof and impermeable to oxygen and CO_2, gas exchange occurs via **lenticels**. These are small patches of powdery cells with air spaces between, through which gases can diffuse. Lenticels are produced by the cork cambium and are often clearly visible in smooth bark. They can also be seen on the surface of apples and potatoes, where they appear as tiny raised dots.

ISBN 9780170191340

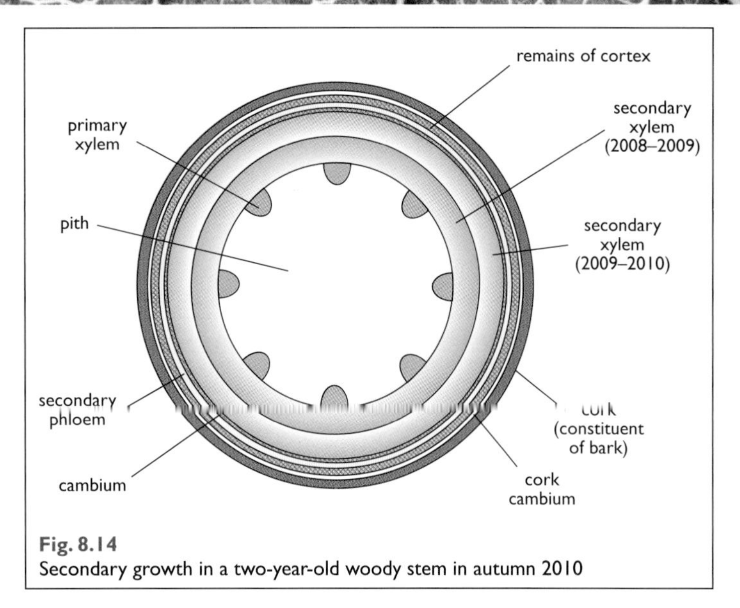

Fig. 8.14
Secondary growth in a two-year-old woody stem in autumn 2010

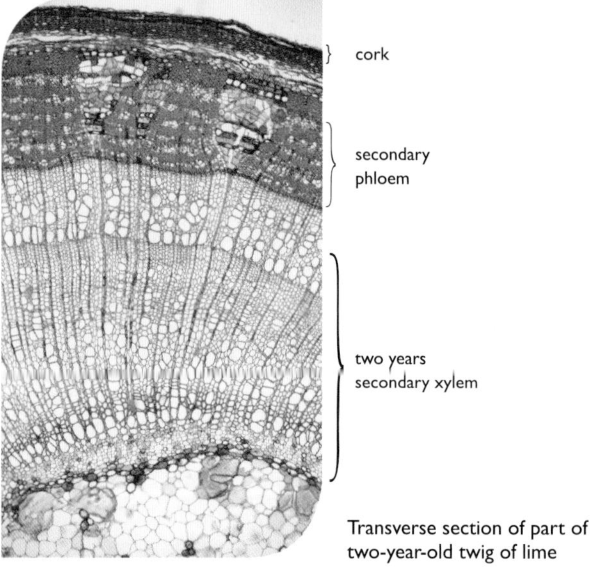

Transverse section of part of two-year-old twig of lime

The xylem produced in spring has wider vessels than the wood produced in late summer, so there is a sharp boundary between the late summer wood of one year and the spring wood of the next. Each year's growth of secondary xylem forms an **annual ring**, so by counting the rings in a tree trunk, you can tell its approximate age (Fig. 8.15).

VEGETATIVE PROPAGATION IN FLOWERING PLANTS

Many plants reproduce by **vegetative propagation**. This is a kind of **asexual reproduction**, in which the offspring are genetically identical to the parents. It is really nothing more than a kind of growth, in which a bud grows out into a shoot, develops roots of its own, and becomes independent. The following are examples of vegetative reproduction:

- Strawberry. In late summer, after fruiting, a strawberry plant produces long, thin lateral shoots called **runners** (Fig. 8.16). Each runner produces tiny *scale leaves* at the first node. At the second node *adventitious roots* and foliage leaves are produced. Eventually the runner rots and the daughter plant is independent.
- Potato. In summer a potato plant produces side shoots that grow *downward* into the soil. The ends become swollen with starch and protein, forming **stem tubers** (Fig. 8.17). You can tell each tuber is a stem because it has tiny buds in the axils of scale leaves. If you were to keep it in the ground, each bud would grow into a shoot that becomes a new potato plant. Because the tuber stores enough energy to survive winter, it also has the function of **perennation**, or survival year after year.
- **Bulbs**, e.g. daffodil (Fig. 8.18). The scales of a bulb are the bases of leaves that have become swollen with stored carbohydrate. The stem is a tiny structure at the base.

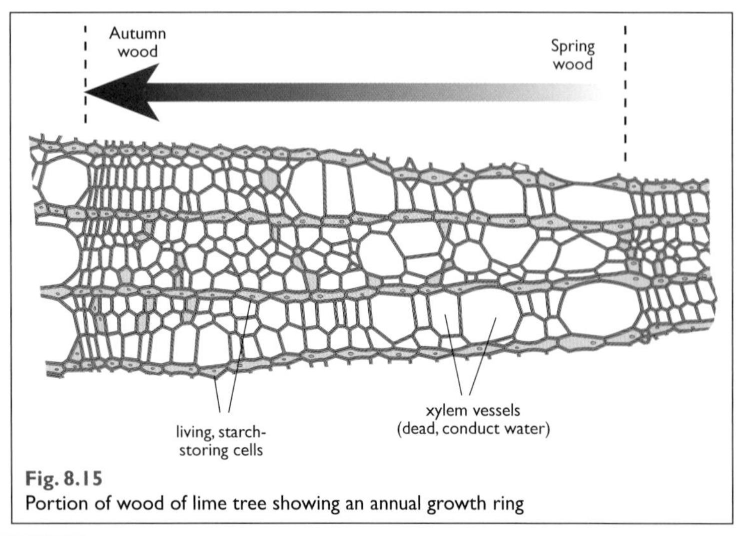

Fig. 8.15
Portion of wood of lime tree showing an annual growth ring

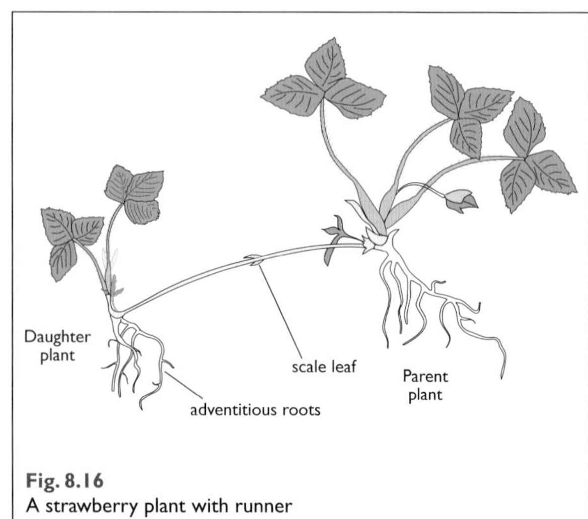

Fig. 8.16
A strawberry plant with runner

ISBN 9780170191340

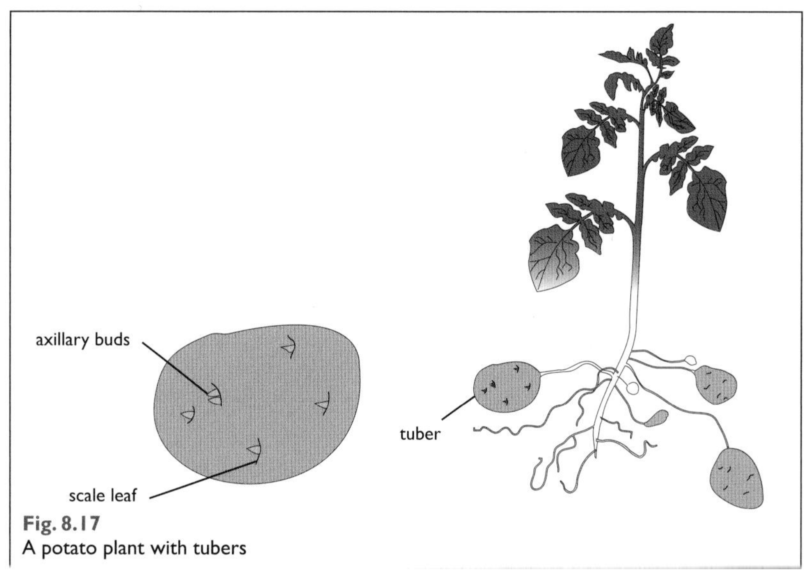

Fig. 8.17
A potato plant with tubers

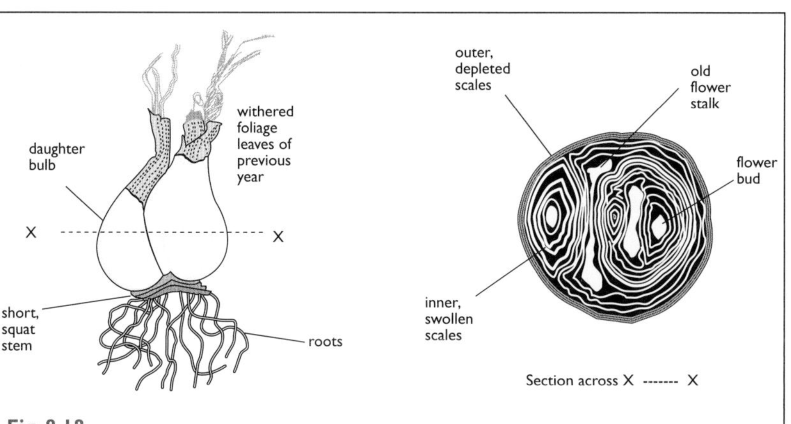

Fig. 8.18
A daffodil bulb with outer scales removed (left) and cut transversely (right)

Stored energy is used not only to survive the winter, but for the growth of one or more buds in spring. As the white, fleshy scales become depleted of energy reserves they become the brown, outer scales. Each bulb usually has several buds, each one becoming a new bulb.

- **Corms**, e.g. crocus (Fig. 8.19). Whereas a bulb stores energy in the bases of foliage leaves, a corm stores it in a squat, vertically growing stem. A corm has several buds, each of which sends up aerial shoots and stores energy in its stem, becoming a new corm.
- **Rhizomes**, e.g. iris (Fig. 8.20). A rhizome is a *horizontally* growing underground stem. In an iris the internodes are very short and growth in length is slow, but in couch grass (the gardener's curse), the internodes are long and the plant spreads rapidly.

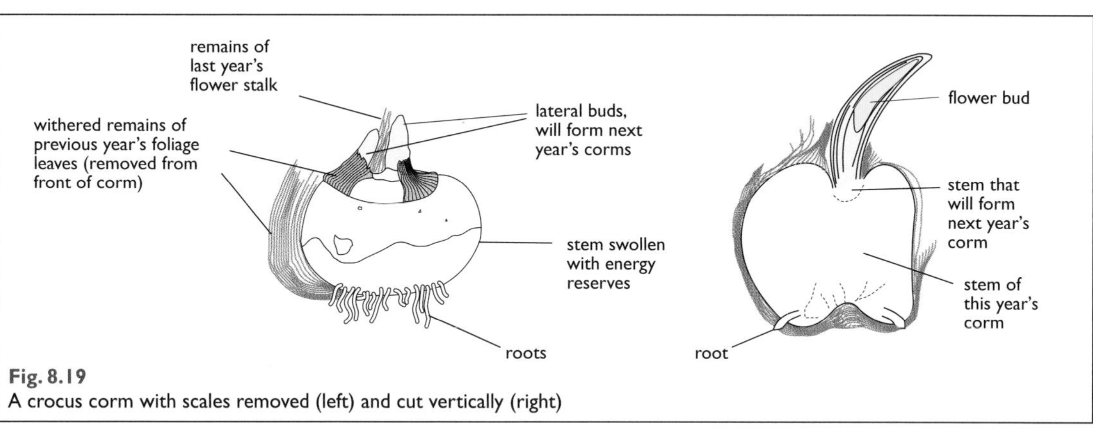

Fig. 8.19
A crocus corm with scales removed (left) and cut vertically (right)

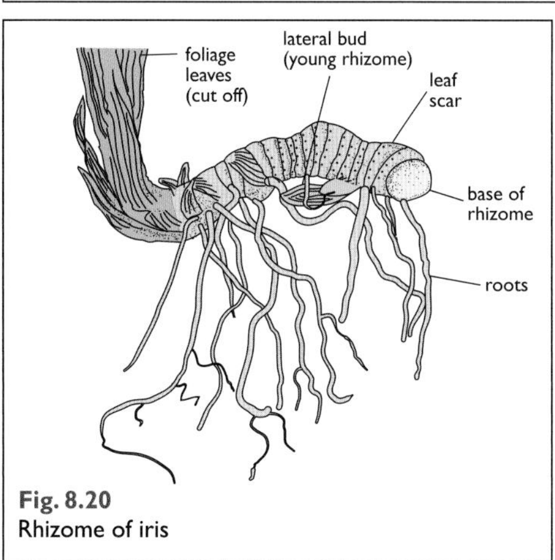

Fig. 8.20
Rhizome of iris

The most important thing about vegetative propagation is that the daughter plants are *genetically identical* to the parent. This is because the cell divisions involved in growth are mitotic (Chapter 1). Vegetative propagation is less wasteful than sexual reproduction for two reasons:

1. Because the daughter plants are initially part of the parent, they obtain energy and nutrients from it until they are independent.
2. Since the parent has proved its fitness by becoming a parent, it has a successful genotype. Because the daughter plants are very close to the parent, they are likely to be growing in similar soil conditions, so are likely to be successful too.

Vegetative propagation is also very useful in horticulture, because once a variety with desirable qualities has been produced (by sexual

ISBN 9780170191340

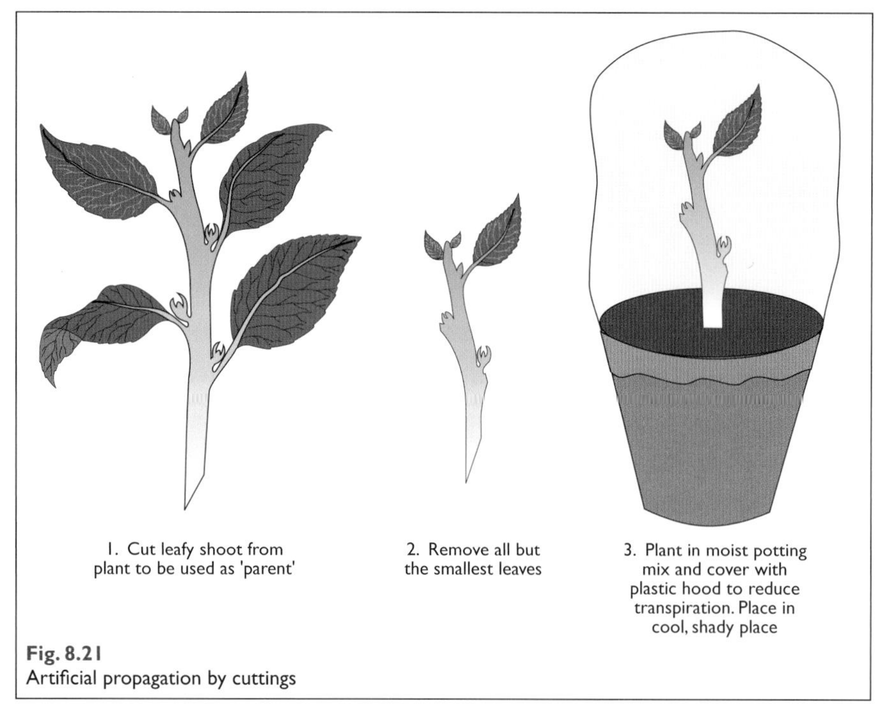

1. Cut leafy shoot from plant to be used as 'parent'
2. Remove all but the smallest leaves
3. Plant in moist potting mix and cover with plastic hood to reduce transpiration. Place in cool, shady place

Fig. 8.21
Artificial propagation by cuttings

reproduction), it can be mass-produced vegetatively. All the plants of a given variety are thus fragments of a single original plant.

As long as the daughter plants experience the same range of environmental conditions as the parent, vegetative propagation is superior to sexual reproduction. But over time, climate and other environmental conditions change. A genotype that has been successful may become less 'fit' in new conditions. This is the advantage of sex; though wasteful, it provides insurance against change because there is a chance that at least some of the offspring will be adapted to new conditions.

Artificial vegetative propagation

Natural vegetative propagation can be rather slow for commercial purposes, and nowadays a variety of artificial methods are used.

Cuttings

Many plants can be propagated by taking a leafy shoot, removing the older leaves, placing in potting mix and covering with a transparent plastic hood (Fig. 8.21). If successful, roots grow from the cut surface of the stem. In some plants root development is helped by first dipping the cut shoot in hormone rooting powder.

Grafting

In this process a shoot of one variety (called the **scion**) is grafted onto another plant (the **stock**). In this way desirable qualities of two plants may be combined. For example, a grape vine that produces desirable fruit may be grafted onto a rootstock of a variety that is resistant to a particular disease.

The essential requirement is that the two varieties are sufficiently closely related (normally of the same species) to be compatible. Grafting depends on the fact that if the cambium of stock and scion are brought into intimate contact, they knit together. Figure 8.22 shows the basic method.

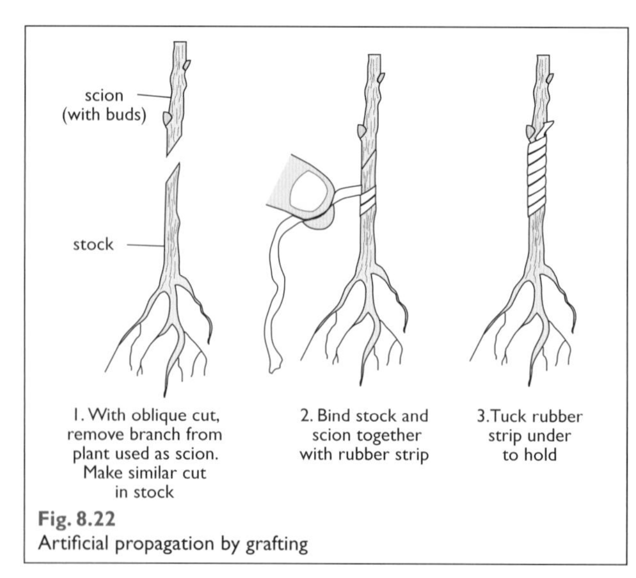

scion (with buds)

3.Tuck rubber strip under to hold

Fig. 8.22
Artificial propagation by grafting

Bud grafting (Fig. 8.23) is a variation in which a bud is used as the scion. First, a bud is cut from a suitable plant in such a way as to include bark and cambium beneath.

ISBN 9780170191340

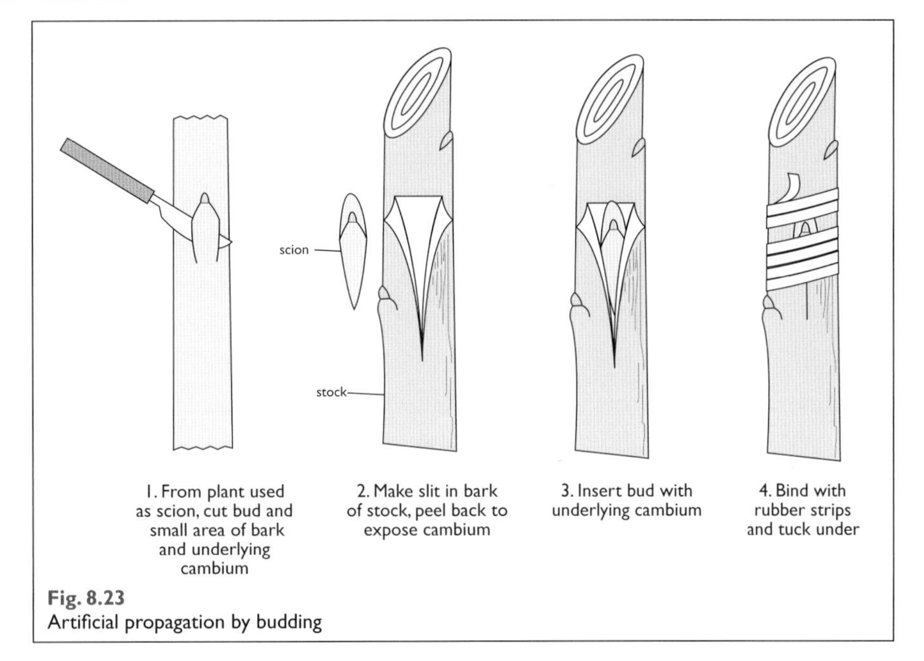

scion
stock
1. From plant used as scion, cut bud and small area of bark and underlying cambium
2. Make slit in bark of stock, peel back to expose cambium
3. Insert bud with underlying cambium
4. Bind with rubber strips and tuck under

Fig. 8.23
Artificial propagation by budding

A 'T'-shaped cut is made in the bark of the stock, and the bark lifted from the cambium just enough to permit insertion of the bud. The graft is then bound with rubber tape to protect against desiccation.

Another method, which produces many more plants, is shown in Fig. 8.24. Cells at the surface of a disc cut from a leaf can be stimulated by hormones to divide, producing a mass of rapidly dividing cells called a **callus**. Further treatment with other hormones induces the callus to develop roots and shoots, and the young plants are eventually planted out.

To produce a new variety, sexual reproduction using flowers and the seeds is used. By carefully selecting the parents, desirable characteristics may (with luck) be produced in some of the offspring. Once a new variety has been produced, it is multiplied by vegetative propagation. A potato seed, for example, is the product of sexual reproduction via flowers, and a 'seed potato' is a tuber that you plant to obtain more of the same variety.

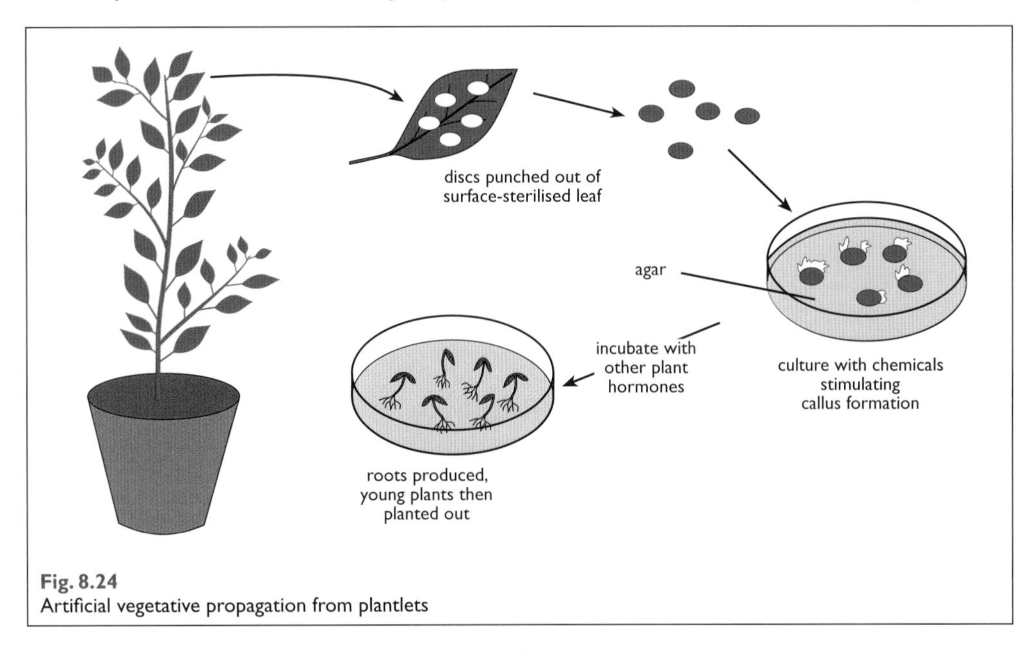

Fig. 8.24
Artificial vegetative propagation from plantlets

ISBN 9780170191340

Summary of key points in this chapter

- Unlike animals, plants use light energy to produce their own organic compounds from CO_2 and water, in photosynthesis.
- Related to this, plants have a *finely divided body*, *cellulose cell walls* and (in some of the cells) *chloroplasts*.
- The flowering plant is typically divided into two parts, an above-ground *shoot system* and a below-ground *root system*.
- The shoot system is adapted for absorbing light and CO_2, and the root system is adapted for absorbing water and minerals and for anchorage.
- Growth of a flowering plant begins with a period of *vegetative* growth, in which energy reserves are produced. Later comes *reproductive* growth, in which flowers, seeds and fruits are produced.
- A flowering plant begins its life with the germination of a *seed*. This is a very young plant (embryo) surrounded by a protective *testa* (seed coat).
- Conditions for germination include *moisture*, *oxygen*, and a *suitable temperature*.
- Germination begins with the uptake of water, followed by the digestion of stored energy (starch in the pea, fat in sunflower), and then the emergence of the *radicle* (young root), and then the *plumule* (young shoot).
- The *dry mass* of the seedling decreases during germination, until the plumule reaches the light and photosynthesis begins.
- During germination the radicle grows downward (*positive gravitropism*) and plumule grows upward (*negative gravitropism*) and later, towards light (*positive phototropism*).
- Light stimulates a number of changes in the growth of the plumule: chlorophyll development, leaf expansion, and slowing down of its rate of elongation.
- Growth involves increase in cell *number*, cell *size*, and cell *diversity*.
- Cell division occurs in regions called *meristems*, e.g. at the root and shoot tips.
- In growth, cells divide *mitotically* (Chapter 1), as a result of which the daughter cells are genetically identical.
- Before a cell can specialise for a particular function, it must first *enlarge*. This occurs mainly by the uptake of minerals and water, plus the enlargement of the cell wall; the volume of the nucleus and cytoplasm remain about the same.
- When a cell *differentiates* (specialises for a particular function), certain genes are activated, and others are 'switched off'.
- The first phase of growth of a shoot or root is growth in length, or *primary* growth. In many plants this is followed by growth in thickness, or *secondary* growth.
- Secondary growth results from division of cells in a meristem called the *cambium*, and gives rise to *secondary xylem* and *secondary phloem*.
- In trees and shrubs, successive years of secondary xylem appear in cross-section as *growth rings*, which enable the age of the tree to be estimated.
- Many herbaceous plants reproduce *vegetatively*, in which the young plants are genetically identical to the parent.
- Compared with sexual reproduction, asexual reproduction has advantages and disadvantages (see p. 72 for details).

ISBN 9780170191340

Test your basics

Copy and complete the following sentences.

a) A terrestrial (land-living) flowering plant is divided into an above-ground ___*___ system and a below-ground ___*___ system.

b) Water and minerals are transported from roots to leaves in the ___*___ tissue, and sugar is transported in the ___*___ tissue.

c) A plant embryo consists of a young root or ___*___, a young shoot or ___*___, and one or two embryo leaves or ___*___.

d) To germinate, a seed needs ___*___ as a solvent in which chemical processes occur, ___*___ for respiration, and a suitable ___*___ for the enzymes that catalyse reactions. In addition, some very small seeds also need ___*___.

e) In growth, cells increase in number by ___*___ divisions. In flowering plants these occur in places called ___*___, for example the ___*___ of roots and shoots.

f) After a cell has been produced, sooner or later it begins to increase in size by developing a large fluid-filled cavity or ___*___. This involves the active uptake of ___*___, accompanied by the uptake of water by ___*___.

g) The final stage in the life of a plant cell is ___*___, in which it begins to ___*___ for a particular function.

h) A structurally distinct part of a cell that is specialised for a particular function is called an ___*___, for example the nucleus. A group of cells working together to carry out a particular function is called a ___*___.

i) A group of tissues that work together to carry out a function that none of the individual tissues can, is called an ___*___.

j) Growth in length of a stem or root is called ___*___ growth. In many plants this is followed by increase in thickness, or ___*___ growth, in which additional xylem and phloem are produced by the activity of a meristem called the ___*___. In trees and shrubs, additional protective tissue is produced in the form of ___*___.

k) Vegetative propagation is a kind of ___*___ reproduction, in which a young shoot develops roots and becomes independent but is genetically ___*___ to the parent.

l) Plants that survive from year to year are called ___*___.

m) In grafting, a shoot (called a ___*___) is grafted onto another plant, called the ___*___.

ISBN 9780170191340

9 How plants feed: Obtaining raw materials and energy

HOW PLANTS OBTAIN THEIR ENERGY: PHOTOSYNTHESIS

Plants are independent of other living organisms for energy (though not for raw materials), because they can harness the energy of the sun in photosynthesis. Water is obtained from the soil; light and carbon dioxide are absorbed from the air. The overall process can be summarised by the following simple equation, in which CH_2O represents sugar (a carbohydrate):

$$\text{light} + CO_2 + H_2O \xrightarrow{\text{chlorophyll}} CH_2O + O_2$$

Investigating the conditions needed for photosynthesis

The need for light, CO_2 and chlorophyll can easily be investigated, as the experiments describe below. Although the organic product of photosynthesis is a sugar, it is rapidly converted into *starch*, and this is used in testing to see if photosynthesis has occurred.

Testing a leaf for starch

Starch is easy to detect because of the intense blue given by the reaction with iodine dissolved in potassium iodide solution. Since this test involves a colour change, the greenness of the chlorophyll must first be removed by boiling the leaf in methylated spirits.

1. The leaf is boiled in methylated spirits using an electrically heated water bath (Fig. 9.1). Dipping the leaf in boiling water for about a minute before boiling in alcohol speeds up the action of the alcohol, though this is not, as frequently stated, 'to kill the leaf' — boiling alcohol does this very effectively.
2. The leaf is placed in a Petri dish and iodine solution added. If starch is present the leaf turns blue-black.

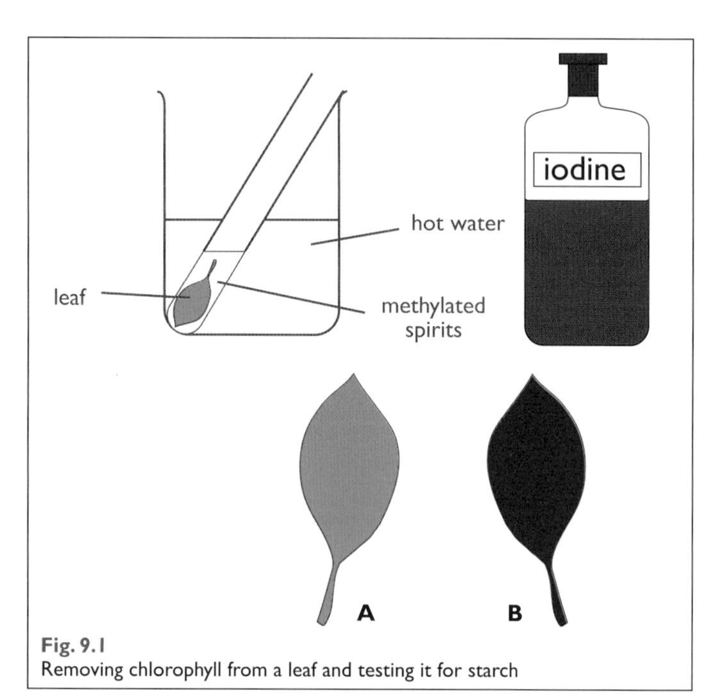

Fig. 9.1
Removing chlorophyll from a leaf and testing it for starch

The need for light

The need for light can be investigated using the apparatus shown in Fig. 9.2. If possible, the stencil should be made of metal foil (which reflects the sun's heat) and should have a thin sheet of transparent plastic glued over it, thus ensuring that the part of the leaf under the 'window' of the stencil has the same (slightly reduced) access to air as the part of the leaf that is in darkness. The stencil is placed on a leaf of a potted plant on a window ledge or other sunny situation. After about two days it is removed and the leaf tested for starch.

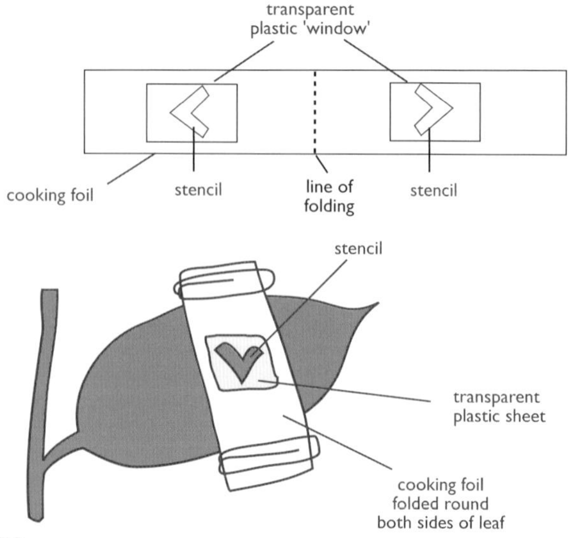

Important: It is essential that the leaf is not 'destarched' by leaving the plant in darkness for 24 hours before attaching the stencil. This would be *assuming the result of the experiment*, which would be *bad science*. If it is done as a *demonstration* rather than an experiment, then prior de-starching is fine.

transparent plastic 'window'
cooking foil
stencil
line of folding
stencil
stencil
transparent plastic sheet
cooking foil folded round both sides of leaf

Fig. 9.2
Apparatus for investigating the need for light in photosynthesis

ISBN 9780170191340

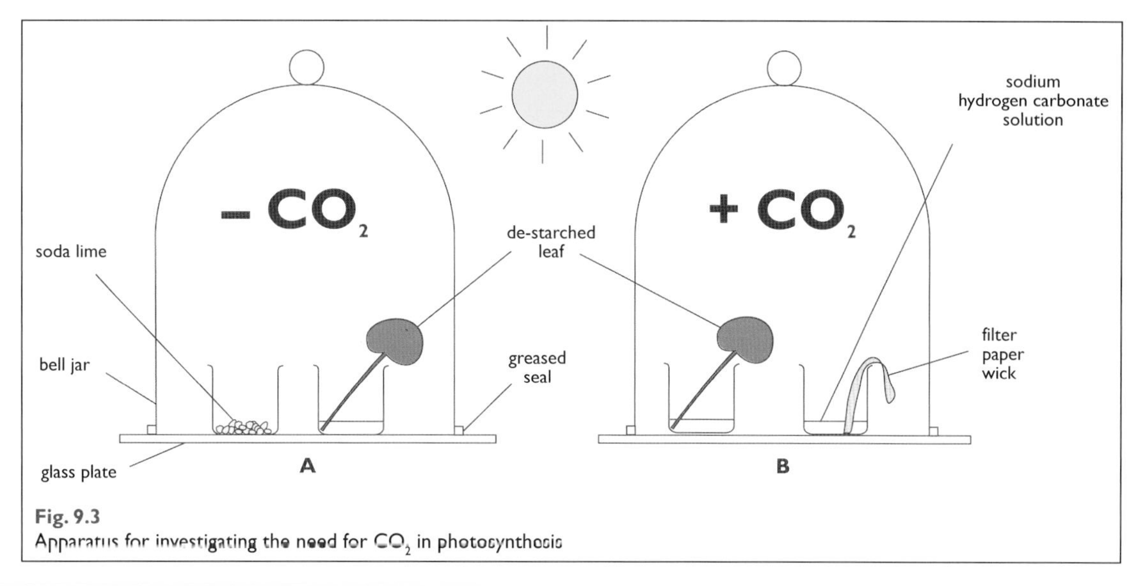

Fig. 9.3
Apparatus for investigating the need for CO_2 in photosynthesis

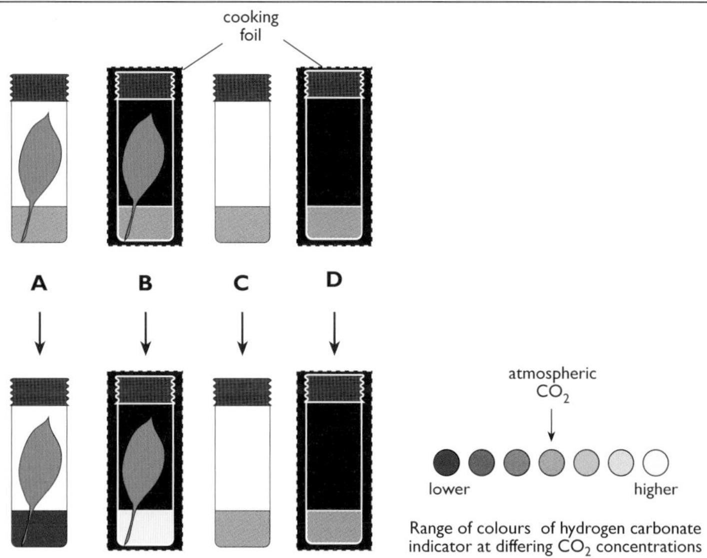

Fig. 9.4
Apparatus for finding out whether CO_2 is absorbed in photosynthesis. McCartney bottles have screw tops and so the interior is sealed from the atmosphere

The need for carbon dioxide

In this case the plant *must* first be de-starched by keeping it in the dark for about 48 hours (and in so doing, assuming the need for light). Next, remove two leaves and set up the apparatus shown in Fig. 9.3.

In A, the beaker contains sodalime to remove CO_2 from the air. Unfortunately, putting a leaf in a bell jar changes the surroundings of the leaf in other ways: the air inside the jar becomes more humid, and in bright sunshine, a lot warmer. It is therefore necessary to set up a *control* (B), identical in every way except for the factor being investigated. The concentrated sodium hydrogen carbonate solution gives off CO_2, replacing any that might be absorbed by the leaf (without this source of CO_2, the amount of CO_2 in the flask would be quite inadequate to produce enough starch for a positive result). After leaving both leaves in bright light for several hours they are tested for starch as described on the previous page.

The results of the above experiment show that CO_2 must be present for photosynthesis to occur, but they do not show whether CO_2 is actually *used up*. This can be put to the test in an experiment using hydrogen carbonate indicator, shown in Fig. 9.4.

When hydrogen carbonate indicator is in contact with normal atmospheric air (i.e. 0.038% CO_2) it is rose pink. If the level of CO_2 rises it turns yellow, and if it falls significantly it turns purple. In the following experiment, four McCartney bottles are set up as shown.

- Tube A subjects the leaf to the presence of light.
- Tube B is to ensure that any colour change is not due to the leaf alone.
- Tube C is to ensure that any colour change is not due to the light alone.
- Tube D is to ensure that the indicator does not change in colour without the influence of light or a leaf.

Tubes B, C and D are controls. If the leaf absorbs CO_2 it will remove CO_2 from the solution, causing the indicator to change colour. If the leaf produces CO_2 the reverse will happen.

The results are shown in Fig. 9.4. Tube A shows that the level of CO_2 has fallen. This can only be due to light acting on the leaf, as tubes C and D remain unchanged, and Tube B shows an *increase* in CO_2. This is due to *respiration* by the leaf.

 ISBN 9780170191340

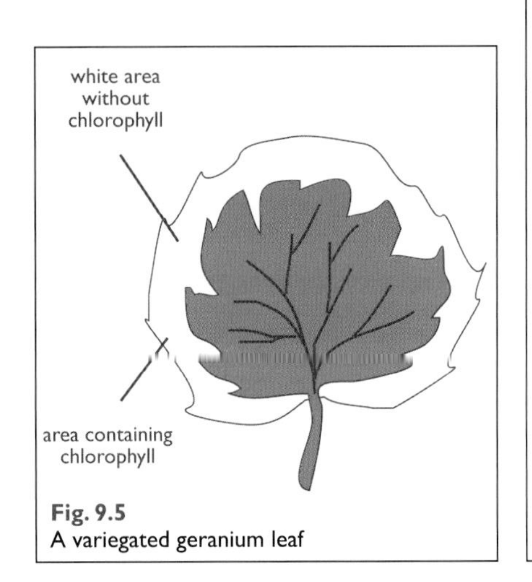

Fig. 9.5
A variegated geranium leaf

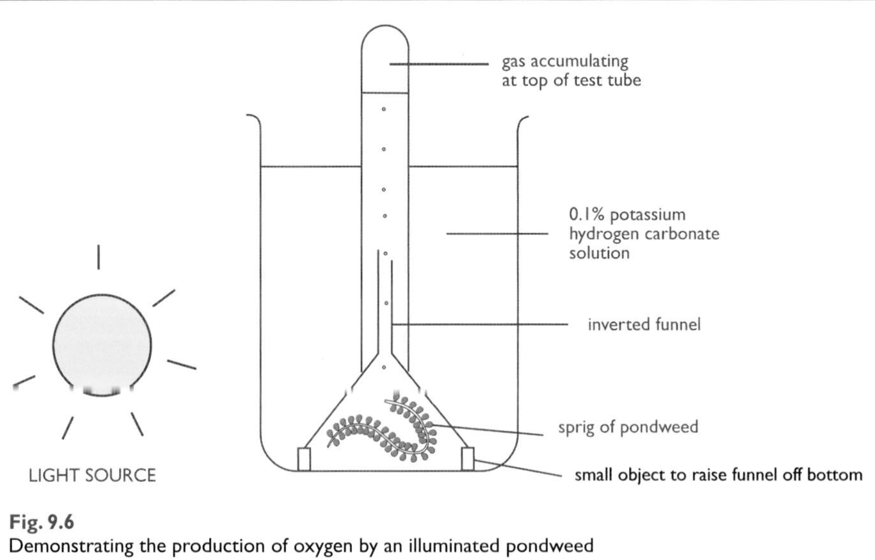

Fig. 9.6
Demonstrating the production of oxygen by an illuminated pondweed

The need for chlorophyll

Chlorophyll cannot be removed from a leaf without killing it. The problem is solved by using *variegated* leaves, in which parts of the leaf are naturally lacking chlorophyll (Fig. 9.5).

A plant with variegated leaves is left in bright light for a few hours (it is pointless to destarch it first, since if chlorophyll is needed, the white parts would not have had chlorophyll before the experiment was set up). Next, a leaf is removed and carefully drawn to show the distribution of chlorophyll, and then tested for starch. The distribution of starch is then compared with the 'map' of the distribution of chlorophyll. If chlorophyll is necessary for photosynthesis, then starch should only be present in the areas that had been green, and this proves to be the case.

The production of oxygen

It is not easy to detect oxygen production by a land plant because the small quantities given off would be difficult to distinguish from the large amount (21%) already present in the air. With pondweed, however, it is a simple matter because oxygen is only slightly soluble in water and during rapid photosynthesis it is given off as bubbles (Fig. 9.6). A trace of potassium hydrogen carbonate added to the water acts as a source of CO_2.

In most pondweeds there are large air spaces extending throughout the plant, so gas produced in the leaves emerges as bubbles from the cut surface of the stem.

Diffusion: A vital process

Plants absorb CO_2 by *diffusion*. You can see diffusion in action by putting a crystal of potassium permanganate in a beaker of *still* water. As the intense purple permanganate dissolves, it very slowly spreads outward and upward (Fig. 9.7).

This process depends on the fact that in any liquid or gas, the particles are moving *randomly* around (Fig. 9.8).

It is important to realise that diffusion is always a *net* movement; particles are moving in all directions, but more are moving away from the area of high concentration than are moving towards it.

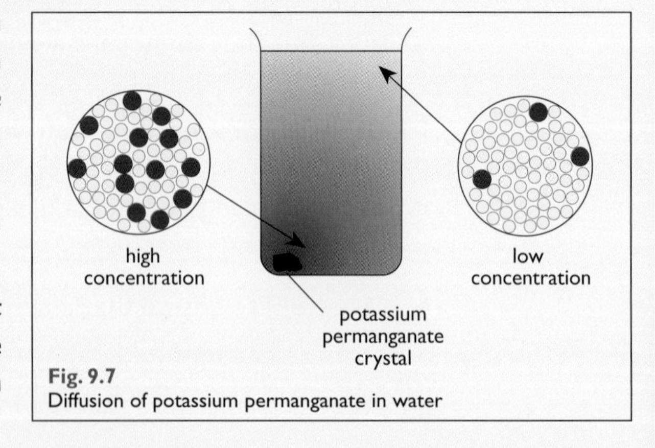

Fig. 9.7
Diffusion of potassium permanganate in water

Just as the rate at which a ball rolls down a hill depends on how steep the hill is, the rate of diffusion depends on the steepness of the *concentration gradient* (Fig. 9.9). A concentration gradient is the rate at which concentration changes with distance. The steeper the gradient, the faster the diffusion.

Because plant cells have large vacuoles, the chloroplasts are near the surface of the cell. This keeps the distance between chloroplasts and air spaces short, and thus the concentration gradient of CO_2 is as steep as possible.

ISBN 9780170191340

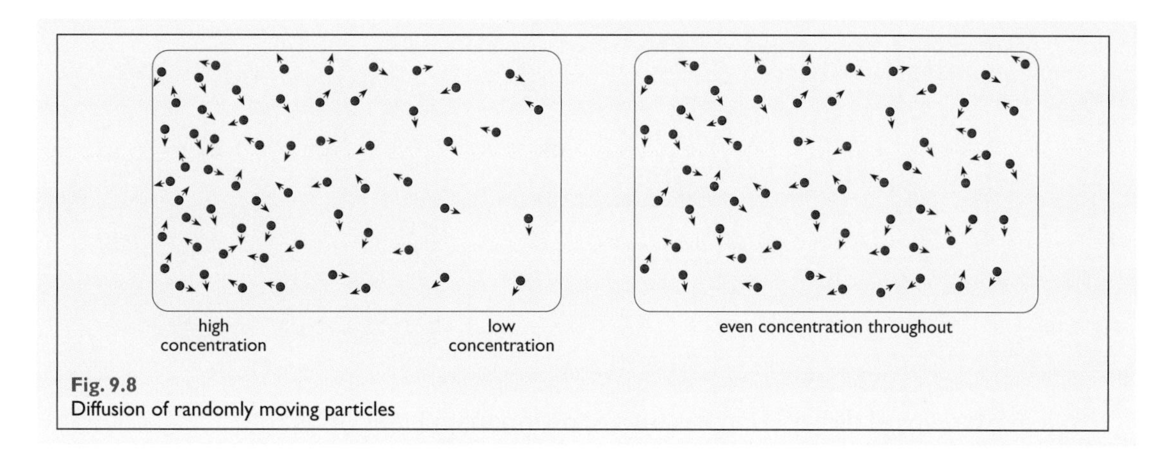

Fig. 9.8
Diffusion of randomly moving particles

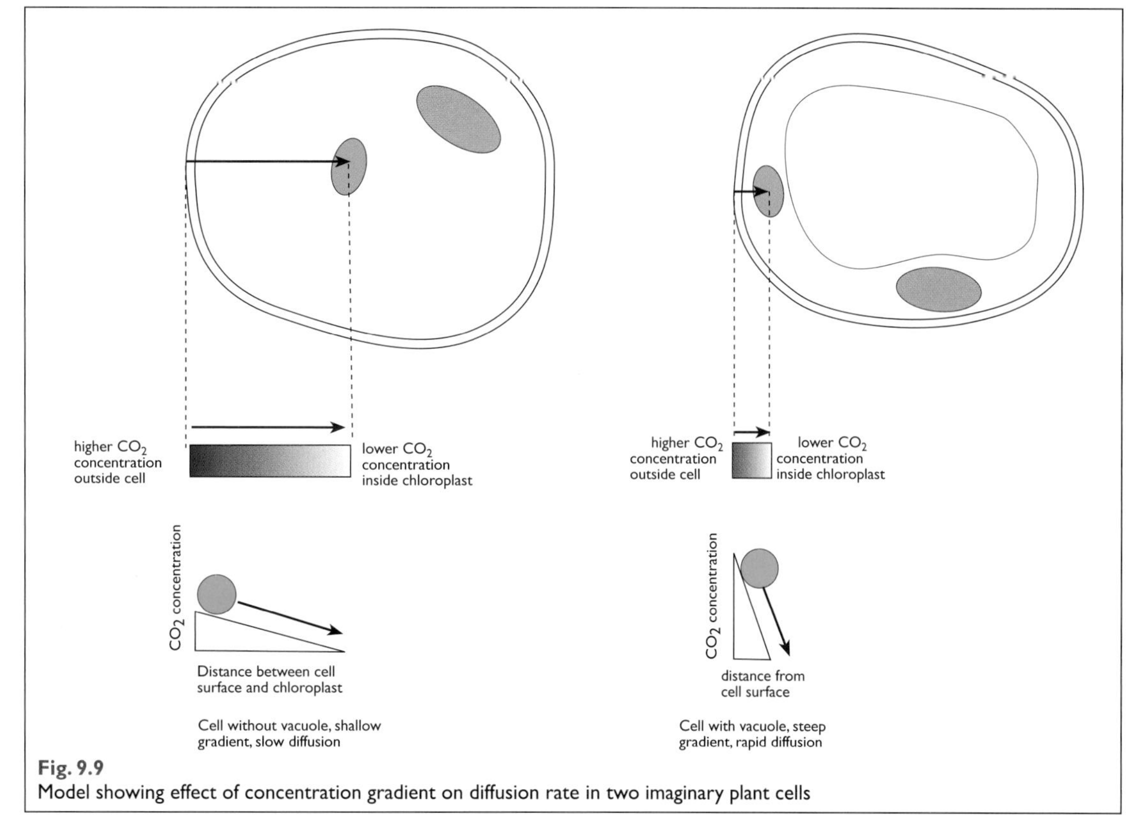

Fig. 9.9
Model showing effect of concentration gradient on diffusion rate in two imaginary plant cells

The leaf as a photosynthetic organ

Most photosynthesis takes place in the *leaves*. To carry out photosynthesis, a leaf must be able to:

- Absorb light.
- Absorb carbon dioxide.
- Restrict and regulate the loss of water.
- Import water.
- Export the carbohydrate product.

A leaf is a highly organised structure. It consists of many different kinds of cell, organised into **tissues**. A tissue is a group of similar cells working together to carry out a particular function, such as protection, water transport, sugar transport, support and photosynthesis.

Of all the tissues in a leaf, only one (palisade mesophyll) is specialised purely for photosynthesis, yet this tissue cannot do its work without the help of the others. When tissues work together to carry out a function that none of the individual tissues can, they are organised into an **organ** (Fig. 9.10). Other examples of plant organs are *roots* and *stems*.

ISBN 9780170191340

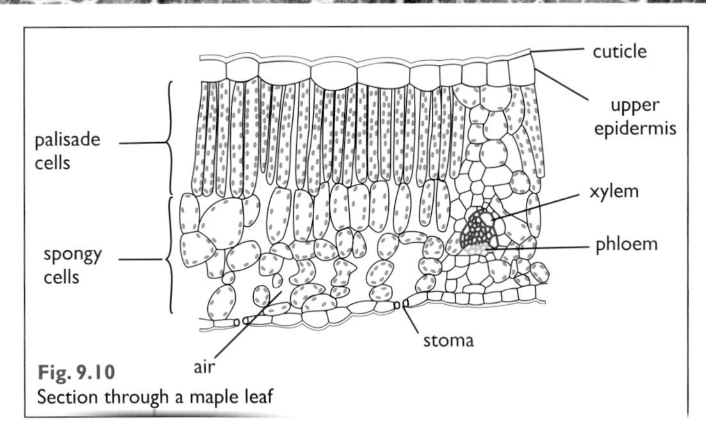

Fig. 9.10
Section through a maple leaf

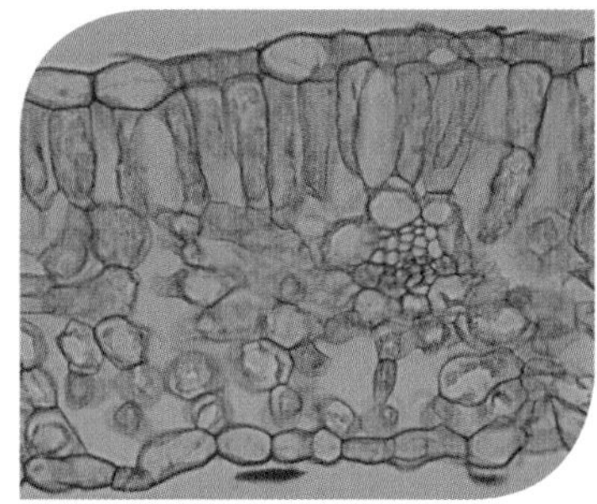
Transverse section of privet leaf

A leaf of a typical dicotyledon has the following main kinds of tissue:

- The **palisade mesophyll**, in which most of the photosynthesis occurs.
- The **spongy mesophyll**, in which some photosynthesis occurs.
- The **xylem**, which delivers water and minerals and also provides support.
- The **phloem**, which exports the sugar produced in photosynthesis.
- The **epidermis**, which regulates the balance between uptake of CO_2 and loss of water.

Absorbing carbon dioxide

Since photosynthesis uses up CO_2, in the light CO_2 is less concentrated inside the leaf than outside. As a result of this concentration difference, CO_2 diffuses into the leaf. The structure of the leaf helps this in several ways (Fig. 9.11):

- The leaf is *thin.* This has two advantages:
  1. It gives it a *large surface area compared with its volume*
  2. It shortens the distance between the atmosphere and mesophyll cells. Diffusion is faster over shorter distances (in which the concentration gradient is steeper)
- The air spaces between the mesophyll cells extend throughout the leaf, including between the palisade cells. This means that most of the diffusion pathway is through gas, in which diffusion is 10,000 times faster than in solution.
- The large vacuole in the mesophyll cells keeps the chloroplasts close to the cell surface, so that the diffusion path through liquid is very short (Fig. 9.11).
- The outer 'skin' of the leaf, the epidermis, is perforated by large numbers of tiny pores called **stomata** (singular: stoma). These allow CO_2 and other gases to diffuse through (see Fig. 9.13).

Absorbing light

The mesophyll is organised into two layers:

1. An upper *palisade* layer, containing abundant chloroplasts.
2. A lower *spongy* layer, containing fewer chloroplasts.

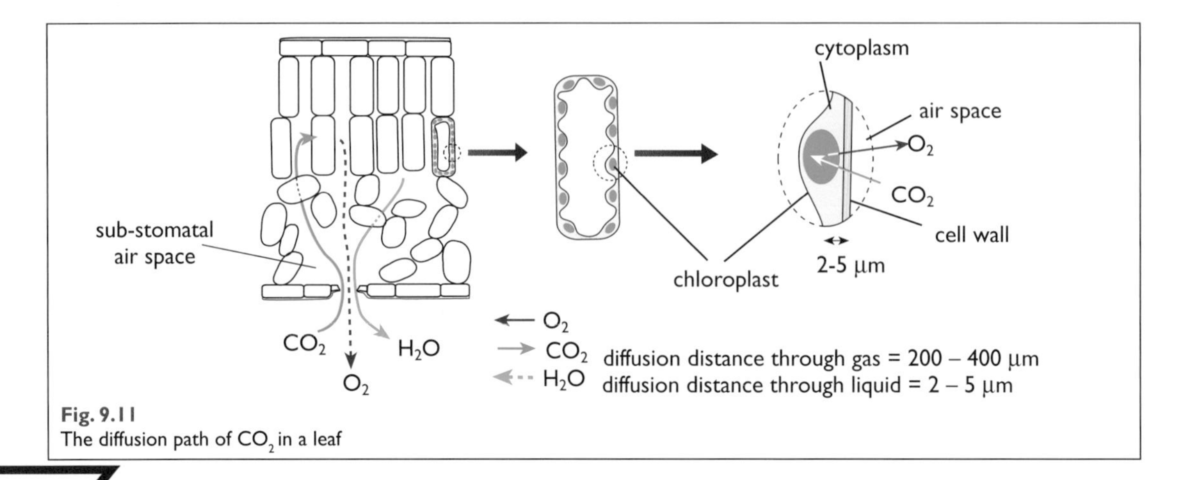

Fig. 9.11
The diffusion path of CO_2 in a leaf

ISBN 9780170191340

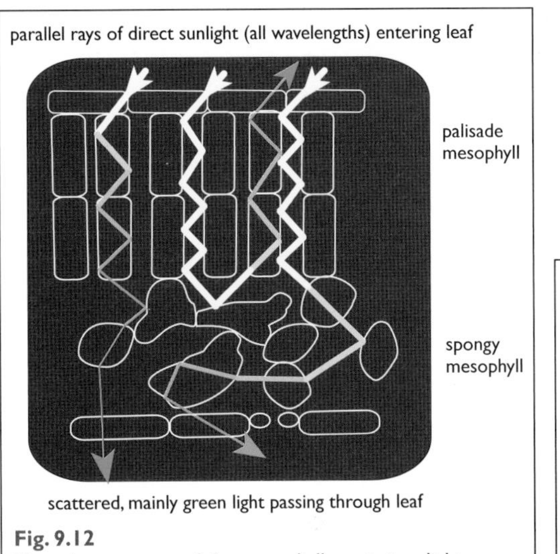

Fig. 9.12
How the structure of the mesophyll maximises light absorption

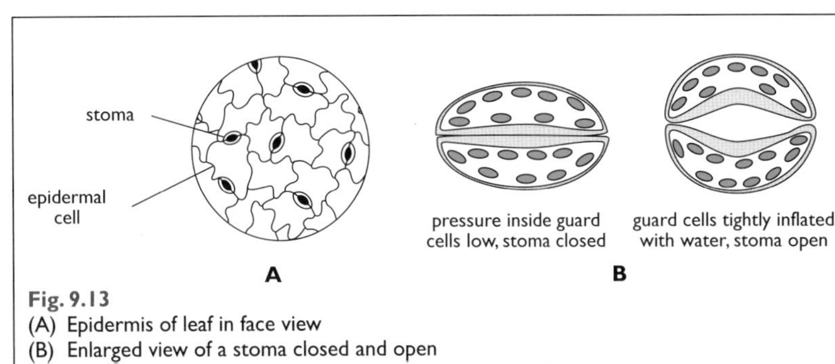

Fig. 9.13
(A) Epidermis of leaf in face view
(B) Enlarged view of a stoma closed and open

Though the palisade cells appear close together, there are narrow air spaces between them (necessary for the absorption of CO_2). As a result about a third of the light entering the leaf passes *between* the palisade cells rather than through them (Fig. 9.12). This light is not wasted, however, because of the spongy mesophyll. Spongy mesophyll cells form a network in which the walls face in all directions. As a result much of the light passing between the palisade cells is reflected back up through the palisade cells for a 'second chance' to be absorbed.

The absorption of light is also helped by the way leaves grow so that they tend to lie at right angles to the prevailing light.

Restricting water loss

The very features of a leaf that promote the inward diffusion of CO_2 also help the outward diffusion of water. This process is called **transpiration** and is the 'price' the plant has to pay to absorb CO_2. The epidermis of a leaf is adapted to reduce transpiration in two distinct ways:

1. It secretes water-resistant, waxy **cuticle**. In plants living in dry environments (xerophytes), the cuticle may be very thick. The thickness of the cuticle is thus an adaptation to long-term conditions, or *climate*.
2. Short-term fluctuations in water supply ('*weather*') are dealt with by *stomata*. Each stoma is surrounded by two **guard cells**. During the daytime, and when water supply is good, these are tightly inflated (turgid) with water and bent, keeping the stoma wide open (Fig. 9.13). In most dicotyledons there are more stomata in the lower epidermis than the upper epidermis, and in most trees and shrubs they are absent from the upper epidermis.

When water is scarce, and also at night, the guard cells lose water and become straight, closing the stoma.

Despite these adaptations for conserving water, a rapidly photosynthesising plant loses water by transpiration over a hundred times faster than it gains CO_2 in photosynthesis. This does not matter when the soil is moist, because water is absorbed by the roots as fast as it is lost by the leaves. It is transported by specialised cells in the *xylem*.

Exporting sugar

The sugar made in photosynthesis is transported in a tissue called the **phloem**. Phloem cells lie adjacent to the xylem cells and together make up the *veins* or **vascular bundles**. On a sunny day sugar is produced by the leaf much faster than it can be exported, and it is temporarily converted into *starch* in the chloroplasts. At night the starch is reconverted to sugar. This is then transported to the other parts of the plant such as roots and developing flowers, fruits and seeds.

ISBN 9780170191340

How the sugar is used

The sugar made in photosynthesis is used in various ways:

- Some is oxidised to release energy in *respiration*.
- Some is used to make other molecules such as starch, cellulose, fatty acids, amino acids and proteins, DNA and RNA.

Net and gross photosynthesis

For a plant to grow, its energy 'income' from photosynthesis must exceed its energy 'expenditure' in respiration. Respiration is the process in which organic matter is oxidised to give useful energy, and occurs in organelles called **mitochondria** (singular: **mitochondrion**).

Of the organic matter produced in photosynthesis, some is later oxidised in the mitochondria. The *net* energy gain is therefore the difference between photosynthesis and respiration (Fig 9.14):

net (apparent) photosynthesis = gross (actual) photosynthesis - respiration

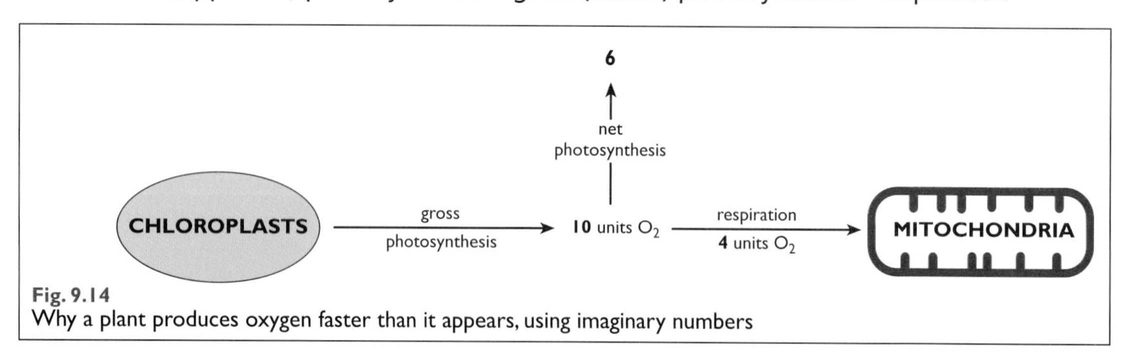

Fig. 9.14
Why a plant produces oxygen faster than it appears, using imaginary numbers

THE EFFECT OF ENVIRONMENTAL FACTORS ON PHOTOSYNTHESIS AND OTHER ASPECTS OF PLANT GROWTH

Photosynthesis is affected by a variety of environmental factors, described below.

Light intensity

Provided other factors are favourable, the rate of photosynthesis is affected by light intensity as shown in Fig. 9.15.

As would be expected, photosynthesis is faster in bright than in dim light, but at higher light intensities it makes little difference. The reason is that other factors are *limiting* (holding up) the rate of photosynthesis. The factor that limits the rate of a process is the one that is in shortest supply, for example:

- CO_2 concentration usually limits photosynthesis on a normal sunny summer's day.
- Temperature usually limits the rate of photosynthesis on a sunny winter's day.

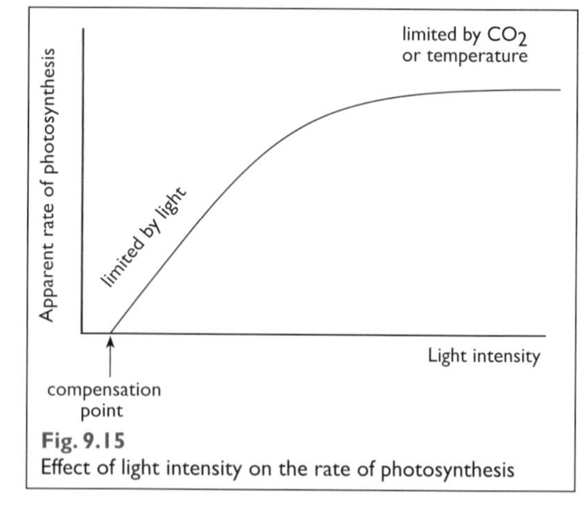

Fig. 9.15
Effect of light intensity on the rate of photosynthesis

Figure 9.15 shows that photosynthetic rate appears to be zero when there is still some light. This is because at this low intensity the rate of photosynthesis is exactly balanced by the rate of respiration. At this light **compensation point** the plant is using carbohydrate as fast as it is making it; the plant is 'marking time'.

It is easy to forget that, while photosynthesis only occurs in the light, respiration occurs all the time. As a result, the organic matter that accumulates in the daytime is actually *less* than the amount the plant actually produces. An easy way to envisage this is to liken a plant to a water tank, as in Fig. 9.16.

The water in the tank represents organic matter in the plant (dry mass). Water entering

ISBN 9780170191340

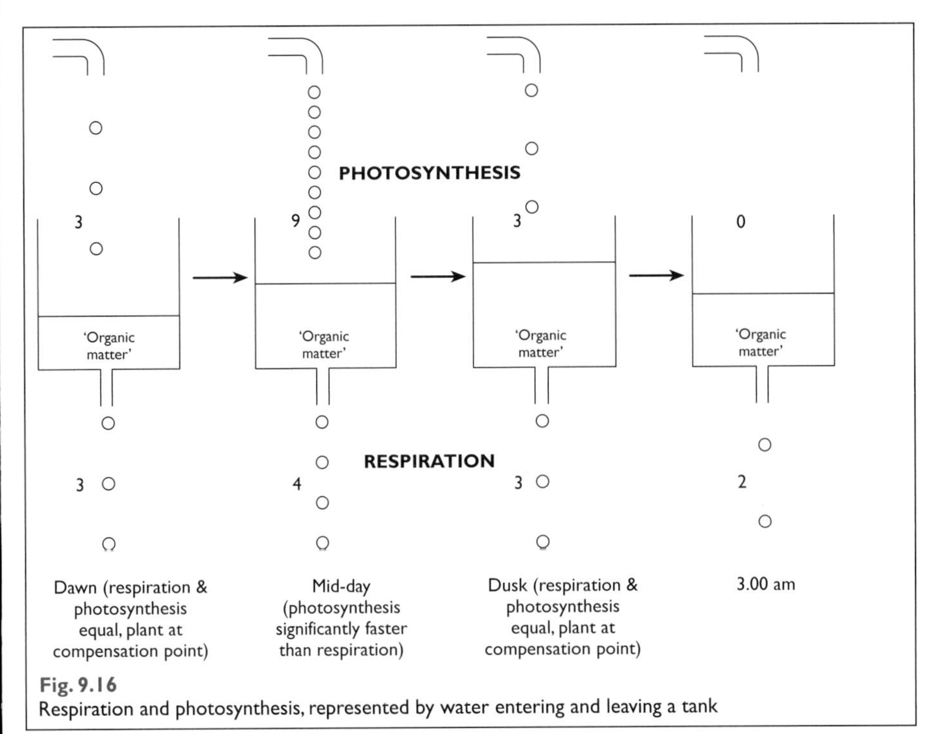

Fig. 9.16
Respiration and photosynthesis, represented by water entering and leaving a tank

the tank represents gain in organic matter by photosynthesis, and water leaving it represents loss of organic matter in respiration. The rate of each process is represented by the rate of water flow, indicated by the number of drops of water.

During the previous night, respiration has continued and the organic matter content has been falling. At dawn, photosynthesis begins and, once it exceeds respiration, the organic matter content begins to rise. By mid-day, photosynthesis is much faster than respiration and the plant is gaining rapidly in organic matter. During the night there is no photosynthesis so the plant is living on its 'savings' until dawn. Although the organic content falls, the plant uses less during the night than it gained the previous day, so over the 24-hour period it has gained dry mass. Notice that respiration is slower at night because it is cooler, so a plant loses less organic matter in a cool night than a warm one.

Extension: Photosynthesis on the forest floor — Shade plants

Plants of the forest floor are called **shade plants** e.g. the Mexican breadfruit plant and most other house plants. They are adapted to make the most of light that is not only dim, but contains a higher proportion of the less useful green light (Fig. 9.17).

Since shade plants have a low 'income', they grow slowly. Figure 9.18 shows why. Notice the following:

- In dim light (e.g. below light intensity 3) the shade plant photosynthesises faster than the sun plant.
- The graph for the shade plant levels off at much lower light intensities than the graph for the sun plant. In bright light, therefore, the shade plant is wasting more light than the sun plant.
- The shade plant has a lower compensation point than the sun plant. The shade plant can therefore gain carbohydrate in light that a sun plant cannot.

To barely survive, a plant needs brighter light than its compensation point because it has to make up all the organic matter it loses during the night.

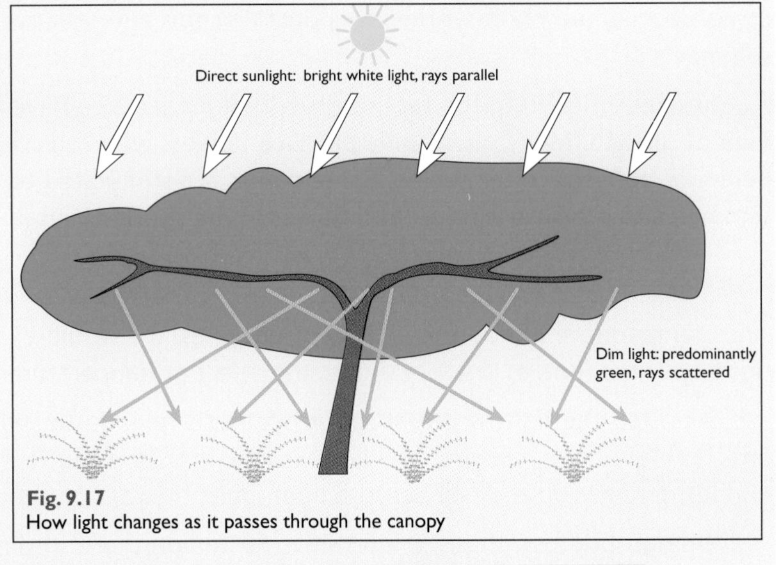

Fig. 9.17
How light changes as it passes through the canopy

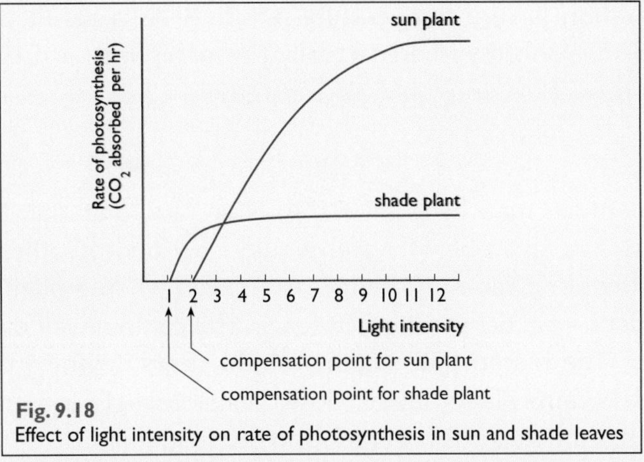

Fig. 9.18
Effect of light intensity on rate of photosynthesis in sun and shade leaves

ISBN 9780170191340

Sun and shade leaves compared

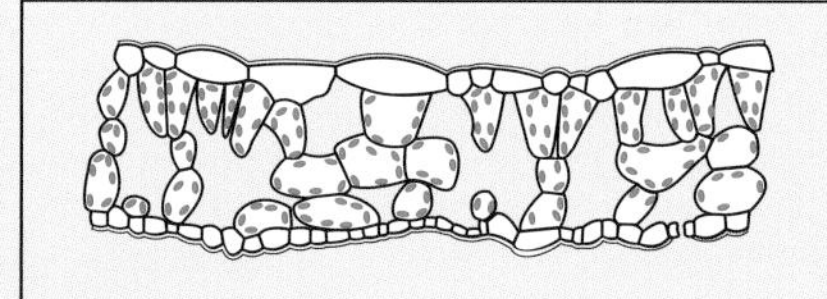

Fig. 9.19
Section through a shade-adapted leaf of maple

Leaves facing south near the bottom of a tree or shrub generally get much less light than those near the top. These leaves are adapted to shade, just like those of plants living in permanent shade. Figure 9.19 shows a shade-adapted leaf of maple. Compare it with the sun-adapted leaf shown in Fig. 9.10. Shade leaves differ from sun leaves in several ways:

- **They have a higher concentration of chlorophyll, so a higher proportion of the light can be absorbed.**
- **They have low respiration rates, and therefore low compensation points.**
- **The leaves are thinner because the mesophyll has little or no distinct palisade layer.**
- **Sun leaves have more stomata per mm² of leaf surface. To use more light, the leaf must be able to absorb more CO_2, requiring more stomata. The cost is a greater rate of transpiration.**

Light quality

Besides the intensity of the light, photosynthesis is strongly influenced by its *wavelength* (that we perceive as colour). Figure 9.20 shows that green light is much less effective than red or blue light. This is to be expected, because chlorophyll transmits green light (meaning that green light passes straight through rather than being absorbed.) A solution of chlorophyll appears green because the eye sees the light, which has not been absorbed, but has passed through it.

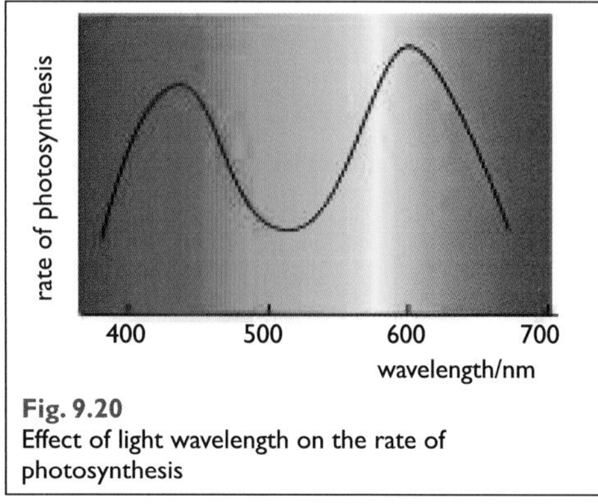

Fig. 9.20
Effect of light wavelength on the rate of photosynthesis

Duration (photoperiod)

Even if the intensity were the same, plants receive a greater *quantity* of light during long summer days than in other seasons, and so they can produce more photosynthetic product. For this and other reasons plants grow faster in summer.

Besides influencing the *rate* of growth, light can also have an important effect on the *kind* of growth. Many plants only produce flowers when the day length or **photoperiod** is below or above a certain value. Some plants are stimulated to flower in the long days of summer, and others are induced to flower in the shorter days of autumn.

Temperature

Since temperature affects the activity of enzymes, we would expect it to affect the rate of photosynthesis. But, as Fig. 9.21 shows, the effect of temperature depends on light intensity.

Provided the light is bright, photosynthesis goes faster with increasing temperature until, at higher temperatures, the enzymes begin to be denatured (inactivated). In dim light, temperature has little effect because light limits the rate of photosynthesis.

Though photosynthesis is insensitive to temperature in dim light, respiration is *very* temperature-sensitive. As a result the apparent or *net* photosynthesis actually *decreases* with a rise in temperature, because respiratory rate increases.

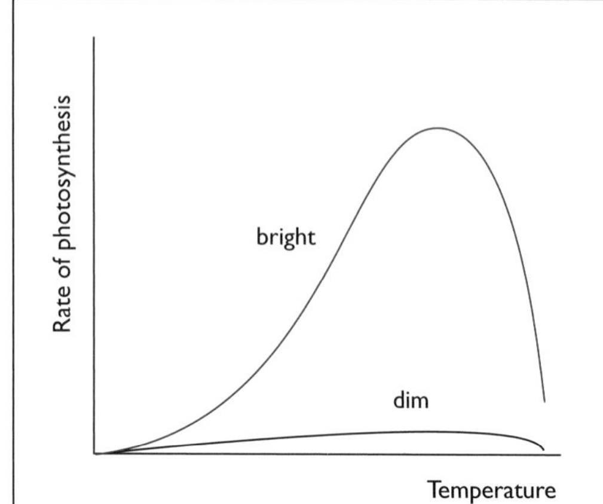

Fig. 9.21
Effect of temperature on the rate of photosynthesis in dim and bright light

Water

You might think that because water is a raw material for photosynthesis, a shortage of water would slow photosynthesis. Actually it does, but for a different reason. Even when a plant is at death's door from dehydration, its leaves still contain well over 50% water. The reason that photosynthesis slows down in dry conditions is because plants have to close their stomata to conserve water. This has the side-effect of reducing the supply of CO_2.

ISBN 9780170191340

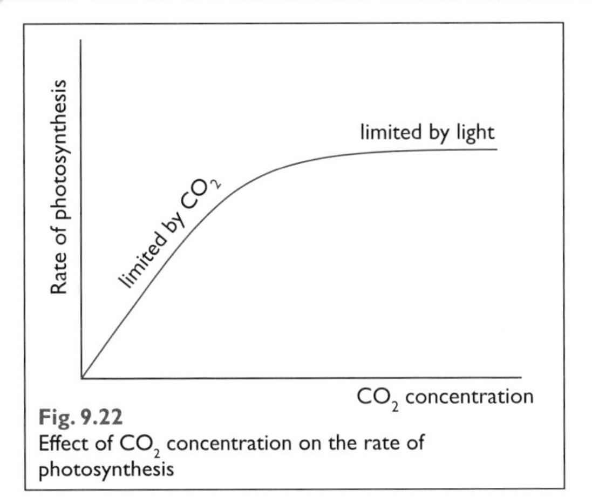

Fig. 9.22
Effect of CO_2 concentration on the rate of photosynthesis

Another way water can affect plant growth is due to the fact that when plant cells enlarge, they do so by taking in water by osmosis. This is why a young shoot elongates more rapidly at night, when the stomata are closed and water is more available.

Too much water can adversely affect plant growth. In a well-aerated soil, oxygen diffuses into the soil through the network of air spaces, but when soil is flooded these air spaces become filled with water. Oxygen is only sparingly soluble in water and is soon used up by the roots in respiration. The reduced energy supply slows the rate of mineral uptake and if the flooding is too prolonged, the roots may die.

Carbon dioxide

In warm, sunny conditions CO_2 is the factor that usually limits the rate of photosynthesis. Some horticulturalists boost plant growth by adding extra CO_2 to the air in their glasshouses. The effect is shown in Fig. 9.22.

Extension: Photosynthesis in hot, dry environments

Most plants native to temperate climates cannot photosynthesise efficiently in high light intensities, dry soils and tropical temperatures. On the other hand most tropical grasses, such as sugarcane, can. Figure 9.23 shows how clover and sugarcane perform in different light intensities and temperatures.

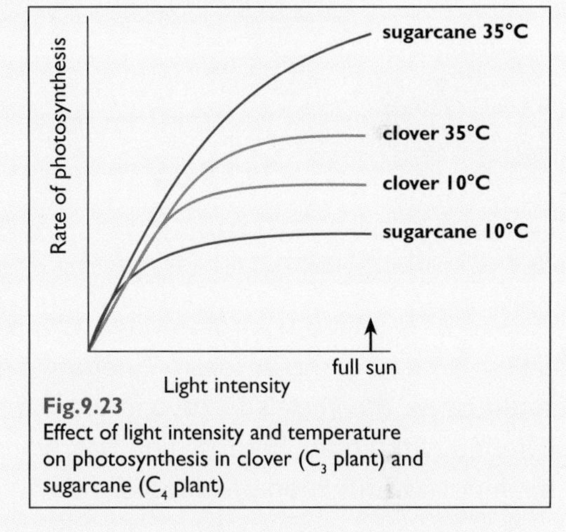

Fig.9.23
Effect of light intensity and temperature on photosynthesis in clover (C_3 plant) and sugarcane (C_4 plant)

These two plants have different photosynthetic mechanisms. Clover is called a 'C_3' plant because the first organic compound produced in photosynthesis has three carbon atoms. Sugarcane is a C_4 plant because the first product of photosynthesis has four carbon atoms.

Note the following:

- **At 10°C the C_3 plant out-performs the C_4 plant.**
- **At 35°C the C_4 plant is more efficient than the C_3 plant.**
- **The curve for sugarcane does not reach a plateau. In other words, sugarcane is not 'light-saturated' even in full sun.**

C_4 plants are also better adapted to dry conditions than C_3 plants. In dry soil, a C_3 plant partly or wholly closes its stomata, which reduces the concentration of CO_2 in the leaf air spaces. C_4 plants are able to absorb CO_2 from much lower concentrations than C_3 plants, so even with their stomata *almost* closed, C_4 plants can still photosynthesise.

ABSORBING WATER AND MINERALS: THE WORK OF THE ROOT

Besides anchoring the plant, the function of the root system is to absorb water and minerals. The structure of a young root is shown in Fig. 9.24 and 9.25.

Just behind the root tip are thousands of **root hairs**. Each root hair is an outgrowth of an individual cell of the epidermis ('skin') of the root. Collectively, the root hairs have a *huge surface area* for the absorption of water and minerals.

Inside the epidermis is the cortex, and in the centre of the root is a strand of *xylem* tissue. The water-conducting tubes of the xylem are called **vessels**. Each vessel is made up a chain of individual cells or *vessel members*, joined end-to-end like the sections of a drainpipe. The vessel members are dead when mature, with no cytoplasm and perforated end walls, so water can flow freely through (Fig. 9.26).

ISBN 9780170191340

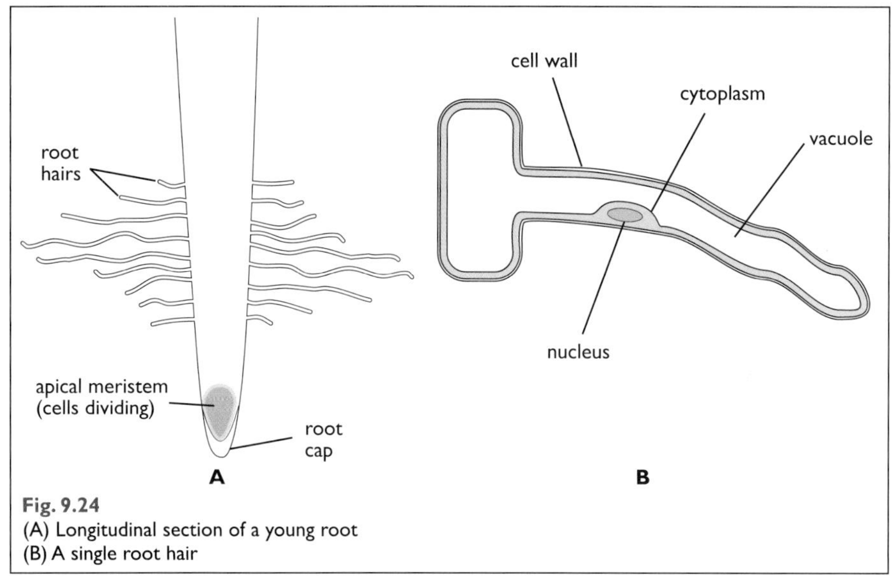

Fig. 9.24
(A) Longitudinal section of a young root
(B) A single root hair

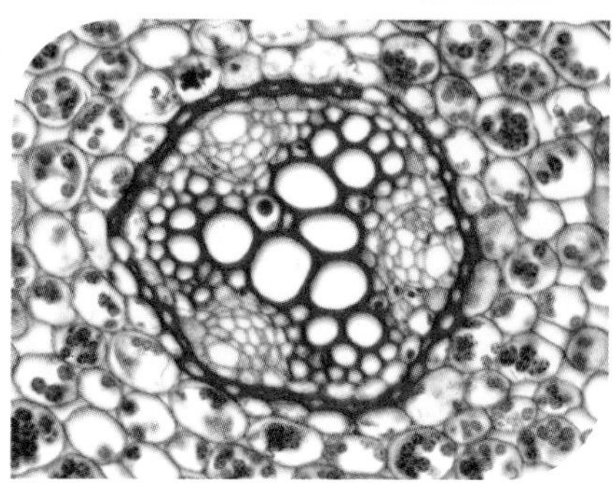
Transverse section of central region of buttercup root. Xylem stained red, starch grains in cortex blue

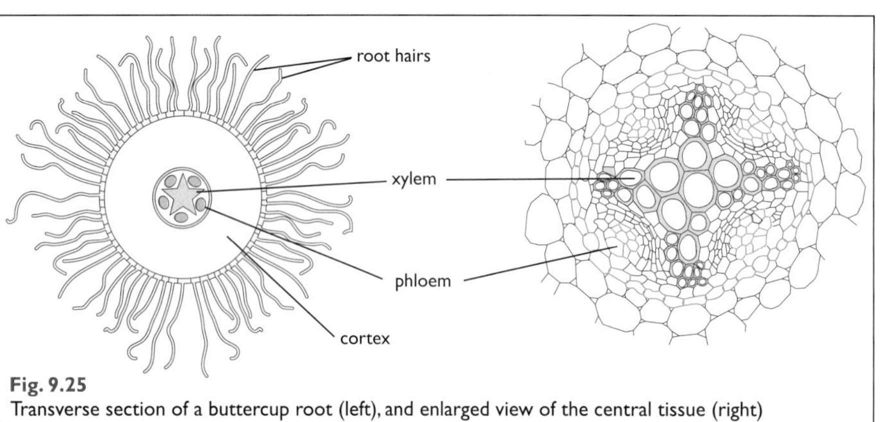

Fig. 9.25
Transverse section of a buttercup root (left), and enlarged view of the central tissue (right)

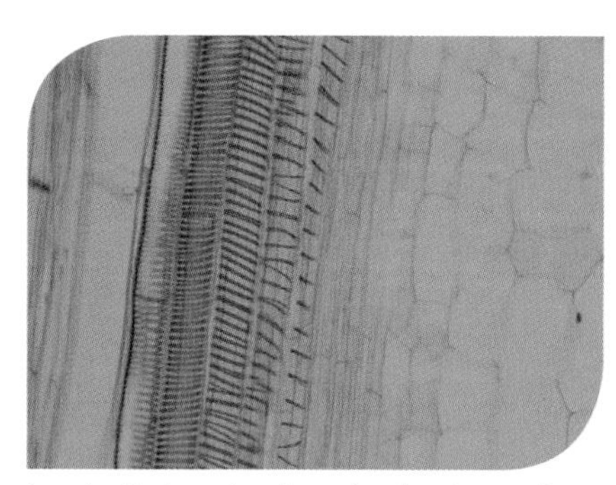
Longitudinal section through xylem in vascular bundle of sunflower, showing lignified spirals and rings stained red

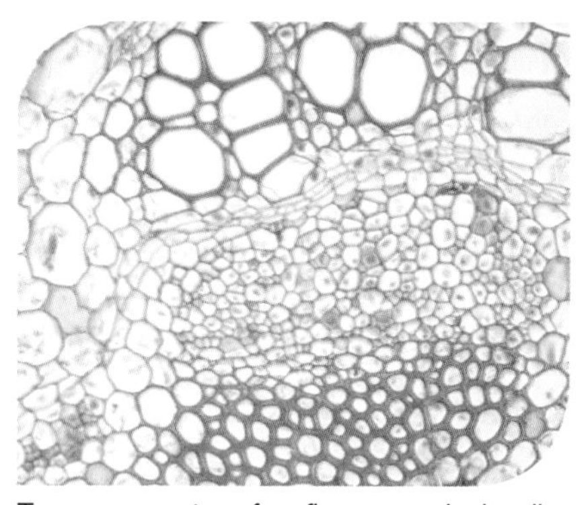
Transverse section of sunflower vascular bundle. The large red-stained cells at the top are xylem vessels. The smaller, red-stained thick-walled cells below are fibres, and the cells in between are the phloem, separated from the xylem by a thin layer of cambium

Water is pulled up from above as a result of evaporation from the leaves. This sets up great tension (negative pressure) in the xylem sap. This would cause the vessels to cave in were it not for the fact that they are stiffened with **lignin**. Because lignified walls are so strong, xylem also has the important function of *support*.

In a young stem, the first xylem vessels to differentiate do so while the stem is still elongating. Vessel members are dead, so they cannot grow in length, but have to passively stretch. To keep the vessels from collapse while allowing them to stretch, the lignin is laid down in spirals or rings.

In stems, the vascular (conducting) tissue is arranged differently. Figure 9.27 shows a transverse (cross) section through a buttercup stem. The vascular tissue is arranged in *vascular bundles*. In a young stem these are typically arranged in a hollow cylinder. As in scaffolding and in the long bones of the human skeleton, a hollow cylinder of supporting tissue gives greater stiffness for a given amount of material.

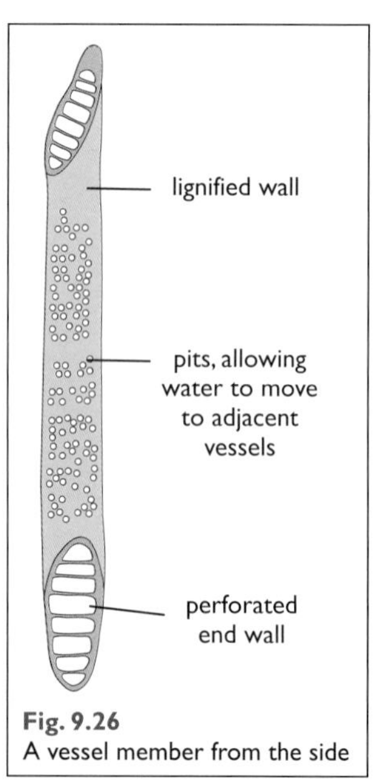

Fig. 9.26
A vessel member from the side

How plants absorb water

Plant cells absorb water by **osmosis** (Fig. 9.28). In osmosis, water moves through a **partially permeable membrane** (also called a *selectively* permeable membrane). This is a kind of molecular sieve — it allows small molecules such as water to pass through, but not larger molecules like sugar.

In Fig. 9.28 the concentration of water in the beaker is higher than it is in the sugar solution (in a sense, the sugar 'dilutes' the water). The dialysis tubing is a partially permeable membrane; water molecules can pass through but sugar cannot.

ISBN 9780170191340

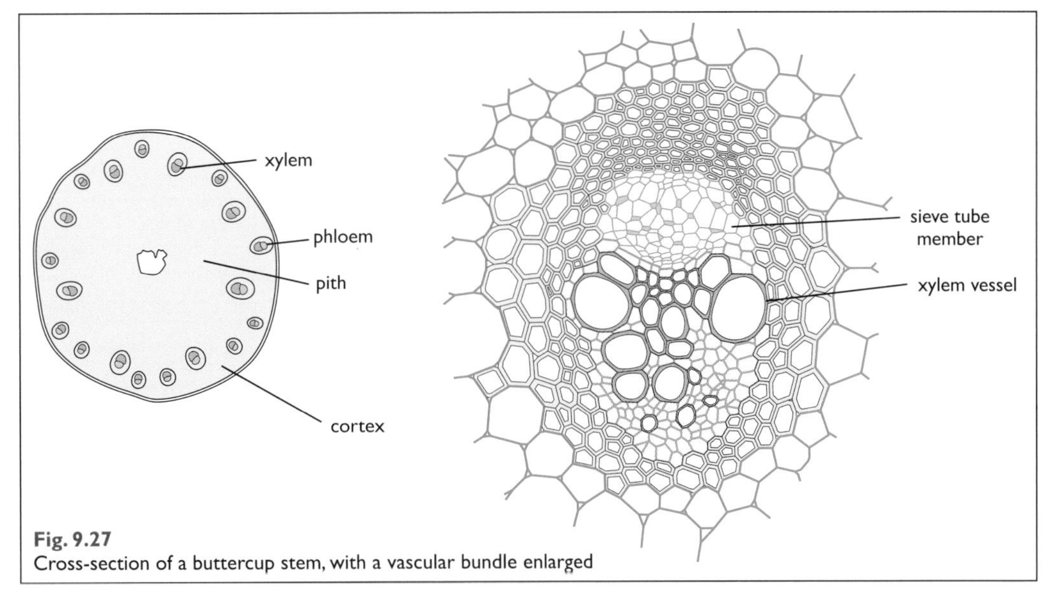

Fig. 9.27
Cross-section of a buttercup stem, with a vascular bundle enlarged

As soon as the apparatus is set up, water begins to pass from the beaker into the sugar solution. This has two effects:

1. As the level of the sugar solution rises, the pressure begins to rise, causing the dialysis tubing to bulge.
2. The water concentration in the dialysis tubing begins to rise (though it never equals that of distilled water).

Eventually (if the tubing were high enough and if the dialysis tubing did not burst), the pressure in the dialysis tubing would be high enough to prevent any further entry of water.

It is important to realise that osmosis is a *net* movement of water. At equilibrium, water molecules are moving at equal rates in both directions.

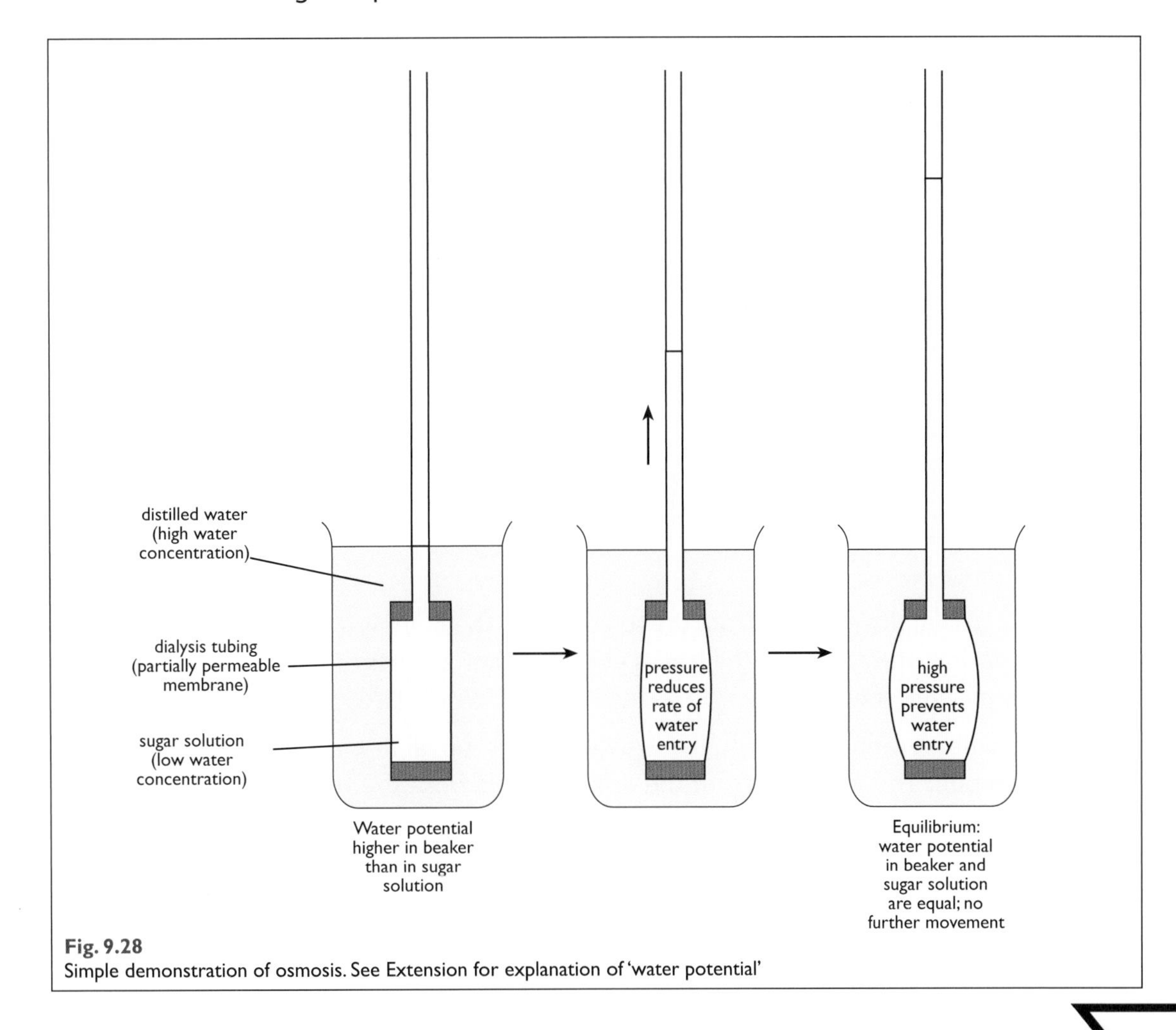

Fig. 9.28
Simple demonstration of osmosis. See Extension for explanation of 'water potential'

ISBN 9780170191340

Osmosis in plants

The process of osmosis in a plant cell is shown in Fig. 9.29. The cellulose wall is fully permeable, meaning that all dissolved substances can pass through. The **plasma membrane,** on the other hand, is partially permeable. The vacuole of a plant cell is a solution of salts, sugars and other solutes. The effect of these is to lower the concentration of the water.

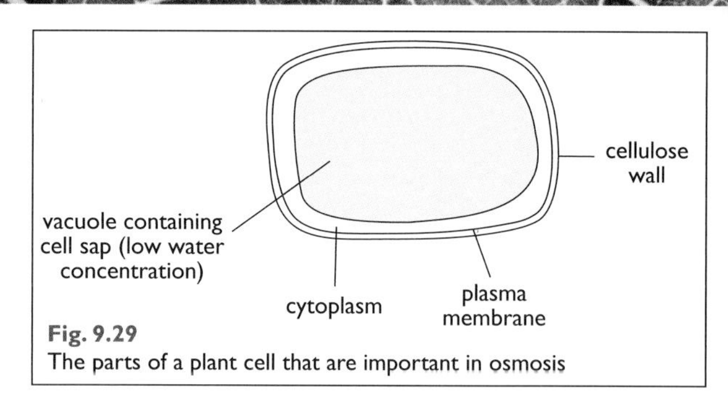

Fig. 9.29
The parts of a plant cell that are important in osmosis

Imagine a plant cell that has lost water and is *flaccid* (limp), like a partially deflated balloon. Suppose such a cell is placed in distilled water (Fig. 9.30). In distilled water the water concentration is higher than it is in the vacuole. Water therefore begins to enter the cell. This causes the pressure inside the cell to rise so it begins to oppose the entry of water. Eventually the pressure of water is high enough to prevent any more water from entering. A cell in this state is tightly inflated or *turgid*. Although the cell is in equilibrium with its surroundings, the water concentration inside and outside are *not* equal.

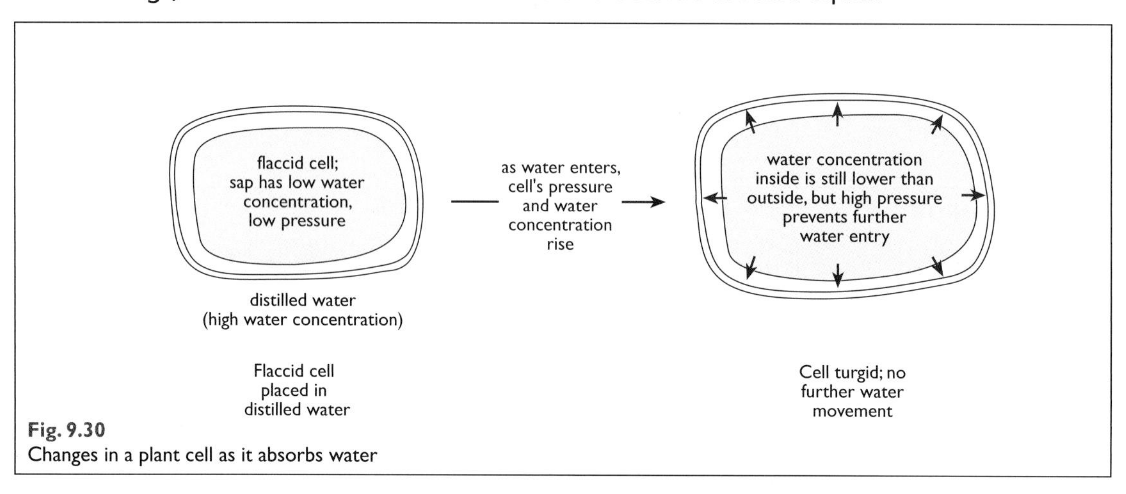

Fig. 9.30
Changes in a plant cell as it absorbs water

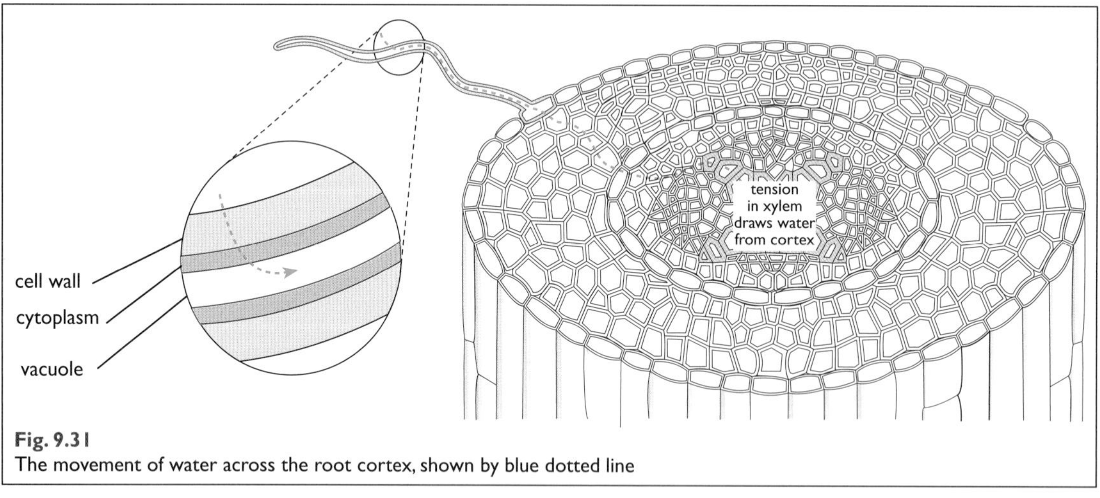

Fig. 9.31
The movement of water across the root cortex, shown by blue dotted line

How water enters roots

The main force that drives water from the soil through the plant is due to **transpiration**. This is the evaporation of water from the leaves and other aerial parts of the plant, and is driven by heat energy from the sun.

When water evaporates from the leaves, it sets up a *tension* in the xylem called **transpiration pull**. This pulls water up the xylem vessels from the roots. The tension in the root xylem draws water from the cells of the cortex. This lowers the water concentration in the cortex cells, causing water to move by osmosis from the root hair cells. This in turn lowers the water concentration in the root hairs, causing water to enter them from the soil water (Fig. 9.31).

ISBN 9780170191340

The overall path of movement through the plant is shown diagrammatically in Fig. 9.32.

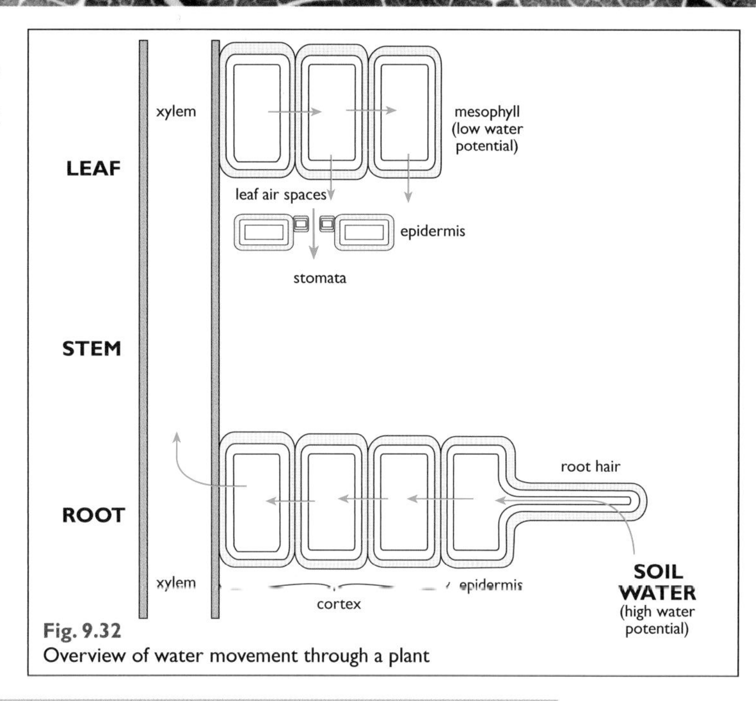

Fig. 9.32
Overview of water movement through a plant

Extension: A closer look at how water moves through plants — water potential

Osmosis in plants is more complicated than described above. To see why, think of a plant cell in equilibrium with distilled water; the cell is turgid, and cannot absorb any more water, even though the water concentration in the cell sap is much lower than the water concentration outside. There is an equilibrium between two tendencies:

1. A tendency for water to enter the cell due to the difference in water concentration.
2. A tendency for water to leave the cell due to the difference in pressure.

Now suppose the cell sap contains solutes equivalent to a 5% sugar solution, and we replace the distilled water with 1% sugar solution; water begins to move out of the cell. This is despite the fact that water is moving from where it is less concentrated to where it is more concentrated (Fig. 9.33). Water leaves the cell until a new equilibrium is established.

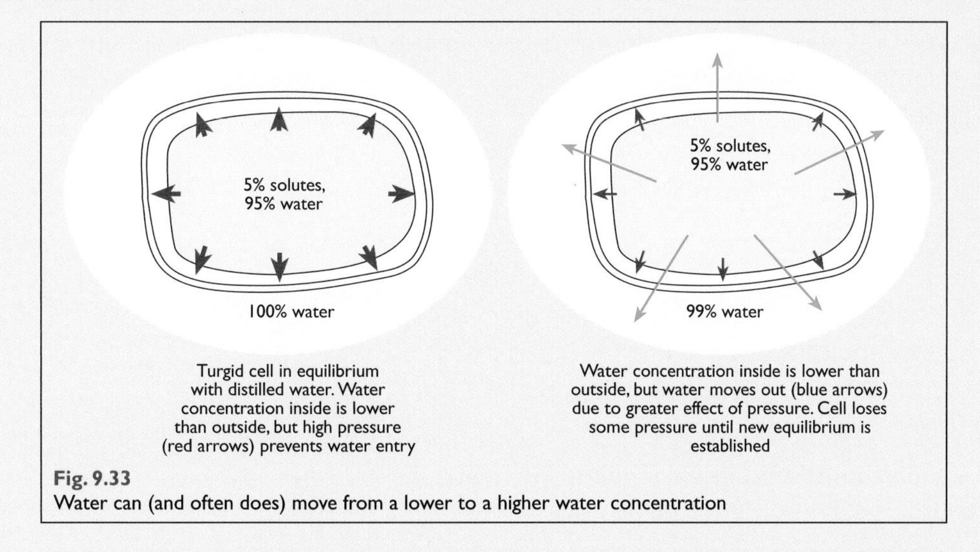

Fig. 9.33
Water can (and often does) move from a lower to a higher water concentration

There are therefore two factors affecting osmosis: water concentration and pressure. Their combined effect is the **water potential** of a solution. In the imaginary example above, water moves out of the turgid cell because its water potential is higher than that of the 1% sugar solution. The movement of water through a plant follows this simple rule.

Water always tends to move into a region with a lower water potential.

Transpiration pull creates a tension (negative pressure) in the xylem, which is transmitted all the way down to the roots. As a result the water potential in the root xylem is lower than the water potential in the soil and consequently water moves into the root hairs, across the cortex and into the xylem vessels.

ISBN 9780170191340

Mineral ions

Besides the carbon, hydrogen and oxygen obtained from CO_2 and water, plants need an additional 14 minerals in the form of inorganic ions. Those required in greatest amount are nitrogen (absorbed as nitrate), phosphorus (absorbed as phosphate), and potassium. Some are required in almost unbelievably minute concentrations; 10 parts per *billion* is sufficient to prevent deficiency of molybdenum.

How plants absorb minerals

Mineral ions cannot pass through cell membranes by simple diffusion, for two reasons:

1. Most are more concentrated inside the root than in the soil, so ions must be absorbed against a concentration gradient.
2. Ions cannot move freely through cell membranes, as they are too large and are electrically charged.

Plants absorb ions by **active transport**. This involves special carrier proteins embedded in the plasma membranes of the root hair cells. Because the ions are moving 'uphill' (from a lower to a higher concentration), the process requires *energy* from *respiration*.

When a mineral ion is in short supply, various *deficiency symptoms* develop. Each kind of deficiency results in characteristic symptoms. Some minerals can be **translocated** (transported) from older to younger leaves, which are more photosynthetically active. In these cases, deficiency symptoms tend to appear in older leaves. On the other hand, deficiency in minerals that cannot be moved around the plant tend to produce symptoms in the younger leaves.

How plants use nitrogen

Nitrogen is absorbed mainly as nitrate but can also be absorbed as ammonium ions. Nitrogen is a constituent of all amino acids and hence proteins (Fig. 9.34), and the nucleic acids DNA and RNA and also of chlorophyll. The effects of deficiency include reduced growth rate and a yellowing of the leaves — a condition called **chlorosis**. Because nitrogen can be re-located round the plant, chlorosis shows up in the older leaves first.

It is important to understand that the effect of increasing the supply of one mineral depends on the supply of other minerals. If, for example, nitrate is in short (limiting) supply, adding extra phosphate will make little difference (Fig. 9.35). The situation is similar to the effect of factors on photosynthesis in which, for example, giving extra light makes little difference if the temperature is too low or CO_2 is in short supply.

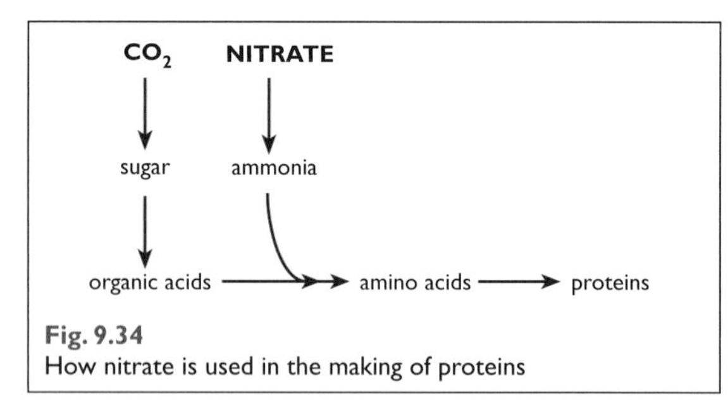

Fig. 9.34
How nitrate is used in the making of proteins

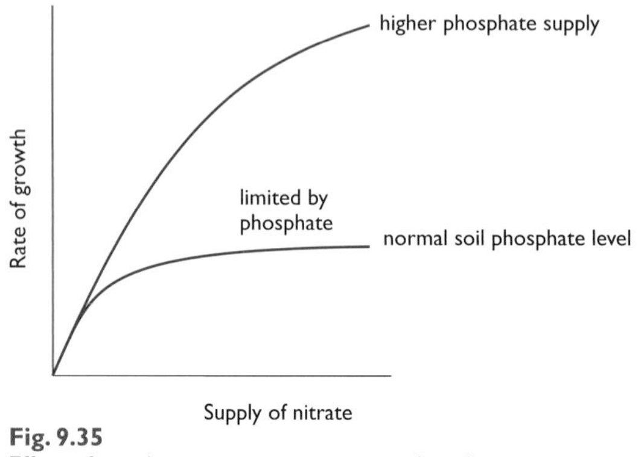

At the lower phosphate level, adding nitrate only benefits growth at lower nitrate levels. At higher nitrate levels the plant is short of phosphate and cannot benefit from extra nitrate. If the phosphate supply is increased, the plant can benefit from higher nitrate levels.

Nitrogen is a **macronutrient**, since it is needed in relatively large amounts. Other macronutrients are:

- Phosphorus: absorbed as phosphate and is a constituent of DNA and RNA.
- Potassium: essential for the action of many enzymes.
- Magnesium: a constituent of chlorophyll and essential for the action of many enzymes.
- Calcium: essential for cell division and many other processes.
- Sulfur: absorbed as sulfate and is a constituent of proteins.

Fig. 9.35
Effect of supplying extra nitrate at two phosphate concentratons on plant growth

ISBN 9780170191340

Wind and plant growth

Apart from occasionally blowing down trees, wind affects plant growth in two ways:

1. It speeds up transpiration which, if severe, may reduce photosynthesis in the way explained above.
2. Evaporation produces cooling, so a rapidly-transpiring leaf is cooler than it would otherwise be. This might be beneficial in hot weather, but in cooler, windy climates such as on mountains, the effect may be to slow photosynthesis.

Summary of key points in this chapter

- The most important physical factors affecting plant growth are light intensity, light wavelength, light duration (photoperiod), temperature, soil pH, soil water supply and wind.
- Different species of plants are adapted to light of different intensities.
- The photosynthetic organ of a plant is the *leaf*.
- A leaf consists of a variety of *tissues* which work together to promote the absorption of light and CO_2.
- The absorption of light is facilitated by the division of the mesophyll into upper palisade and lower spongy layers.
- The absorption of CO_2 by a leaf is promoted by its thin shape, internal air spaces, and position of the chloroplasts in the mesophyll cells.
- The absorption of CO_2 is accompanied by the inevitable loss of water (*transpiration*). Balancing the need to absorb CO_2 and conserve water is the function of the *stomata* in the epidermis.
- Roots absorb water by *osmosis*, which is *the movement of water across a partially permeable membrane*.
- Plants need certain *mineral ions*, which they obtain from the soil. Each ion is essential; none can substitute for any of the others. If any is in short supply, *deficiency symptoms* develop.
- Roots absorb mineral ions by *active transport*, which requires energy from respiration.
- Minerals needed in relatively large amounts are called *macronutrients*; those required in minute amounts are called *micronutrients* or trace elements.

ISBN 9780170191340

Test your basics

Copy and complete the following sentences.

a) In photosynthesis, __*__ from the air and __*__ from the soil are combined to make __*__ and oxygen, using __*__ energy from the __*__.

b) Before testing a leaf for starch, the __*__ must first be removed by boiling it in __*__.

c) A leaf can be deprived of CO_2 by confining it in a glass vessel containing __*__. To supply a leaf with CO_2, concentrated __*__ __*__ __*__ solution is used.

d) In a leaf such as privet or sunflower, the tissue specialised for photosynthesis is the __*__ mesophyll, in which the cells are elongated at right angles to the epidermis. Below this is the __*__ mesophyll, in which the cells are more randomly oriented. This serves to __*__ light that has passed between the __*__ mesophyll cells, giving it a second chance to be absorbed.

e) Diffusion is the movement of a substance from a region of __*__ concentration to a region of __*__ concentration by __*__ movement of its particles.

f) The function of the __*__ in the epidermis of a leaf is to balance the loss of __*__ and the uptake of __*__.

g) The uptake of CO_2 is promoted by the __*__ shape of a leaf, which gives it a large __*__ compared with its __*__, and also by the extensive system of __*__ spaces inside the leaf (diffusion is much faster in __*__ than in __*__).

h) Photosynthesis occurs in the __*__ of the cell. Because of the large __*__, the cytoplasm and its organelles are displaced close to the cell surface. Therefore CO_2 has to diffuse a very short distance through __*__.

i) During the day, the __*__ made in photosynthesis accumulates much faster than it can be exported by the __*__, and it is temporarily stored in the __*__ as __*__.

j) The rate of photosynthesis may be affected by several factors in the environment of the leaf: __*__ intensity, __*__ concentration in the air, and the __*__.

k) The rate of photosynthesis is also affected by the water content of the soil. In dry conditions, the __*__ close, restricting the supply of __*__.

l) Water is transported to the leaf in xylem __*__, which consist of dead cells with perforated end walls, situated end to end to form continuous tubes. The walls are stiffened with __*__ which prevents them collapsing under the __*__ set up as a result of __*__.

m) Osmosis is the movement of water through a __*__ __*__ membrane.

n) Roots absorb water by __*__.

o) The wall of a plant cell is fully __*__ but is strong enough to resist internal __*__. A cell that is tightly inflated with water is said to be __*__; in this state the internal pressure is high enough to oppose entry of more water.

p) Roots absorb mineral ions by __*__ __*__, which requires __*__ from __*__.

q) Nitrogen is absorbed mainly in the form of __*__ ions. In the plant these are used to make __*__ acids which are then built up into __*__. Nitrogen is also a constituent of the green pigment __*__ and of nucleic __*__ such as DNA.

ISBN 9780170191340

10 Reproduction and life cycle of flowering plants

Reproduction in plants can be of two fundamentally different kinds:

1. *Sexual* reproduction, which involves the re-shuffling of genes in the processes of **meiosis** and **fertilisation** (Chapter 1). In fertilisation, male and female *gametes*, each with one set of chromosomes, join together to form a **zygote** or *fertilised egg*, which is *diploid*, with two sets of chromosomes. The essential result of sexual reproduction is that the offspring are genetically different from each other and from the parents.
2. *Asexual* reproduction, which involves mitotic cell divisions only, with the result that the offspring are genetically identical to the parent.

SEXUAL REPRODUCTION IN FLOWERING PLANTS

In flowering plants, some shoots are highly modified for sexual reproduction and form *flowers*. The male gametes are produced inside haploid cells called **pollen** grains. The female gametes are produced inside **ovules**. After fertilisation the zygote develops into a young plant or **embryo**. This is inside a **seed**, which is protected inside the ripened ovary or **fruit**. The function of the fruit is usually to enable the seed to be dispersed.

Before fertilisation can occur the pollen grains must be transferred from the male part of a flower to the female part, a process called **pollination** (Fig. 10.1).

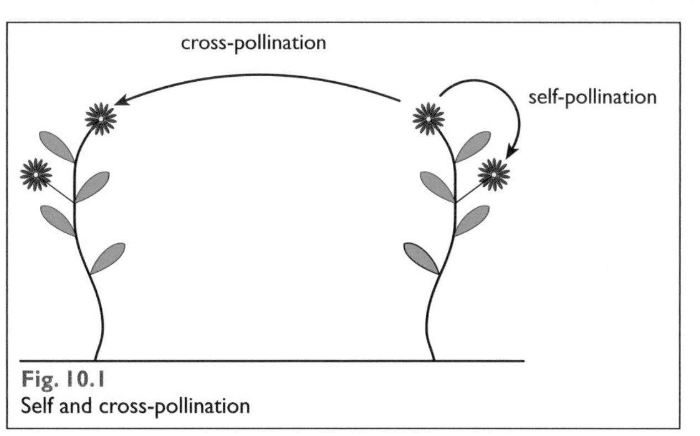

Fig. 10.1
Self and cross-pollination

In *cross-pollination* the pollen is transferred between *different* plants, and leads to **outbreeding**. In **self-pollination** the pollen is transferred to a stigma on the *same* plant (though often in a different flower), and leads to **inbreeding**. Each kind of pollination has costs and benefits, and the balance of advantage depends on circumstances.

In New Zealand, pollination may be carried out by animals (mainly insects), wind or, in a few cases, by water.

Insect-pollination

Successful insect-pollination requires three key features:

1. Insects must be 'rewarded' with some form of *food*. This is usually **nectar**, a solution containing various kinds of sugar. Some flowers provide food in the form of pollen, which is rich in protein. By far the most effective pollinators are bees, because they have to collect protein-rich pollen for their larvae as well as nectar for themselves. Some bee-pollinated flowers make no nectar, relying entirely on pollen as a lure, for example broom, gorse and poppies.
2. The flowers must *advertise* their presence, usually by being conspicuous and/or by producing some kind of odour.
3. The pollen must be *sticky* so that it adheres to the insect's body. This is of course quite accidental so far as the insect is concerned.

A simple flower: the buttercup

Figure 10.2 shows half of a buttercup flower, cut vertically. The tip of the flower stalk is the **receptacle**, to which the floral parts are attached.

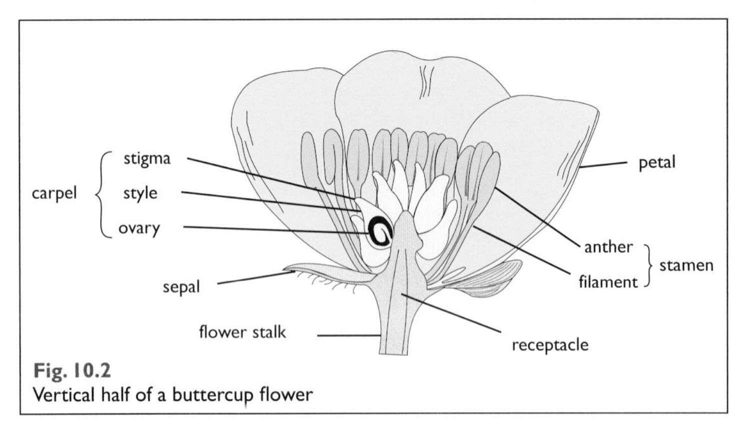

Fig. 10.2
Vertical half of a buttercup flower

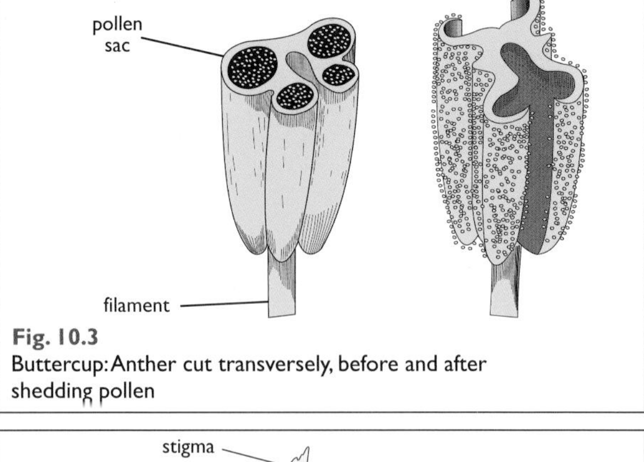

Fig. 10.3
Buttercup: Anther cut transversely, before and after shedding pollen

Outermost are the **sepals**, which protect the inner parts in the bud stage, and collectively form the *calyx*. When the flower opens the sepals are bent back to reveal the five large **petals**, collectively called the *corolla*. To humans they look a uniform yellow, but there are lines radiating from the base of each petal that reflect UV light, which most insects can see. These lines are called *nectar guides*. They help to guide the insect to the *nectary* at the base of each petal. It secretes nectar, which is easily accessible to insects with a short 'tongue' or, more correctly, the *proboscis* (plural: proboscides), such as small beetles and flies.

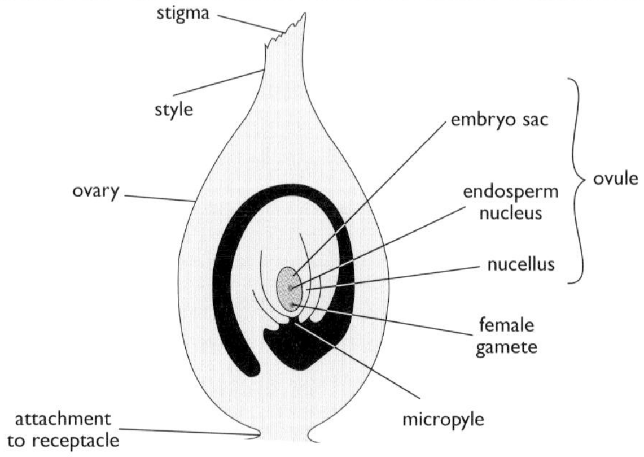

stigma
style
ovary
embryo sac
endosperm nucleus
nucellus
ovule
female gamete
micropyle
attachment to receptacle

Fig. 10.4
Buttercup: Vertical section through a carpel

Inside the corolla is the male part of the flower, the *androecium*, consisting of many **stamens**. Each stamen consists of an **anther**, held up by a long stalk or *filament* (Fig. 10.3). The anther contains four **pollen sacs**, in which pollen grains are produced by **meiosis**. When the anther is ripe it splits down the middle to release the pollen, some of which sticks to the bodies of insects feeding on the nectar.

The centre of the flower forms the *gynoecium*, or female part of the flower. It consists of many **carpels**, which are the female equivalent of stamens. Each consists of three parts: *stigma*, *style* and *ovary* (Fig. 10.4). The **stigma** not only receives pollen sticking to the bodies of insects, but also secretes a sugar (sucrose), which stimulates pollen grains to germinate. The **style** is a short stalk holding the stigma up. Each **ovary** houses a single **ovule**, which develops into a *seed*.

Preventing inbreeding

In the buttercup the flowers are *hermaphrodite* (having both male and female parts), but the male part ripens before the female part so that at any given time the flower is unisexual. This reduces the chances of self-pollination, but does not prevent it since pollen can still be transferred from a young flower to an older flower on the same plant.

In some plants unisexual flowers are produced, each flower producing either male or female parts. If these are on the same plant, self-pollination is still possible. If they are on separate plants, as in *Coprosma* and kiwifruit, inbreeding is impossible.

In some plants with hermaphrodite flowers inbreeding is prevented by *self-incompatibility*, in which the pollen cannot grow on a stigma of the same plant, for example apple.

Wind-pollination

Like pines and other conifers, many flowering plants are wind-pollinated, for example grasses. In contrast to the 'showy' animal-pollinated flowers, wind-pollinated flowers are usually small, green and inconspicuous, and never produce nectar or scent.

The energy saved by not producing nectar, scent or large petals is offset by the fact that pollen has to be produced in huge amounts. This is because wind carries pollen randomly, so only a tiny proportion of the pollen grains find their target. Since the pollen grains are very small they have a high surface to weight ratio, so they fall slowly and are easily caught

ISBN 9780170191340

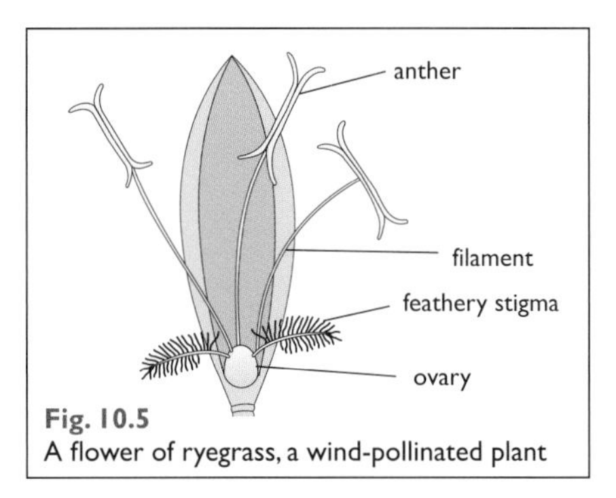

Fig. 10.5
A flower of ryegrass, a wind-pollinated plant

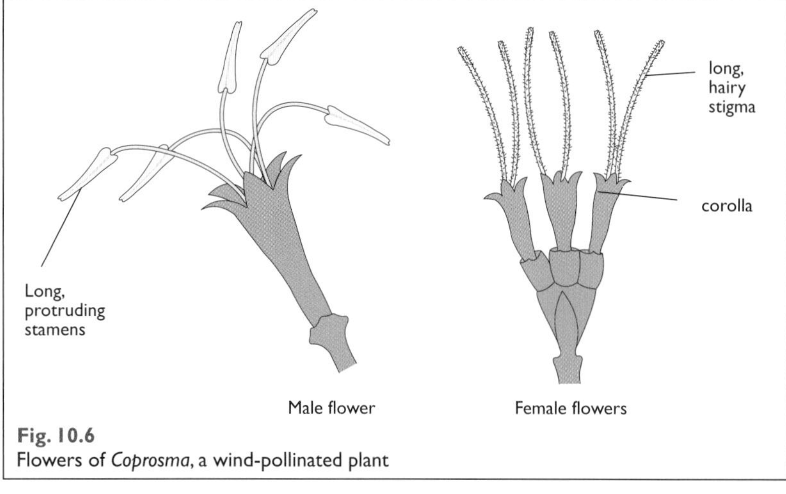

Fig. 10.6
Flowers of *Coprosma*, a wind-pollinated plant

by the wind. They are also non-sticky so they do not adhere to leaves, branches or other obstacles. Many wind-pollinated trees produce their flowers before the leaves open, thus taking advantage of free air movement.

Figure 10.5 shows a flower of ryegrass, a wind-pollinated flower. The clusters of tiny flowers are produced on long stems, well above the leaves, so the wind is able to catch them. A grass flower has three stamens, each with its filament attached mid-way along the anther so that it rocks in the wind, shaking out the pollen. The stigmas are feathery, exposing a large surface to the wind. Even so, the chances of a stigma intercepting more than one pollen grain are not high, and like most wind-pollinated flowers, grasses have only one ovule in each ovary. Another example of a wind-pollinated flower is *Coprosma* species, a genus of native trees (Fig. 10.6). In *Coprosma* all the flowers on one plant are of the same sex, so only female plants produce berries.

Differences between wind and insect-pollinated flowers are summarised in Table 10.1.

Typical insect-pollinated flower	Typical wind-pollinated flower
Large and conspicuous	Small and green
Usually produce nectar and scent	Never produce nectar or scent
Anthers do not pivot on filaments	Anthers usually pivot on filaments
Pollen sticky	Pollen non-sticky
Pollen grains relatively large	Pollen grains very small
Smaller amounts of pollen	Larger amounts of pollen
Stigmas not usually feathery	Stigmas feathery or hairy

Table 10.1 Comparison between insect- and wind-pollinated flowers

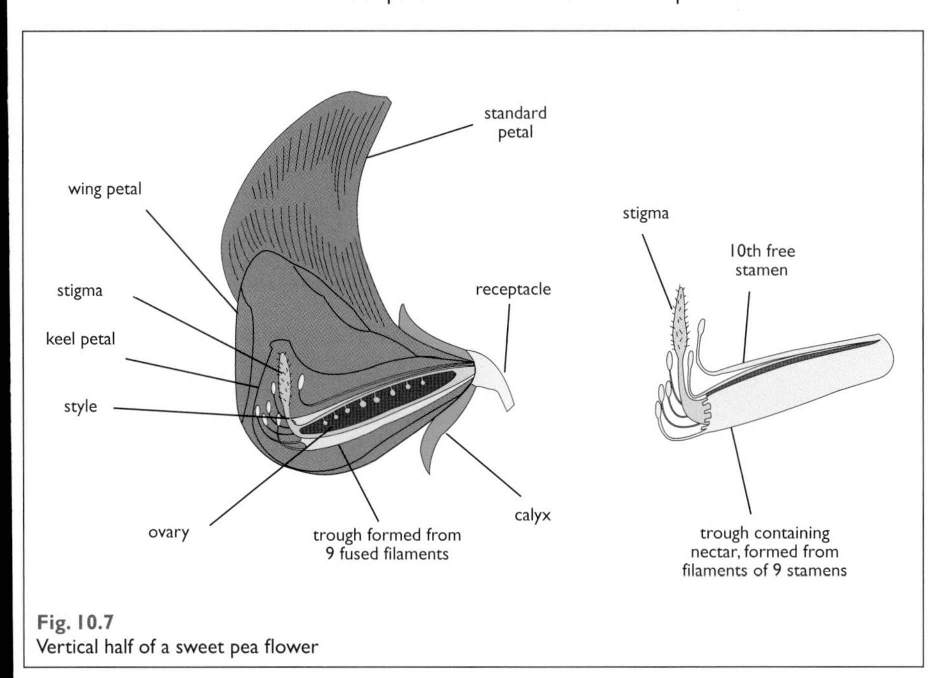

Fig. 10.7
Vertical half of a sweet pea flower

Floral variety

The buttercup is often said to be a 'typical' flower, but in reality there is no such thing, as flowers vary enormously in almost all their parts.

Many flowers are *bilaterally symmetrical* or **zygomorphic**, meaning that the flower can only be cut into two equal halves in one plane. In effect the flower has left and right halves, for example the sweet pea (Fig. 10.7) and other members of the legume family such as clover and gorse. An insect visiting these flowers has to take the

ISBN 9780170191340

same position relative to the flower, so the same part of the insect's body touches the stigma and stamens, wasting less pollen.

Concealing the reward

Many flowers have their nectaries situated at the end of long tubes, so that only insects with a long 'tongue' (e.g. bees, butterflies and moths) can reach it. An example is *Nicotiana* (Fig. 10.8). It pays for bees to concentrate on flowers with concealed nectar, since they have an advantage over insects with short tongues. It is also advantageous to the plant since bees visiting them are likely to carry pollen of the same plant species. This is an example of *co-evolution*, in which two species influence each other's evolution.

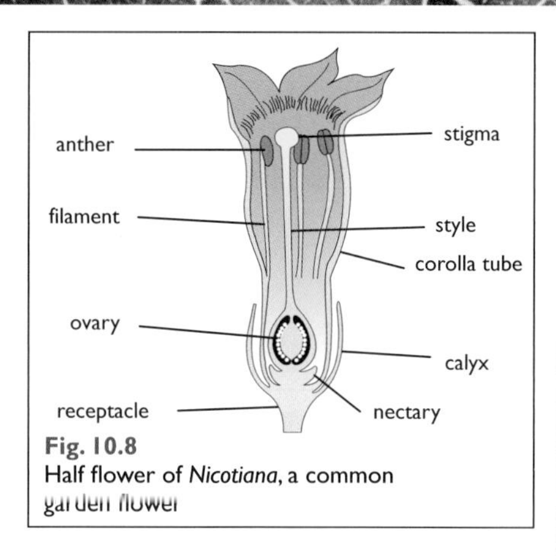

Fig. 10.8
Half flower of *Nicotiana*, a common garden flower

Some flowers restrict the range of insect visitors even more. The sweet pea and many other members of the legume family secrete their nectar into a trough formed from the filaments of nine of its ten stamens. To reach the nectar an insect needs more than a long proboscis; having landed on the wing petals, it must be heavy enough to depress the keel petals to which the wing petals are attached. Only bumblebees can do this.

Flower parts may be separate (as in the buttercup) or various parts may be joined together. Some flowers are clustered into **inflorescences**, such as the dandelion (Fig. 10.9). In this plant each 'petal' is actually the corolla of a tiny flower or *floret*, the five petals being fused into a single strap-shaped structure. The sepals are highly modified as tiny hairs that function as a parachute, dispersing the fruit.

In *dicotyledons* the floral parts are usually in multiples of four or five, while in *monocotyledons* they are usually in threes, for example the tulip (Fig. 10.10).

In the tulip and many other monocotyledons there appear to be six petals in two layers or 'whorls' of three. Only the inner whorl is the equivalent of the petals of a buttercup. The outer three are the equivalent of the sepals, but they resemble petals. These two whorls together make up the *perianth*. In most dicotyledons the two whorls of the perianth are different and are given different names (*calyx* and *corolla*). In the tulip and in many other monocotyledons the two whorls are similar and function as petals.

The formation of pollen

Pollen grains are produced within the four pollen sacs of each anther. Although they are produced by meiosis, *they are not male gametes*, but produce the gametes inside them. Each pollen grain is *haploid* because it has only one set of chromosomes.

The wall of a mature pollen grain consists of an inner, continuous *intine* and an outer *exine* that is perforated by pores. The exine is waterproof and very resistant to decay. Its shape and pattern is so

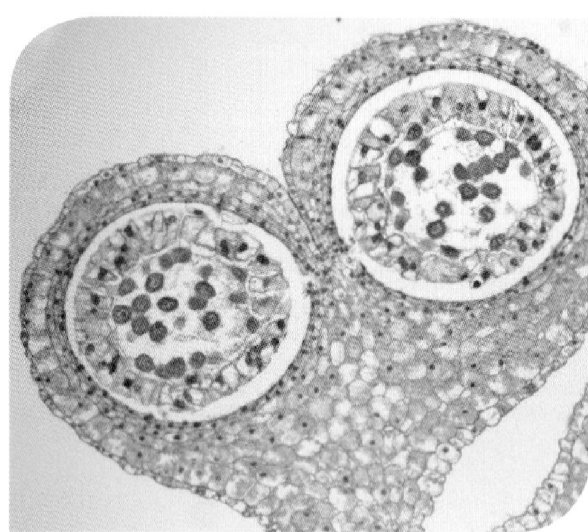
Transverse section of pollen sacs of lily

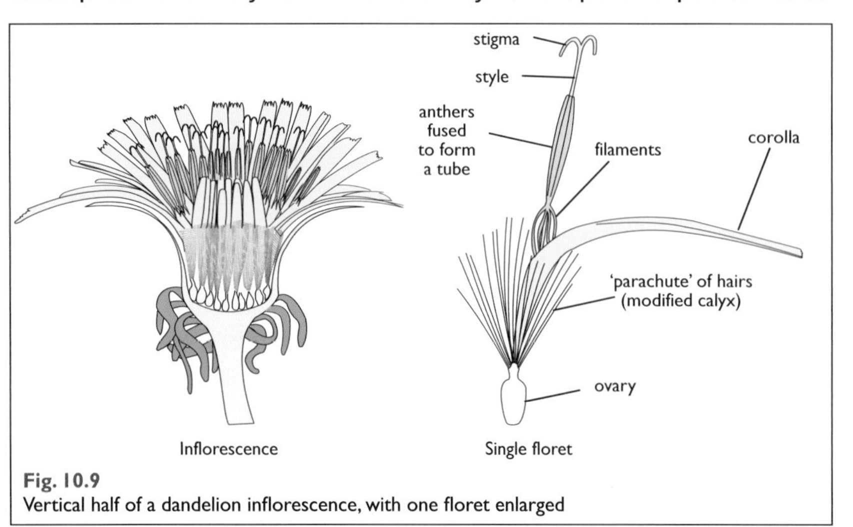

Fig. 10.9
Vertical half of a dandelion inflorescence, with one floret enlarged

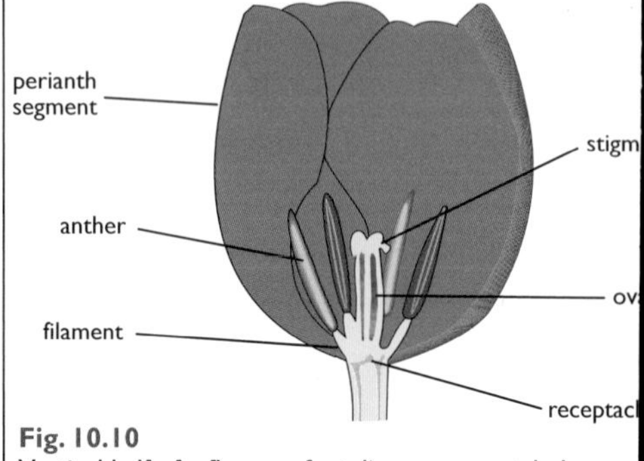

Fig. 10.10
Vertical half of a flower of a tulip, a monocotyledon

ISBN 9780170191340

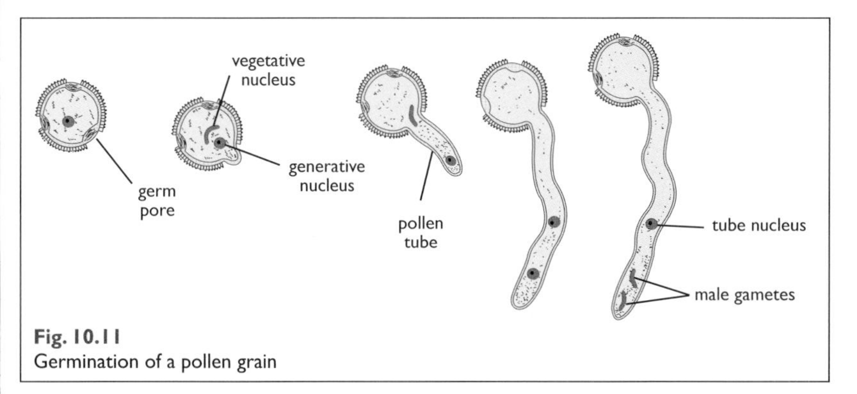

Fig. 10.11
Germination of a pollen grain

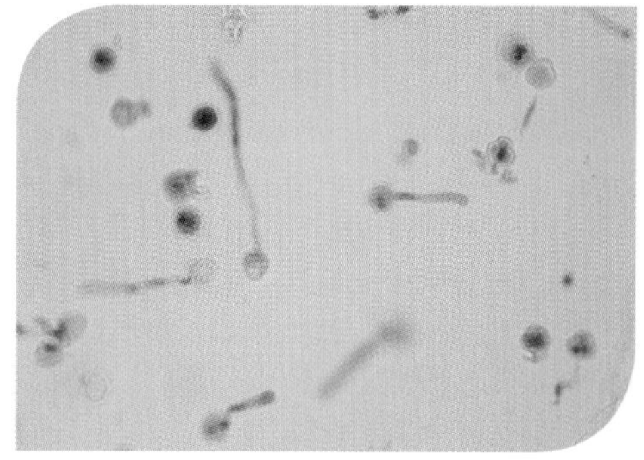
Pollen grains germinating in sugar solution

characteristic of the species that experts can identify the plant that produced it.

As the anther ripens, tensions develop as the wall dries, causing it to split lengthwise, releasing the pollen.

At about this time the nucleus of the pollen grain divides *mitotically* into two nuclei, a *tube nucleus*, and a *generative nucleus.* The generative nucleus later divides again to form two male gametes (Fig. 10.11).

The ovule and egg

An ovule is a potential seed, and develops on a part of the ovary wall called the **placenta**. It is partially surrounded by two collars of tissue called **integuments**, with a pore or **micropyle** (Fig. 10.12). Inside the ovule are eight haploid nuclei, only two of which are shown. One of these is the *female gamete* or egg.

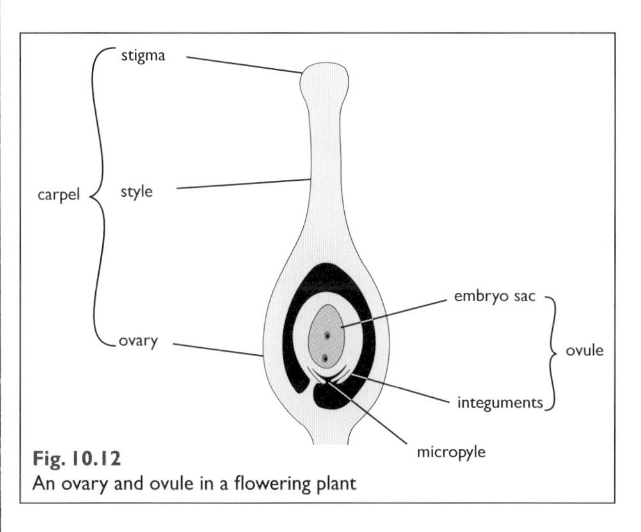

Fig. 10.12
An ovary and ovule in a flowering plant

Fertilisation

Pollination merely brings the pollen to a stigma; the egg remains some distance away inside an ovule. The next stage is the growth of the **pollen tube**, which brings the sperm to the egg — a process that can take several days.

Stimulated by sugar secreted by the stigma, the pollen tube grows out through one of the pores in the outer layer of the pollen grain. This tube grows down into the stigma and style, secreting digestive enzymes as it goes (Fig. 10.13). The simple products of digestion (such as sugars and amino acids) are absorbed by the pollen tube and used for growth. The length to which the pollen tube grows depends on the length of the style. In a buttercup the stigma is little more than a millimetre away from the ovule, but in some species it may be as much as several centimetres.

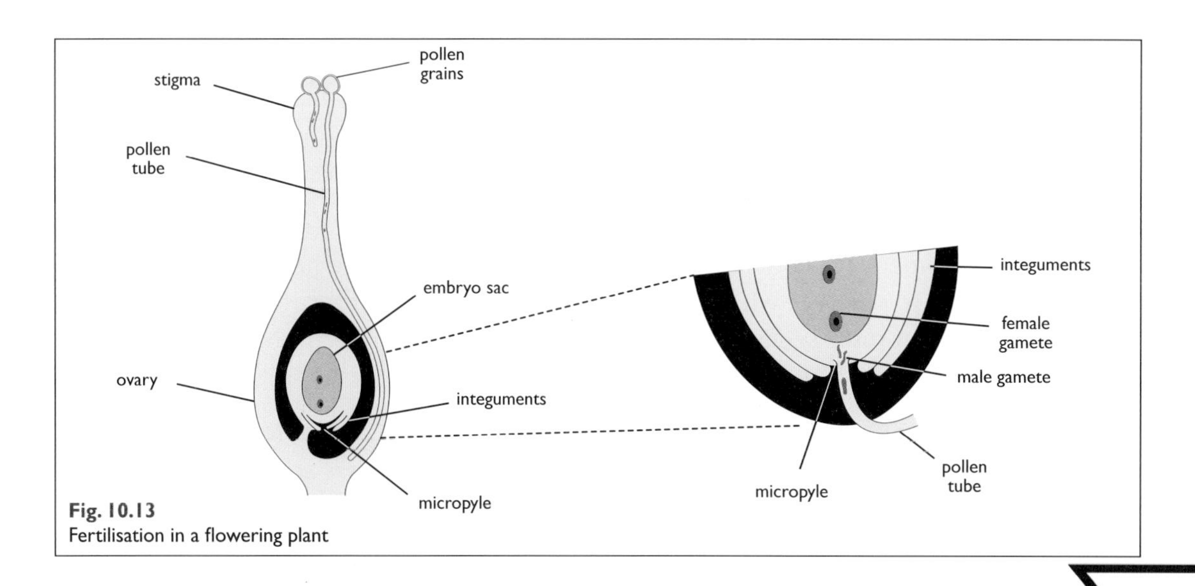

Fig. 10.13
Fertilisation in a flowering plant

ISBN 9780170191340

During its growth down the style, the generative nucleus divides mitotically into two *male gametes*, each of which retains a thin envelope of cytoplasm.

When the pollen tube reaches the ovary it grows toward an ovule, attracted by chemicals produced by the egg. The pollen tube usually grows through the micropyle, and the tip breaks down on reaching the egg, releasing the two sperm. One joins with the egg to form a *zygote*, and the other fuses with the other nucleus to form an *endosperm nucleus*. In most plants the ovary contains many ovules, so many pollen grains will be needed if they are all to develop into seeds.

The formation of fruit and seed

After fertilisation the endosperm nucleus divides many times to form a spongy storage tissue called the **endosperm**. Though all seeds have an endosperm to begin with, in *non-endospermic seeds* it disappears early in development, to be replaced by an energy store in the embryo itself. In *endospermic seeds*, such as cereals (e.g. rice, corn), the endosperm continues to develop and forms a large part of the mature seed.

The zygote meanwhile has been dividing mitotically and develops into the *embryo* that has three parts: the *radicle* or embryonic root, the two embryonic leaves or *cotyledons* (one in monocotyledons) and the embryonic shoot or *plumule*.

Pollination and fertilisation trigger a number of other changes:

- The integuments continue to grow and form the seed coat or **testa**.
- The ovary grows into the **fruit**, and the ovary wall becomes the fruit wall or **pericarp**.
- In most flowers the sepals, petals and stamens fall off.

The developing seed lays down large amounts of nutrient in the form of protein, and either fat or starch. In most seeds the final stage of development is marked by a large decrease in water content to about 5–20% and the complete cessation of any detectable metabolic activity. These changes are a necessary preparation for dispersal, in which the seed has to travel through conditions hostile to growth.

Fruits

A fruit is usually defined as a ripened ovary, and in many plants it plays an important part in the dispersal of the seeds. According to the way the pericarp develops, fruits fall into two categories:

1. In *succulent fruits* the pericarp becomes soft and juicy and is eaten by animals.
2. In *dry fruits* it becomes dry and tough and is not eaten.

Dry fruits containing more than one seed are usually **dehiscent** — that is, they open to release the seeds. In these fruits the seeds are dispersed individually. *Indehiscent fruits*, which do not open, usually contain only one seed. Some fruits are formed from parts of the flower other than the ovary and are called 'false fruits', for example a strawberry is a greatly swollen receptacle, the actual fruits being the little seed-like structures on the surface. An apple is formed by the ripening of a deeply cup-shaped receptacle, the actual ovary being the 'core'.

Dispersal of fruits and seeds

In most plants the survival chances of seeds are increased by *dispersal*. This always involves a considerable element of chance, and many do not reach a suitable habitat. The successful seeds do not have to compete with each other or with the parent for resources, and there is always a chance that some seeds may reach a better environment than that of the parent.

Dispersal by wind

Wind-dispersed fruits and seeds take diverse form but have one essential feature in common; a large surface/weight ratio (Fig. 10.14). In most fruits and seeds this is achieved by a complex shape in the form of a wing (e.g. sycamore) or feathery outgrowths (e.g. dandelion). Although

ISBN 9780170191340

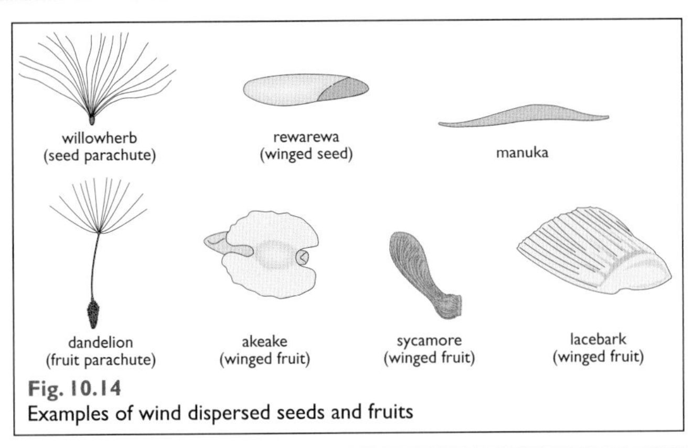

Fig. 10.14
Examples of wind dispersed seeds and fruits

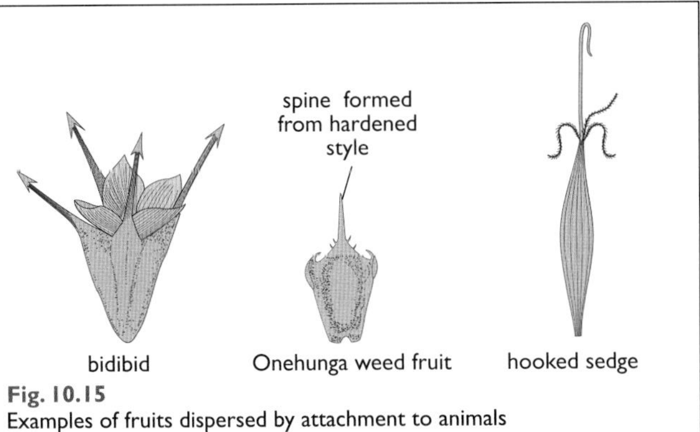

Fig. 10.15
Examples of fruits dispersed by attachment to animals

these outgrowths may be superficially similar, in some cases they have evolved by modification of different structures. In willowherb, for example, the parachute is an extension of the testa (seed coat), and in dandelion it is a highly modified calyx called a *pappus*.

In orchids the seeds are microscopic, are blown about like dust and carry insufficient stored energy to get established by themselves. Instead they obtain energy provided by a saprobic fungus, with which the orchid lives in a mutualistic relationship. Pohutukawa and manuka seeds are not as small as orchid seeds, but their elongated shape further increases their surface area/weight ratio.

Wind dispersal of seeds and fruits may appear wasteful, but this is offset by the fact that because the seed or fruit has to be light, larger numbers can be produced without great cost.

Dispersal by attachment

Dispersal by attachment to animals is somewhat less wasteful than wind-dispersal because animals tend to frequent areas supporting plant growth. In pre-human New Zealand, flightless birds such as kiwi, weka and moa were probably important in the dispersal of a number of kinds of fruits and seeds.

Hooks can be developed in various ways. In hook grass (not actually a grass but a sedge) the hook is the tip of the tiny stem to which each flower is attached. Bidibid has barbs projecting from the base of the joined sepals. In Onehunga weed, the style hardens to form a spine that becomes embedded in animals' feet (Fig. 10.15).

Dispersal by being eaten

Many fruits are eaten by birds, but pass unharmed through the gut to be deposited when the bird defaecates several hours later (Fig. 10.16). For example, the fruits of the puriri, *Coprosma,* fivefinger, supplejack and nikau palm are all favourite food of kereru (native pigeon). These fruits advertise themselves either by their bright colours or, in certain fruits, being shiny blue-black. Once eaten the seeds are protected from the digestive juices of the animal by a hard coating. In berries such as cabbage tree, fuchsia and tomato this is the testa. In puriri and other fruits with a 'stone', it is the inner layer of the pericarp. Since the food content of succulent fruits must be sufficient to act as a lure, this represents a cost to the plant and so puts a limit on the number of fruits it can produce.

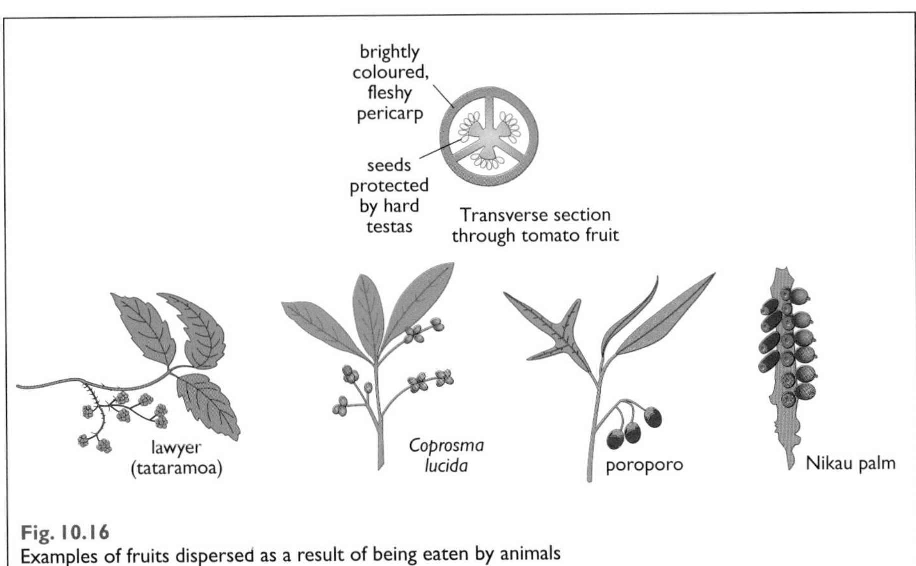

Fig. 10.16
Examples of fruits dispersed as a result of being eaten by animals

Water dispersal

The native mangrove is dispersed by water (Fig. 10.17). Its seeds begin to germinate before they are released from the fruit and float in the water, eventually coming to rest and beginning growth. In the coconut, the pericarp consists of many fibres which contain air, so the fruit floats. The outer layer of the pericarp is waterproof, so the fruit does not become waterlogged and

ISBN 9780170191340

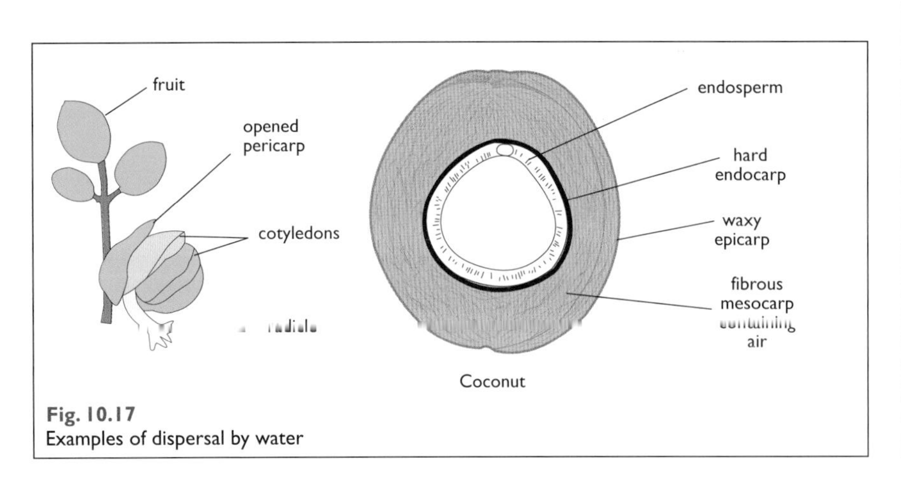

Fig. 10.17
Examples of dispersal by water

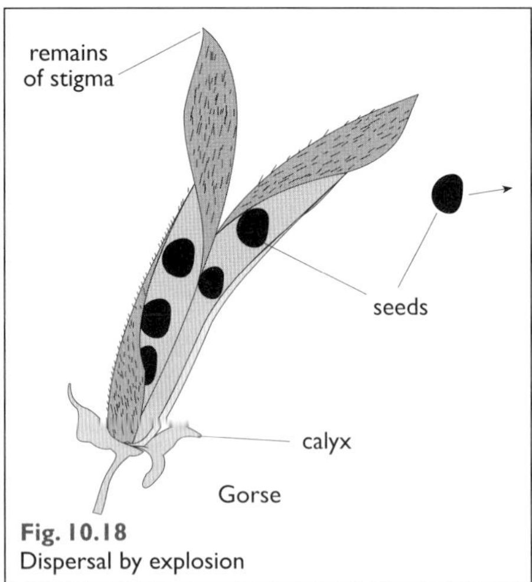

Fig. 10.18
Dispersal by explosion

can be carried for hundreds of kilometres (the coconut you buy has had its thick layer of fibres removed).

Self-dispersal

In some dehiscent fruits the pericarp is constructed so that it warps as it dries out. Tensions build, which are released suddenly with the violent opening of the fruit, jerking out the seeds, up to several metres in lupin (Fig. 10.18). Explosive dispersal is rare among native plants, known to occur only in matagouri, a spiny shrub.

Pepper-pot dispersal

Another form of dispersal is the censer, or 'pepper-pot' mechanism, in which seeds are jerked out of a fruit when knocked by a passing animal (e.g. a poppy, Fig. 10.19). Wind action is probably too gentle to jerk the stems sharply enough.

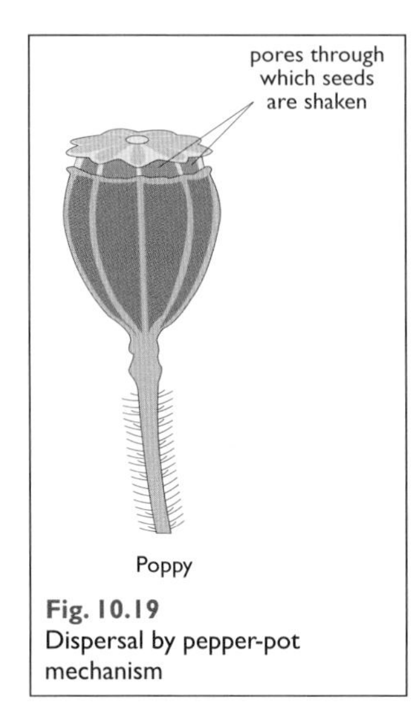

Fig. 10.19
Dispersal by pepper-pot mechanism

Kinds of plant life cycle

Based on the duration of the life cycle, plants can be divided into the following three groups:

1. **Annuals,** for example the garden pea. These have a life cycle lasting a year or less. Some annuals have life cycles measured in weeks and are called **ephemerals**.
2. **Biennials,** for example the carrot. These need two years to complete their life cycle. The first year is devoted to vegetative (non-reproductive) growth, in which energy reserves are stored. In the second year flowers and then seeds are produced, after which the plant dies.
3. **Perennials**. These have a life cycle lasting many years. In **herbaceous** perennials the aerial parts (shoots) die down at the end of each season, and the plant survives using energy stored below ground. In **woody** perennials the aerial parts persist, new growth resuming each spring from the previous year's shoots. In these plants energy is stored above as well as below ground, in the secondary xylem and phloem.

ISBN 9780170191340

Summary of key points in this chapter

- Flowering plants reproduce sexually by highly modified shoots called *flowers*.
- Before fertilisation can occur, pollen must be transferred from male to female parts of a flower on the same plant (self-pollination) or a different plant (cross-pollination).
- Pollen is transferred by external agents such as *insects* or other animals, or by *wind*.
- Insect-pollinated flowers invest much energy in advertising their presence by bright colours, and rewarding visitors by nectar and/or pollen.
- Wind-pollinated flowers are dull-coloured and do not produce nectar, but produce massive quantities of pollen.
- Pollen grains produce male gametes within them, and each ovule produces a female gamete within itself.
- After pollination, a pollen grain develops a pollen tube that grows down to and penetrates an ovule, releasing two male gametes, one of which fertilises the female gamete to form a *zygote*.
- After fertilisation the zygote develops into an embryo plant, the ovule develops into a seed, the integuments develop into the seed coat or *testa*, and the ovary develops into a *fruit*.
- Dispersal of fruits and seeds helps to avoid competition and makes it possible for the offspring to find better conditions.
- Seed and fruit dispersal may occur by animals, wind, water, or by self-dispersal.
- Plants may have life cycles extending over a year or less, two years, or many years.

Test your basics

Copy and complete the following sentences.

a) Sexual reproduction differs from asexual reproduction in that the offspring show genetic ___*___, whereas the offspring of asexual reproduction are always genetically ___*___ to the parent. This is because in sexual reproduction, chromosomes are 'reshuffled' into new combinations. This occurs at two stages in the life cycle; in meiosis, when the chromosome number is ___*___, and at ___*___, when it is doubled.

b) In flowering plants, fertilisation is preceded by ___*___, which is the transfer of ___*___ from an ___*___ to a ___*___ of a flower of the same species.

c) Insect-pollinated flowers are usually brightly coloured and scented, have ___*___ pollen, and most produce a sugary solution called ___*___. Wind-pollinated flowers are inconspicuous, produce ___*___ quantities of ___*___ pollen grains and never produce ___*___.

d) The male parts of a flower are the ___*___, each of which has a pollen-producing ___*___.

e) The female part of a flower is called the ___*___. In a buttercup this consists of many separate ___*___. The base of each carpel is the ___*___ and contains an ___*___ which contains the female ___*___ or egg.

f) The top of each carpel is the ___*___ on which pollen grains are deposited. The ___*___ secretes ___*___ that stimulate pollen grains to ___*___ by developing a ___*___ ___*___. This grows down into the ___*___ and ___*___, towards the ___*___.

g) The pollen tube contains two male ___*___ or sperm. When it reaches the ___*___, one of the ___*___ joins with the egg to form a ___*___. The other joins with one of the other nuclei in the ovule to form the ___*___ nucleus. The zygote undergoes many ___*___ divisions and eventually develops into the ___*___ plant. The endosperm nucleus divides repeatedly, forming a storage tissue called the ___*___. In many seeds this disappears before the seed is mature.

h) A fruit is a ripened ___*___, and in many plants is important in the ___*___ of seeds. The wall of the fruit is the ___*___.

i) Wind-dispersed fruits and seeds have a large ___*___ compared with their ___*___. This is achieved either by being very ___*___, or by having thin or feathery outgrowths.

j) Animal-dispersed seeds and fruits either stick to animals by means of ___*___, or are eaten and resist ___*___.

k) In ___*___ plants the life cycle is completed in less than a ___*___. ___*___ have a life span of two years; the first year being devoted to developing an ___*___ store, and the second to the production of ___*___.

ISBN 9780170191340

Plant Processes: Assesement Practice

QUESTION ONE

The potato and many other flowering plants reproduce both sexually and asexually.

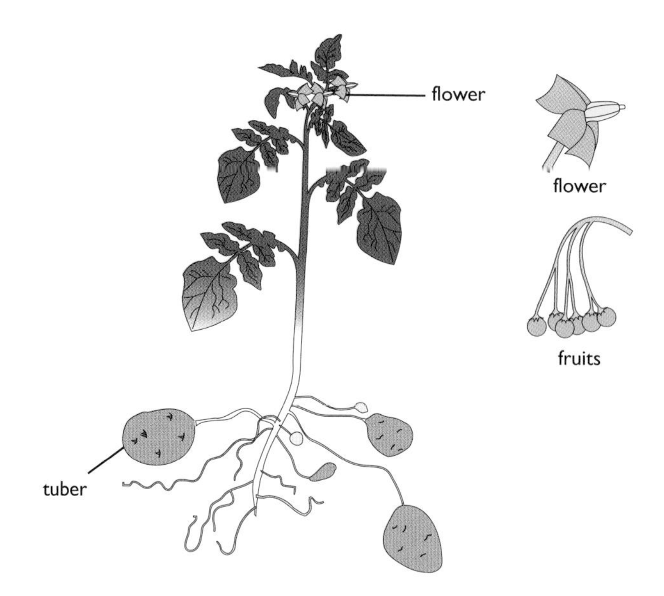

With reference to the potato, **discuss** the **advantages** and **disadvantages** to the plant of both sexual and asexual reproduction.

In your answer:

- **Describe** how the plant reproduces both sexually AND asexually.
- **Explain** how both sexual AND asexual reproduction can affect the dispersal AND variation in plants.
- **Explain** how the above ideas are related to reproductive success.

(Guidance: Your answer should be approximately 30 lines.)

ISBN 9780170191340

QUESTION TWO

In a demonstration of the conditions necessary for photosynthesis, the leaves of a variegated kumara plant were destarched by keeping the plant in darkness for 48 hours. One of the leaves was then partly covered with a stencil cut from cooking foil as shown below. The stencil was covered on both sides with transparent sellotape (sticky side away from the leaf surface) to ensure that access to air was equal in all parts of the leaf under the stencil. The plant was then left in bright sunlight for six hours. The leaf was then tested with iodine for starch and the results are shown below.

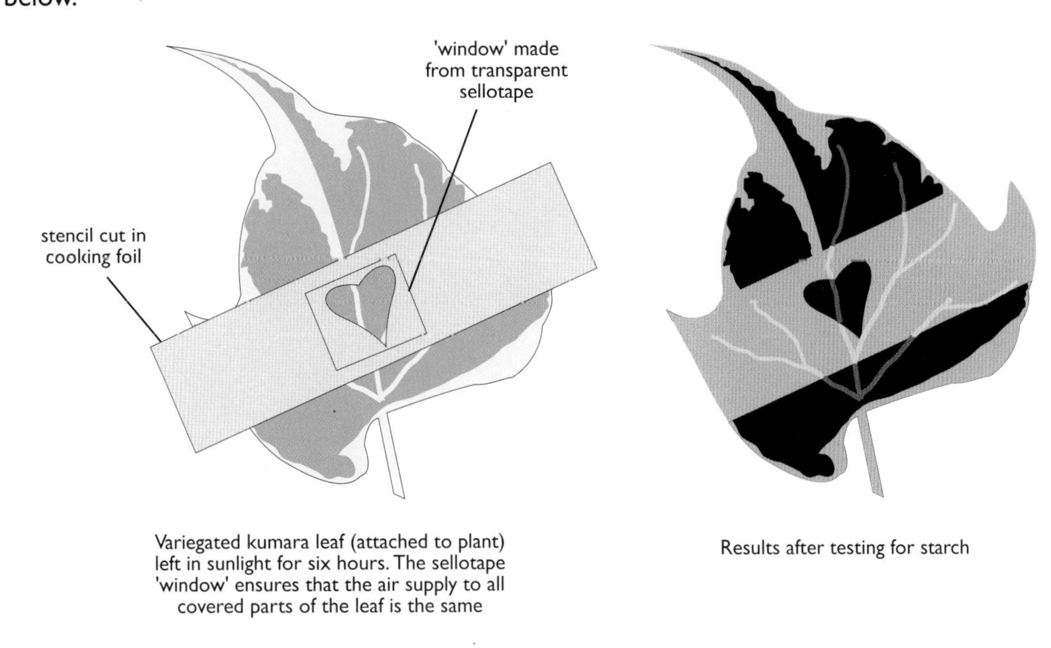

Variegated kumara leaf (attached to plant) left in sunlight for six hours. The sellotape 'window' ensures that the air supply to all covered parts of the leaf is the same

Results after testing for starch

(a) Using the information provided, describe the conditions needed for photosynthesis.

(Guidance: Your answer should be approximately 3 lines.)

(b) **Discuss** (i.e. *explain*) the results in part (a), **justifying** the resulting colour changes in **all** parts of the leaf.

(Guidance: Your answer should be approximately 28 lines.)

QUESTION THREE

Many flowering plants undergo secondary growth in addition to primary growth.

Discuss (i.e. *explain*) how both primary AND secondary growth occur in the stem of a plant.

In your answer you should consider:

- The role of the **apical meristem** in the growth of plants.
- **Where** primary and secondary growth occur in a stem.
- The **role** of primary and secondary growth in the life of the plant.

(Guidance: Your answer should be approximately 15 lines.)

ISBN 9780170191340

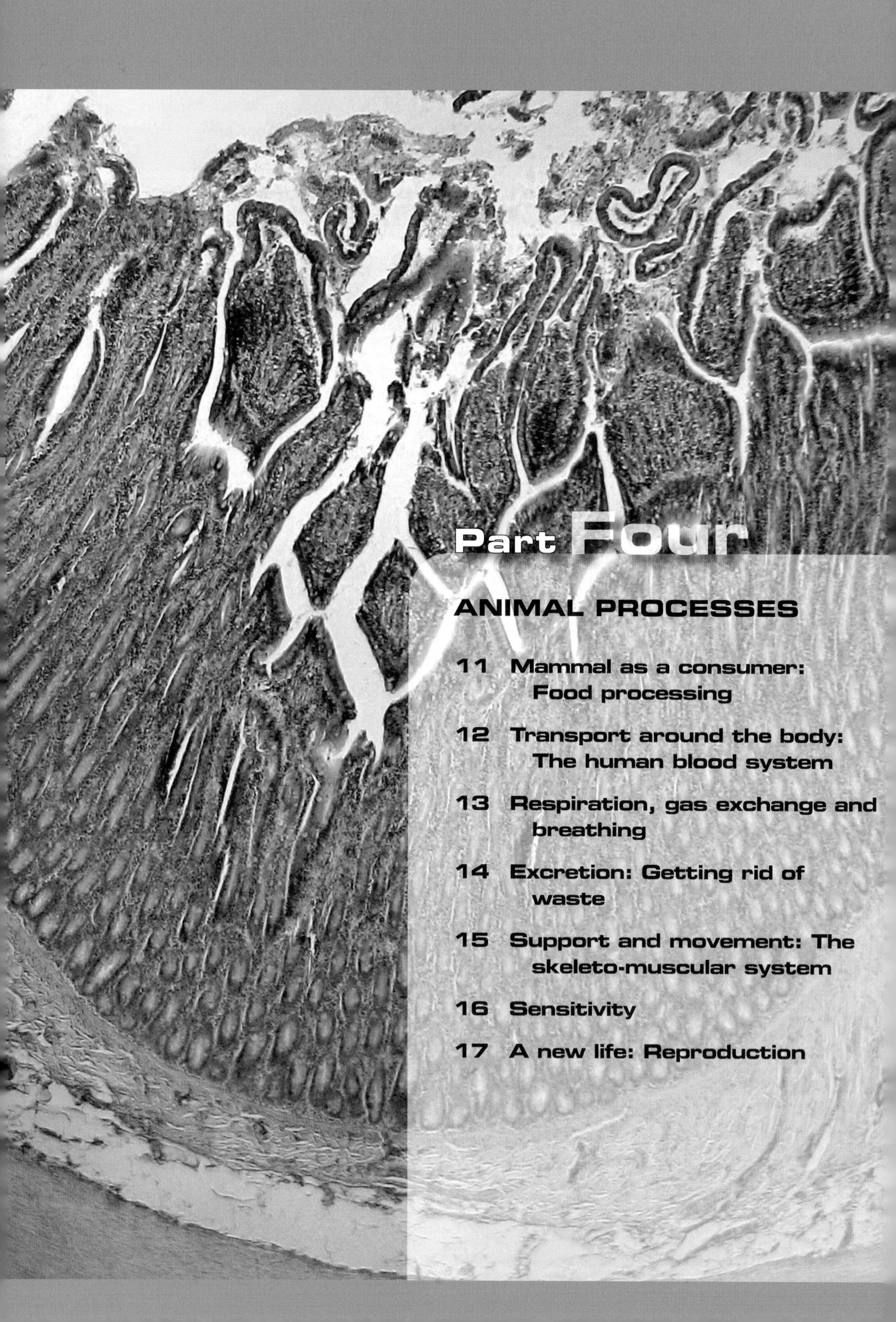

Part Four

ANIMAL PROCESSES

11 Mammal as a consumer: Food processing

Unlike plants, animals cannot make their own organic materials from carbon dioxide and water. They are **heterotrophic**, meaning that they obtain their food from other organisms (Latin: *heteros* = other, *trophic* = feeding). In most animals the food consists largely of organic substances such as starch, proteins and fats, together with water, vitamins and minerals.

Starch, proteins and fats are complex substances with large molecules. Before they can be used by the body they must be broken down into smaller molecules. This process is called *digestion* and is brought about by **enzymes**.

ENZYMES

The enzymes that bring about digestion are secreted by various *digestive glands* in the walls of the alimentary canal.

Each enzyme is *specific* for a particular action, so many are needed to digest a meal. Figure 11.1 summarises the essential changes.

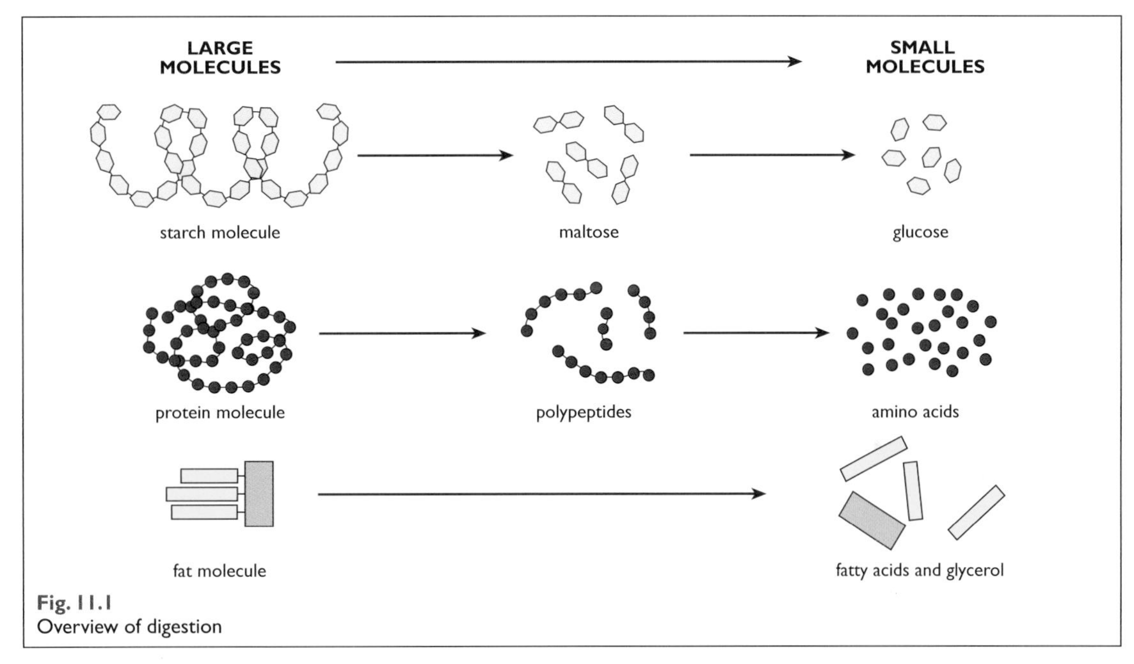

Fig. 11.1
Overview of digestion

Enzymes are very sensitive to temperature, and each enzyme has a 'best' or *optimum* temperature, at which it works fastest. In mammals this is body temperature, which in humans is about 37°C.

Enzymes are also sensitive to pH. Most enzymes work best at the pH inside cells, which is about 7.2 (just on the alkaline side of neutral). Digestive enzymes, however, work outside the cells and some work at pH values far from neutral. The pH of gastric (stomach) juice is about 1.5, which is extremely acid.

THE HUMAN ALIMENTARY CANAL

Digestion occurs in the **alimentary canal** or *gut*, which is a long tube beginning at the mouth and ending at the anus. You may think that food in your stomach is inside you, but as Fig. 11.2 shows, the cavity of the gut is actually an extension of the outside world. In theory, it would be possible to thread a long piece of string all the way from mouth to anus without causing injury!

What this means is that the food in the gut is strictly speaking *outside* the body. To get inside the body it must pass through the gut lining and into the blood, which carries it to the cells where it is used.

ISBN 9780170191340

Digestion is one of five stages of food processing. These are:

1. **Ingestion**, or the taking in of food into the gut at the mouth. This often involves physical breakdown of the large food lumps into smaller bits, by *chewing*.
2. **Digestion**, or the chemical breaking down of the complex foods into simpler, soluble substances.
3. **Absorption**, or the passing of the simple products of digestion through the walls of the gut.
4. **Egestion**, or the passing out of the gut of indigestible remains as *faeces*. Egestion is quite different from *excretion*, since faeces have not been produced in metabolism and have never actually been inside the body.
5. **Assimilation**, or the absorption of the digested foods by the cells.

The human alimentary canal is shown in more detail in Fig. 11.3.

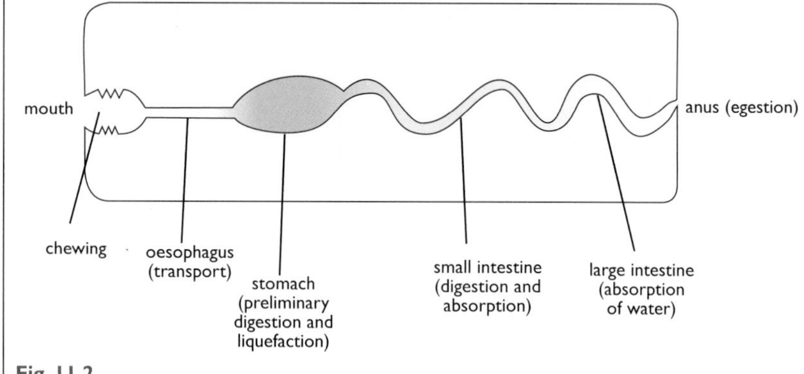

Fig. 11.2
Digrammatic view of the alimentary canal showing different activities and changes in pH (red = acid, blue = alkaline)

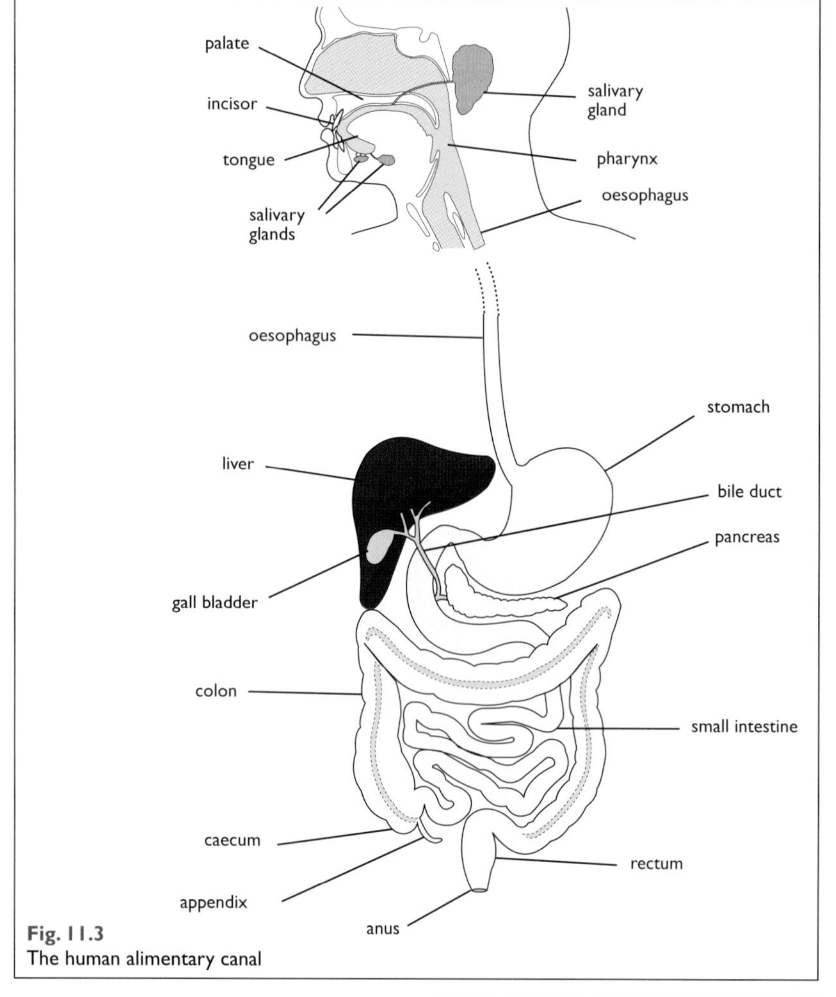

Fig. 11.3
The human alimentary canal

Teeth

The function of teeth is to break the food up into smaller pieces. As a result of chewing, the surface area of the food is increased, so digestive enzymes can work faster (Fig. 11.4).

Humans and other mammals have different kinds of teeth, adapted for special tasks. From front to back these are:

- **Incisors**, with chisel-like edges for cutting off pieces of food.
- **Canines**, only slightly pointed in humans and have a similar function to incisors.
- **Premolars** have one root and two cusps (projections on the surface of the tooth), and are used for crushing food into smaller pieces.
- **Molars** differ from premolars in that they are *first* teeth — they do not replace any earlier 'milk' or deciduous teeth. They have several roots and four or five cusps. They are larger than the premolars and, like them, crush the food. The last molars are called 'wisdom teeth' because they erupt (emerge from the gum) late — in the teens and early adulthood.

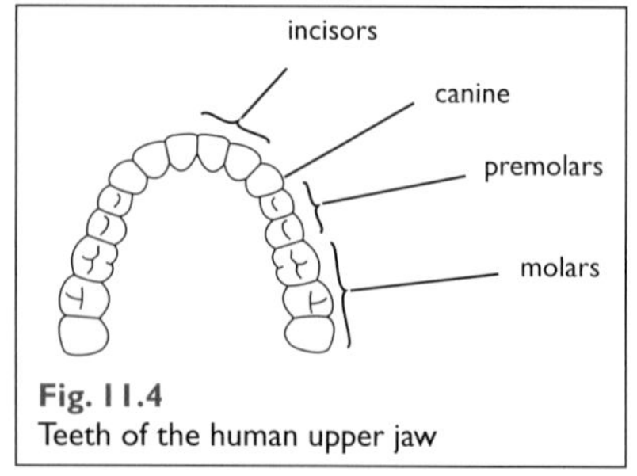

Fig. 11.4
Teeth of the human upper jaw

Teeth are anchored in sockets in the jawbone by tough collagen fibres of the **periodontal ligament**. Each tooth consists of a number of parts (Fig. 11.5):

- **Enamel** is the hardest substance in the body and covers the *crown*, forming the biting surface. It consists almost entirely of calcium phosphate.

ISBN 9780170191340

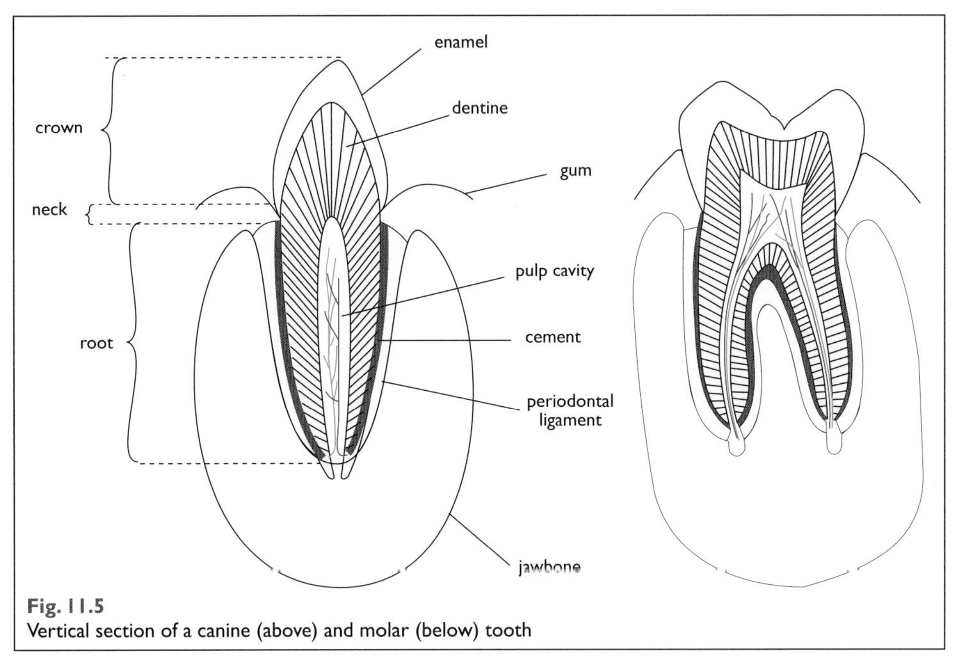

Fig. 11.5
Vertical section of a canine (above) and molar (below) tooth

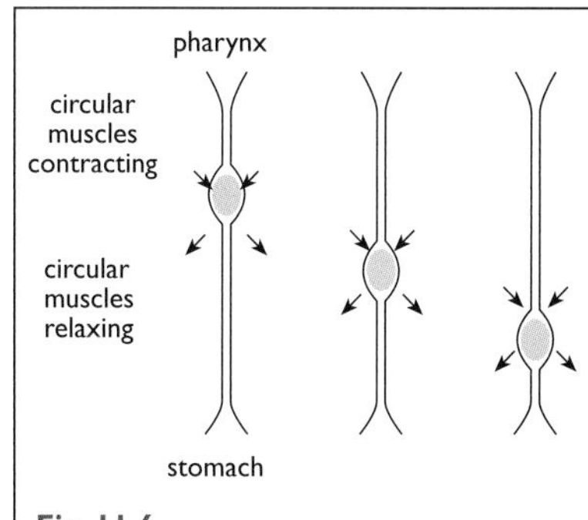

Fig. 11.6
How food passes down the oesophagus by peristalsis

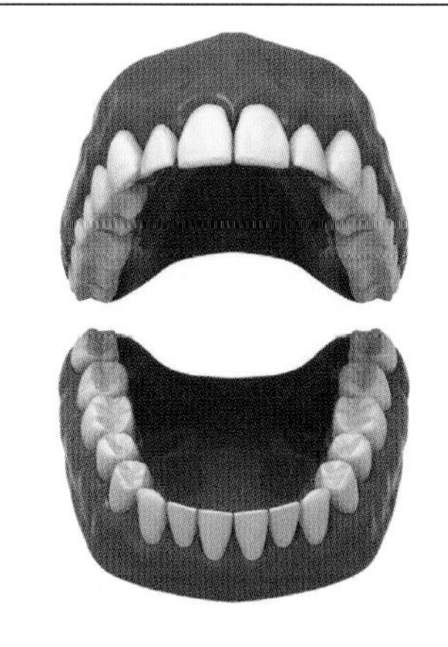

- **Dentine** forms the bulk of the tooth.
- **Pulp cavity** contains blood vessels and nerve endings.
- **Cement** covers the root of the tooth and resembles bone.

There are 32 teeth in the adult set, but these are often lost or damaged due to dental *caries* or decay. This is caused by bacteria attacking the dentine. In healthy teeth the dentine is protected by enamel, but sugar in the mouth is converted by bacteria into lactic acid which slowly eats away at the enamel, finally exposing the dentine. This allows a different group of bacteria to feed on the dentine (which, unlike enamel, contains protein).

Chewing

Aided by saliva secreted by **salivary glands**, chewing breaks the food up into a soggy lump or *bolus*. At the same time the enzyme *salivary* **amylase** converts starch into a complex sugar, *maltose*.

Chewing is sometimes called 'physical digestion'. It is, however, quite different from digestion, which is a *chemical* process.

Swallowing

Swallowing is the muscular propulsion of food from mouth to stomach through the **oesophagus**. First, the tongue pushes the bolus to the **pharynx** in which air and food passages merge. The food is prevented from entering the trachea (windpipe) by the flap-like **epiglottis**. The bolus is squeezed down the oesophagus by waves of muscular contraction called **peristalsis** (Fig. 11.6).

Digestion in the stomach

Gastric juice is secreted onto the food, and at the same time it is mixed by the peristaltic action of muscle in the **stomach** wall. The enzyme **pepsin** converts proteins into **polypeptides**. Gastric juice contains *hydrochloric acid*, which kills most bacteria and provides the low pH needed by pepsin. Over a period of 2–3 hours the food is slowly converted into a creamy white liquid called **chyme**.

The exit to the stomach, the *pylorus*, is guarded by a ring-like **sphincter** muscle, which regulates the rate that the stomach empties. As the chyme enters the duodenum it is neutralised by alkaline juices secreted by its walls.

Why doesn't the stomach digest itself?

The stomach is protected from its own digestive juice in several ways:

- Pepsin is secreted in an inactive form as *pepsinogen*, which is converted to pepsin

ISBN 9780170191340

by hydrochloric acid. The cells that secrete the acid are different from the cells that secrete pepsinogen, so pepsin is not activated until it reaches the stomach cavity.

- The stomach lining is protected by a thick layer of mucus.

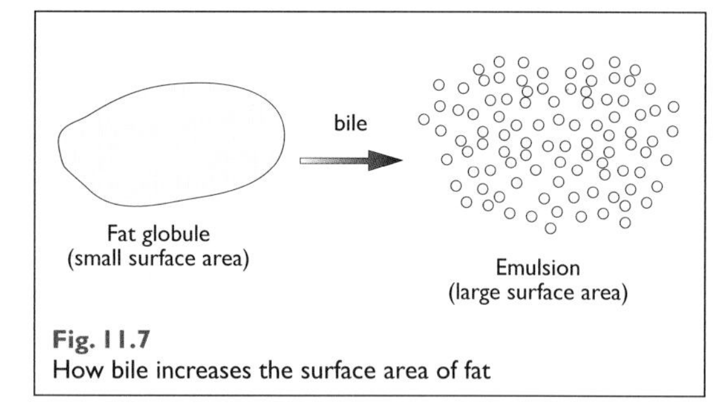

Fig. 11.7
How bile increases the surface area of fat

Digestion in the small intestine

As the food is liquefied it is slowly released into the **duodenum**, in which it receives two digestive juices:

1. **Bile** is a yellow-green, alkaline liquid. It is secreted by the **liver** (the largest gland in the body) and temporarily stored in the **gallbladder**. It contains no enzymes but contains salts which break large fat globules up into an *emulsion* of millions of tiny droplets (Fig. 11.7). This greatly increases its surface area, speeding up the digestion of fat. It also contains *sodium hydrogen carbonate*, which makes it alkaline.
2. *Pancreatic juice*, secreted by the **pancreas**, contains many enzymes that digest fat, proteins and starch:
   - *Amylase* breaks down starch to the complex sugar maltose.
   - **Lipase** breaks down fat to fatty acids and glycerol.
   - *Trypsin* breaks down proteins into polypeptides. This is secreted as inactive *trypsinogen*, which is activated by an enzyme secreted by the duodenum, thus protecting the pancreas from its own juice.
   - *Peptidases* convert polypeptides to dipeptides (consisting of two amino acids).

The rest of the small intestine, or **ileum**, is about 6–7 metres long in an adult, and is where the digestion of proteins and carbohydrates is completed by other enzymes, for example:

- *Lactase* breaks down lactose (a complex sugar found in milk) to the simple sugars glucose and galactose.
- *Maltase* converts maltose (a complex sugar) to glucose, a simple sugar.
- *Dipeptidases* split dipetides into amino acids.

The result is a mixture of *simple sugars* (mainly glucose), *amino acids*, *fatty acids* and *glycerol*. While the food is in the intestine it is mixed and propelled by muscular action.

Absorption of the digested food

The surface area of the intestinal lining is increased in three ways:

1. The lining has folds running around it.
2. It has many finger-like **villi**, each about 1 mm long.
3. The area is further increased by millions of **microvilli** on the individual cells lining the

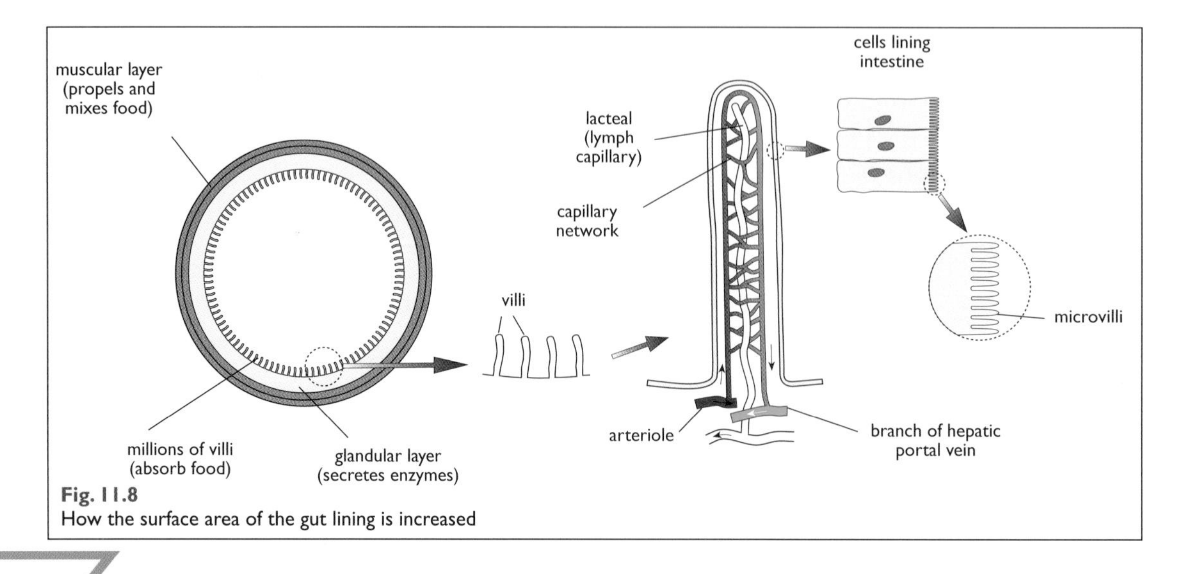

Fig. 11.8
How the surface area of the gut lining is increased

ISBN 9780170191340

The lining of the duodenum, showing transverse folds

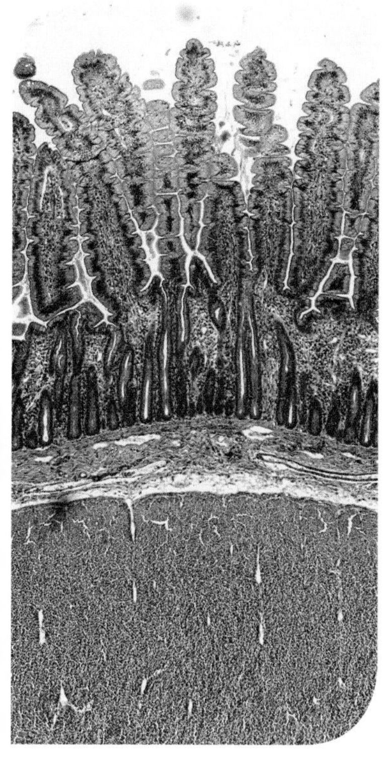

Section through the wall of the small intestine, showing villi

intestine (Fig. 11.8). Each villus contains a network of microscopic blood vessels called **capillaries**. These eventually join to form the **hepatic portal vein**, which takes blood to the liver. Unlike almost all other veins it begins and ends as capillaries.

Also inside each villus is a lymphatic capillary called a **lacteal**. The lacteals are the blind endings of the 'twigs' of the *lymphatic system*, which extends all over the body.

Glucose and amino acids

Glucose and amino acids are absorbed into the blood by *active transport*, which is one reason why the gut uses more energy after a meal. The energy for active transport is supplied by *respiration* in the abundant *mitochondria* in the cells lining the small intestine.

Fatty acids and glycerol

Fatty acids and glycerol get into the blood in a roundabout way via the lymphatic system. They cross the lining of the intestine and recombine to form microscopic fat droplets. These enter the tiny lymph capillary in each villus as an extremely fine suspension of fat droplets called an *emulsion*. This gives the lymph a milky appearance, hence the name 'lacteal'. The lymphatic vessels eventually open into the veins at the base of the neck.

The absorption of water

Water is absorbed by *osmosis*. The absorption of soluble food such as glucose makes the gut contents more dilute than the blood, so water follows by osmosis.

This is just as well, because the digestive juices contain large amounts of water (roughly six litres per day), which is too valuable to waste. About 95% is reabsorbed in the small intestine and most of the rest is reabsorbed in the **colon**, which forms the greater part of the large intestine.

The large intestine

By the time undigested food has reached the large intestine, all that remains is cellulose and some water, together with huge numbers of bacteria. The bacteria in the colon convert the material into **faeces**, which are temporarily stored in the rectum before being **egested** via the anus.

Many of these bacteria also synthesise certain vitamins, such as vitamin K, and people who are on prolonged courses of antibiotics may become deficient in this vitamin.

In rabbits and rodents an important part of the large intestine is the **caecum** and its extension, the **appendix**. These are large and contain cellulose-digesting bacteria. These bacteria also synthesise essential amino acids from non-essential ones.

In humans the caecum is a small pouch and the appendix is reduced to a worm-like extension.

The assimilation of the digested food

Once absorbed into the blood, the food is delivered to the cells, where it is used in one of two ways:

1. It may be broken down and used for energy (*catabolism*).
2. It may be built up into larger molecules (*anabolism*).

The energy needed to drive anabolism is supplied by the catabolic reactions of respiration.

ISBN 9780170191340

The fate of glucose

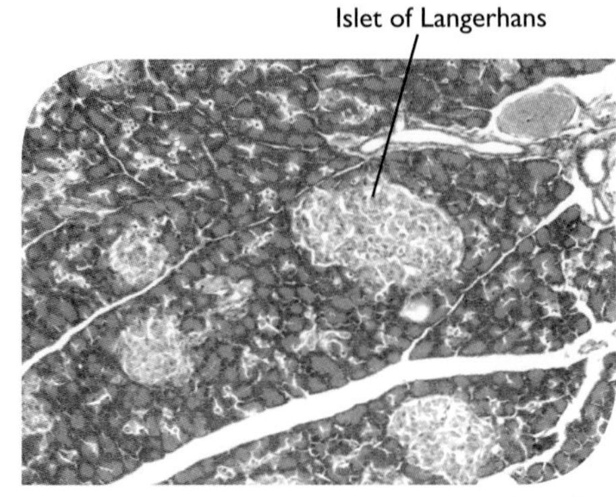

Section through pancreas tissue, showing Islets of Langerhans (seen as clusters of paler cells)

Glucose can be used in three possible ways:

1. Some is used to supply energy in *respiration* (Chapter 12).
2. Some is stored as **glycogen** (a complex carbohydrate resembling starch) in the liver and skeletal muscle.
3. Some is converted to *fat* and stored under the skin.

The use of glucose is under the control of **insulin** and other hormones. Insulin is secreted by tiny clusters of cells in the pancreas called **Islets of Langerhans**. After a meal the level of glucose in the blood rises, and this stimulates certain cells in the Islets of Langerhans to raise their secretion of insulin. Insulin has several effects, all acting to lower the concentration of glucose in the blood:

- It stimulates the uptake of glucose by cells.
- It stimulates the liver to convert glucose to glycogen.

In between meals the level of glucose in the blood falls and so does the output of insulin. Other hormones then help to maintain the blood glucose concentration by stimulating the conversion of glycogen into glucose.

In this way the concentration of glucose in the blood is regulated and does not vary much. This is one of many examples of **homeostasis**, or the regulation of the 'internal environment'.

If the pancreas cannot produce enough insulin, or if the cells become unable to respond to it, the cells cannot use glucose. The glucose level in the blood consequently rises — a condition known as **diabetes**.

Another hormone secreted by the Islets of Langerhans (but by different cells) is **glucagon**. This is antagonistic to insulin, promoting the conversion of glycogen into glucose and thus raising the concentration of glucose in the blood. After a meal the output of glucagon falls, and during a fast it rises.

The fate of amino acids

Amino acids are used in one of three ways:

1. Some are joined together to make proteins and used in *growth*.
2. Most are **deaminated** in the liver. This is the removal of *ammonia*, leaving an organic acid that is then used in respiration to supply energy. Ammonia is extremely poisonous and is immediately converted to *urea*, which is much less toxic. The urea is then excreted by the kidneys.
3. Some are converted to other amino acids; of the 20 kinds of amino acid used in making proteins, only nine are essential in the diet; the rest can be made from other amino acids.

The fate of fat

Fat is used in respiration, yielding more than twice as much energy per gram as carbohydrate. Fatty acids are also used in making cell membranes, which consist partly of fatty material.

ISBN 9780170191340

Summary of key points in this chapter

- Animals are *heterotrophic*, meaning that they obtain their energy by feeding off other organisms.
- Before food can be used by the cells, it must be *digested*, or broken down into simpler substances. This occurs in a tubular extension of the outside world called the *alimentary canal* or *gut*.
- Digestion is brought about by protein catalysts called *enzymes*, which are contained in juices secreted by glands lining the gut.
- In the first stage of food processing, the food is broken up by the *teeth* into smaller pieces, thus increasing its surface area for attack by enzymes.
- Mammals have different kinds of teeth specialised for different actions.
- The various enzymes are *specific* for different chemical reactions, and operate best in particular pHs.
- As a result of digestion starch is converted into glucose, proteins are converted into amino acids, and fats are converted into fatty acids and glycerol.
- Glucose and amino acids are absorbed by *active transport* and transported to the liver.
- Fatty acids and glycerol are absorbed and immediately reconverted to fat droplets, which reach the blood system via the *lymphatic system*.
- Most of the water is absorbed from the small intestine, while most of the remainder is absorbed via the large intestine.
- Digested products are either built up into larger molecules again, or broken down into smaller molecules with the release of energy.

Test your basics

Copy and complete the following sentences.

a) Digestion is a process in which __*__ food molecules are broken down into __*__ ones by protein catalysts called __*__.

b) Enzymes are very sensitive to conditions, especially __*__ and pH. Each enzyme is __*__ in its action, meaning that it only __*__ one kind of reaction.

c) As a result of digestion, starch is eventually broken down into __*__, proteins are broken down into __*__ __*__, and fats are converted to __*__ __*__ and __*__.

d) The chemical breakdown of food is assisted by the physical breakup of the food by __*__ by the action of the __*__.

e) A tooth consists mainly of __*__, capped by __*__, which consists of __*__ __*__ and is the hardest substance in the body. Inside the tooth is the __*__ __*__, containing __*__ vessels and __*__.

f) Digestion begins in the mouth with the breakdown of __*__ into __*__ by __*__ in the __*__, secreted by the __*__ glands. This continues until stopped by the acid in the __*__ juice.

g) In the stomach, proteins are broken down into __*__ by the enzyme __*__, activated by the __*__ acid in the __*__ juice. Digestion is assisted by the mixing of the food by __*__ action of the stomach walls. The food is slowly converted into a creamy white liquid called __*__.

h) The partially digested food is slowly allowed to leave the stomach by relaxation of the __*__ muscle guarding its exit. As it enters the __*__, it is neutralised by alkaline juices secreted into it.

i) Two alkaline juices acting in the small intestine are bile, secreted by the __*__, and __*__ juice. Bile contains no enzymes, but contains salts that break up fat globules into a fine __*__, which greatly increases its __*__ area.

j) The pancreas secretes a juice containing a number of enzymes. These include __*__ which breaks down fat into __*__ __*__ and glycerol, __*__ which converts __*__ into maltose, and trypsin, which converts __*__ into shorter chains of amino acids called __*__.

ISBN 9780170191340

k) The absorption of the digested food is helped by the presence of minute, finger-like ___*___ in the intestinal lining. Each ___*___ contains a network of microscopic blood vessels called ___*___, together with a lymphatic capillary called a ___*___. The surface area of the gut lining is further increased by millions of ___*___, which are submicroscopic extensions of the plasma membranes of the cells lining the intestine.

l) Glucose and amino acids are absorbed by ___*___ ___*___, which requires ___*___ from respiration. The glucose and amino acids are transported to the ___*___ by the ___*___ ___*___ vein.

m) The absorption of digested food lowers the ___*___ concentration in the intestine, causing water to enter the blood by ___*___.

n) Most of the remaining water is absorbed in the ___*___.

o) In the liver, surplus glucose in converted to ___*___ and stored in the liver cells. This is promoted by the hormone ___*___, secreted by the ___*___. Any glucose that cannot be stored in the liver is converted to fat and stored under the skin. Glucose that is not stored is used in ___*___ to supply energy.

p) Amino acids are either built up into ___*___, or are ___*___ in the liver. In this process the nitrogen is removed as ___*___, which is immediately converted to ___*___, which is much less toxic.

q) Fats are used in ___*___ to supply energy.

ISBN 9780170191340

12 Transport around the body: The human blood system

The blood vascular system is the body's *transport system*. It carries food (e.g. glucose and amino acids), oxygen, wastes (carbon dioxide and urea), hormones, defence proteins and defence cells, and heat.

PARTS OF THE BLOOD SYSTEM

The human blood system consists of five components:

1. **Blood**; the material in which substances are transported.
2. **Heart**; provides the driving force that propels the blood.
3. **Arteries**; carry blood *away from* the heart and distribute it around the body.
4. **Veins**; return blood from all parts of the body *toward* the heart.
5. **Capillaries**, microscopic vessels in which substances enter and leave the blood, and connect the smallest arteries and veins.

What is blood made of?

Blood consists of blood cells (about 45% of the blood volume), suspended in a pale straw-coloured liquid called **plasma** (Fig. 12.1).

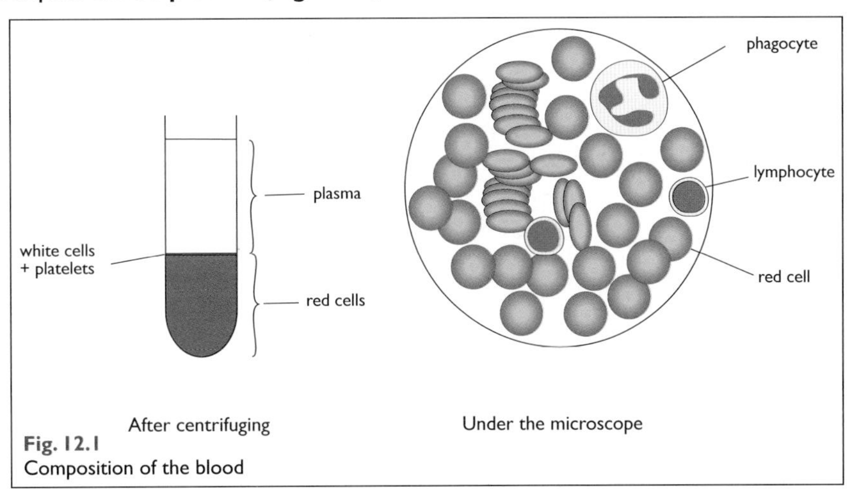

Fig. 12.1
Composition of the blood

Plasma

This is a complex liquid consisting of about 95% water and a wide variety of solutes, including:

- Inorganic ions, e.g. sodium, calcium and chloride.
- Nutrients, e.g. glucose, amino acids and vitamins.
- Waste products, e.g. CO_2 and urea.
- **Hormones**, e.g. insulin, adrenaline, oestrogen and testosterone.
- Antibodies.

Blood cells

These are all produced in the *bone marrow* and are of two general types: red cells and white cells.

Red cells (erythrocytes)

These carry oxygen from the lungs to all other parts of the body and are packed with **haemoglobin**, a red oxygen-carrying protein containing iron. They are biconcave discs and have *no nucleus*, making more room for haemoglobin (Fig. 12.2).

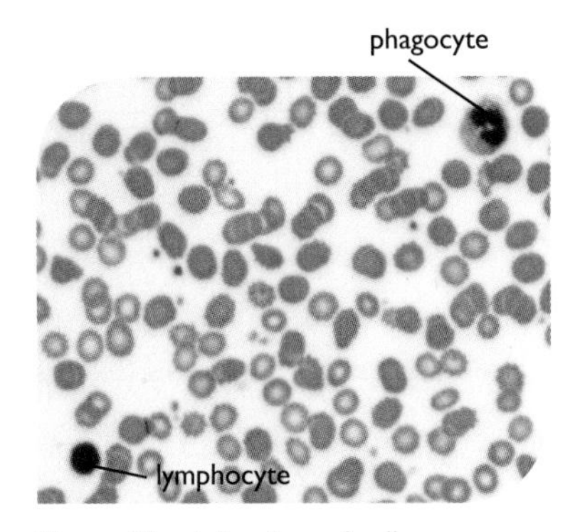

Human blood showing red cells, a phagocyte and a lymphocyte

ISBN 9780170191340

At the high oxygen concentrations in the lung capillaries, haemoglobin combines with oxygen to form *oxyhaemoglobin*. At the same time the blood changes from dark red to scarlet. At the low oxygen concentrations in all other capillaries, oxyhaemoglobin splits up into haemoglobin and oxygen, and the blood becomes dark red again:

$$\text{haemoglobin + oxygen} \underset{\text{rest of body}}{\overset{\text{lungs}}{\rightleftharpoons}} \text{oxyhaemoglobin}$$

Most organs remove less than half the oxygen from the blood flowing through them, so 'deoxygenated' blood still contains a reserve of oxygen.

Fig. 12.2
A red blood cell from above (left) and section through (right)

The biconcave shape of the red cells gives them a *high surface area compared with their volume*. It also causes them to take up a bell-shape as they pass along the capillaries (see below).

Red cells are constantly being produced by cell division in the bone marrow. They only live about four months, after which they are destroyed in the *liver*. In the breakdown of haemoglobin, green-yellow *bile pigments* are produced. These are excreted in the bile and give its characteristic colour. The iron is not excreted but stored in the liver and 'recycled' to make haemoglobin in new red cells.

Bile pigments are not always formed in the liver. As a result of a mechanical blow, capillaries are damaged and blood leaks out into the tissues. White blood cells attack the red cells and break down the haemoglobin, forming the characteristic yellow or green of a bruise.

Despite the recycling and storage of iron, some is slowly lost from the body, so you need to replenish your iron from your diet. Good sources are liver, kidney and meat, or if you are a vegetarian, fresh green vegetables.

Carbon dioxide transport

Whereas oxygen is carried in the red cells, most CO_2 is carried in the plasma in the form of *hydrogen carbonate* ions:

$$CO_2 + H_2O \underset{\text{lungs}}{\overset{\text{rest of body}}{\rightleftharpoons}} HCO_3^- \text{ (hydrogen carbonate)} + H^+$$

White blood cells (leucocytes)

These defend the body against **pathogens,** or disease-causing organisms. They are produced in the *bone marrow* and are of two main kinds (Fig. 12.3):

1. **Lymphocytes** can grow into **plasma cells** that make defence proteins called **antibodies**. These combine with 'foreign' substances or **antigens** on the surfaces of pathogens.
2. **Phagocytes** feed on bacteria, digesting them *intracellularly* (within the cells). Bacteria are especially eaten if they have first been covered with antibody.

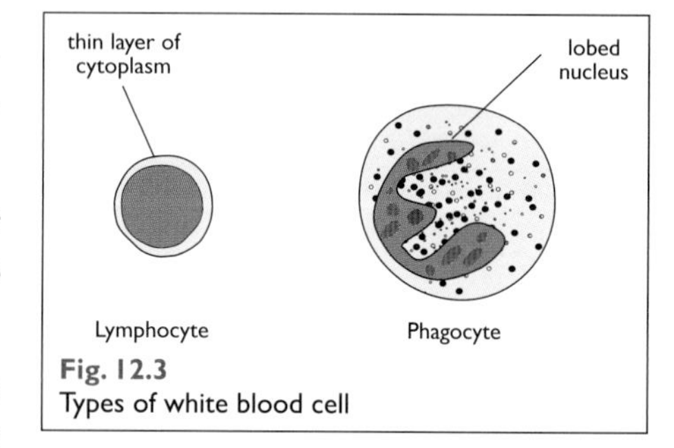

Fig. 12.3
Types of white blood cell

The heart

The mammalian heart is a *double pump*; the right side pumps deoxygenated blood through lungs, and the left side pumps oxygenated blood through the rest of the body. Each side consists of a thin-walled receiving chamber, the **atrium**, and a thicker-walled pumping chamber, the **ventricle**. There are thus two atria and two ventricles (Fig. 12.5). Humans and other mammals are said to have a *double circulation* because blood passes through the heart *twice* in each complete circuit.

- In the *pulmonary circuit* the right side of the heart pumps blood through the **pulmonary arteries** to the lungs and back to the left side of the heart via the **pulmonary veins**.
- In the *systemic circuit* the left side of the heart pumps blood through the **aorta** and its branches to the rest of the body and back to the right side of the heart via two huge veins, the **venae cavae**.

ISBN 9780170191340

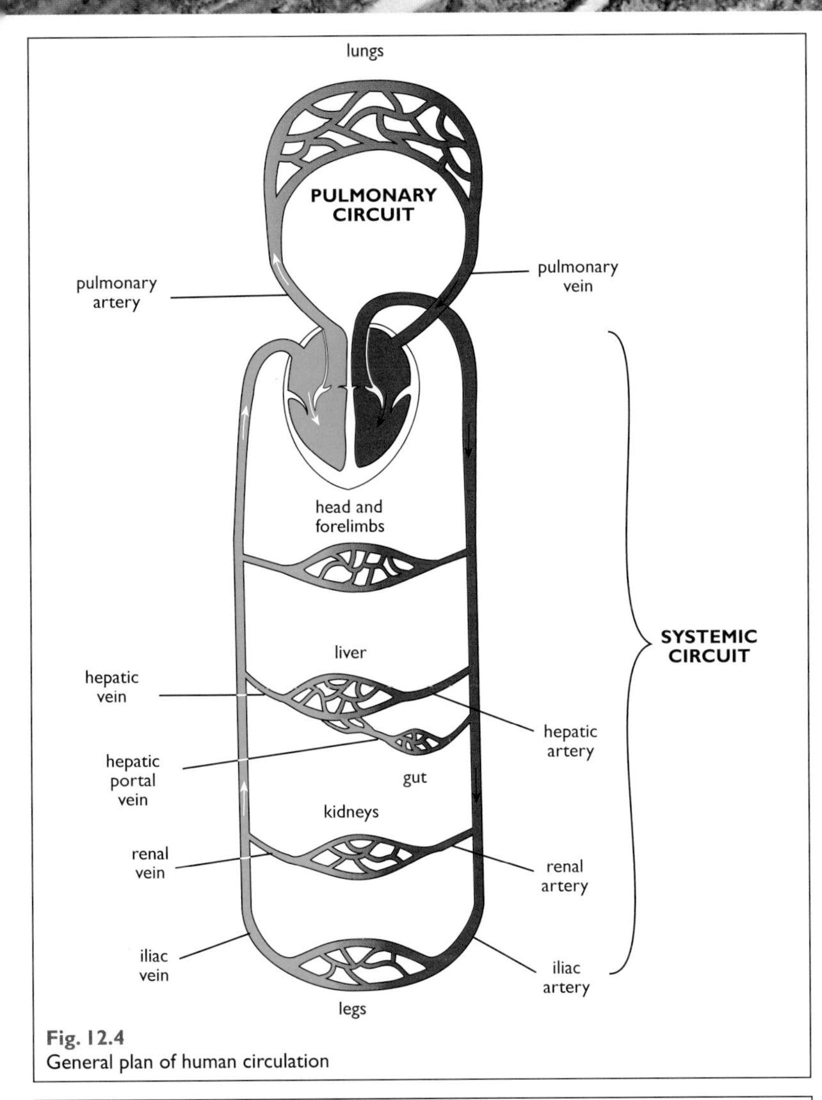

Fig. 12.4
General plan of human circulation

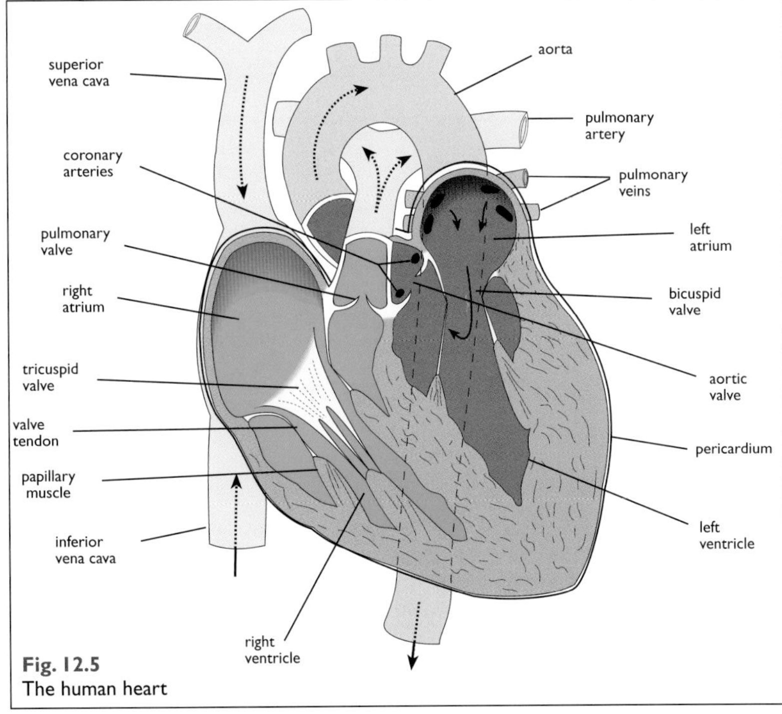

Fig. 12.5
The human heart

The heart lies in a bag called the **pericardium**, separated from the heart by a thin layer of lubricating fluid.

Since all the blood that leaves the left side of the heart comes back to the right side, the two sides of heart are actually in *series* rather than in *parallel*. The two halves must therefore pump *equal volumes* of blood. However, the left ventricle is much more muscular than the right because it has to exert a much higher pressure, for reasons explained later.

Each chamber alternately contracts and relaxes; when the two ventricles are contracting, the two atria are relaxing, and vice versa. The blood is prevented from flowing backward by four valves, of two kinds. Between each atrium and ventricle is an **atrio-ventricular valve** (A–V valve). Its flaps are prevented from turning 'inside out' by tendons attached to the ventricle wall. The A–V valve on the right side is called the **tricuspid valve** because it has three flaps. The valve on the left side is called the **bicuspid valve** (with two flaps). At the base of the aorta is a **semilunar valve**, consisting of three half-moon shaped flaps. Another semilunar valve lies at the base of the pulmonary artery. The heart valves cannot move themselves, but are pushed open or closed by blood pressure. The cycle of events in the heart is shown in Fig. 12.6.

The walls of the heart consist of **cardiac muscle**, which differs from skeletal muscle in two important ways:

1. It contracts *spontaneously* (without outside stimulation via nerves).
2. After each contraction it *has to relax*, so it cannot make a sustained contraction and cannot get cramp or fatigue.

The heart muscle has its own internal blood supply — the two **coronary arteries**, which are the first branches of the aorta, and the **cardiac veins**.

The heart of an average adult beats about 70 times a minute at rest, and pumps about 5 dm^3 ('litres') per minute (both values vary considerably). During exercise the heart output increases because it beats faster and also pumps more out each beat. Heart output is regulated partly by the hormone **adrenaline** (which speeds it up), and partly by the two sets of nerves that supply it. The *sympathetic* nerves speed it up while the *parasympathetic* nerves act to slow it down.

ISBN 9780170191340

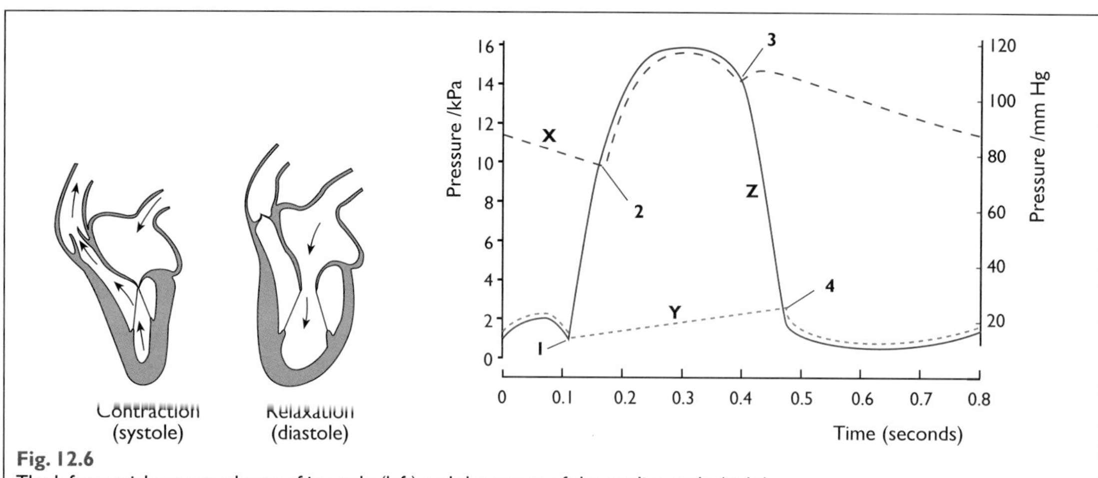

Fig. 12.6
The left ventricle at two phases of its cycle (left) and the events of the cardiac cycle (right).
X = pressure in aorta, Y = pressure in left atrium, Z = pressure in left ventricle, 1 = closing of bicuspid valve, 2 = opening of aortic valve, 3 = closure of aortic valve, 4 = opening of bicuspid valve.

Heart sounds

The clapping shut of the heart valves give rise to the *heart sounds* that can be heard with a *stethoscope*. Closure of the A–V valves produce a soft ('lub') sound, and the semi-lunar valves causing a sharper ('dup') sound. An abnormal heart sound may indicate a defective heart valve.

Arteries

These carry blood *away* from the heart. All arteries *except* the pulmonary artery carry oxygenated blood. To resist the high pressure, the walls are *thick* and *strong* due to abundant fibres of **collagen**, which is the main protein in tendons and ligaments (Fig. 12.7).

The larger arteries have very *elastic* walls. This helps to smooth out the surge of the pulse, so by the time the blood reaches the smallest arteries the pressure and flow are steady (Fig. 12.8). Arteries also have *smooth muscle* in their walls, which is not under our conscious control. Along the arterial tree, the proportion of elastic tissue decreases and the proportion of muscular tissue increases.

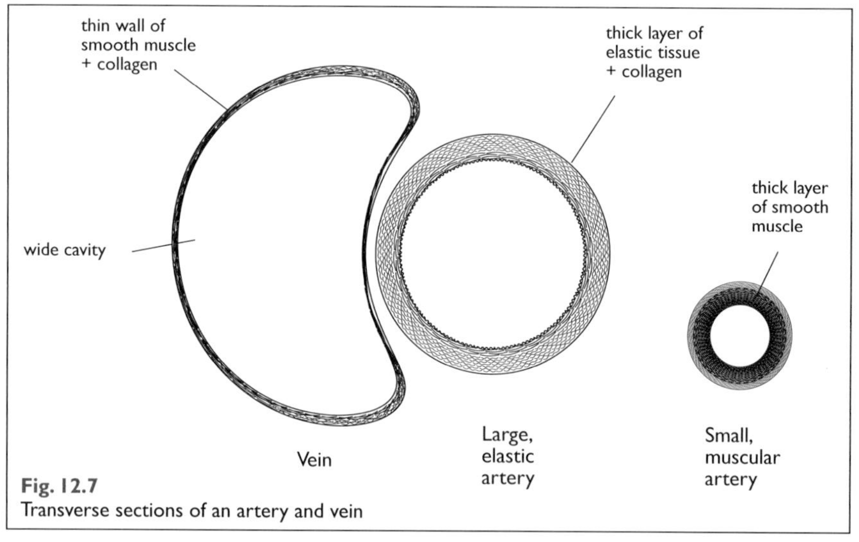

Fig. 12.7
Transverse sections of an artery and vein

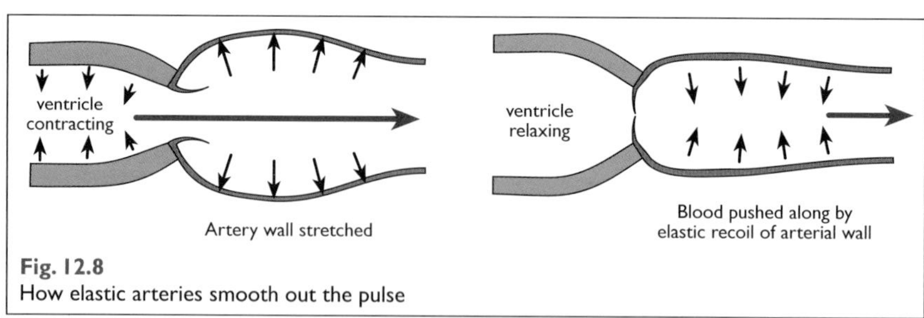

Fig. 12.8
How elastic arteries smooth out the pulse

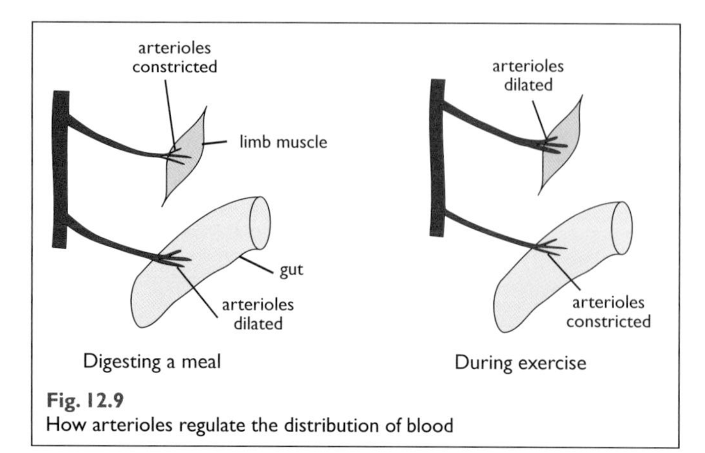

Fig. 12.9
How arterioles regulate the distribution of blood

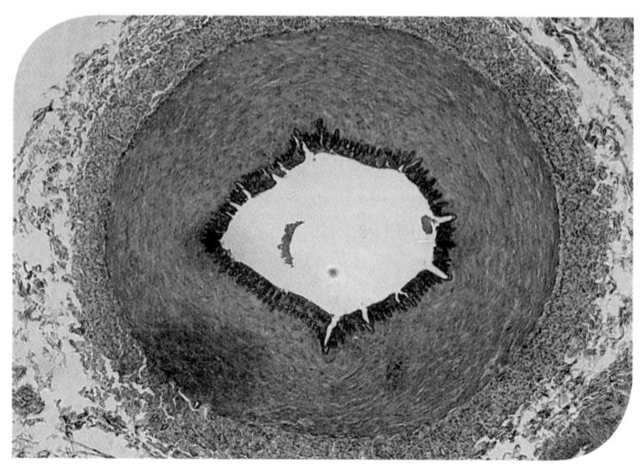

Transverse section through a smaller, muscular artery

ISBN 9780170191340

The smallest arteries are **arterioles**, and their walls contain a high proportion of smooth muscle. Arterial muscle does not *propel* the blood, but regulates its *distribution*.

For example, during exercise blood is diverted to the limb muscles from other parts, such as the gut. The muscle in arterioles supplying the limb muscles relaxes, so blood pressure forces them to get *wider*, allowing more blood to get through (Fig. 12.9). At the same time the muscle in the arterioles supplying the gut contract, causing them to become *narrower*, so less blood gets through.

Why is the left side of the heart more muscular than the right?

It is often stated that the left ventricle is more muscular than the right ventricle because 'it has to pump blood all the way around the body, whereas the right ventricle only has to pump it through the lungs'.

This is quite incorrect, for the following reason. By far the greatest resistance to blood flow is in the arterioles and capillaries, which only account for the last few millimetres of the journey from heart to tissues. The distance from heart to the largest arterioles is therefore of little consequence.

So, what is the explanation? There are three reasons:

1. **Whereas blood pressure forces tissue fluid out of the *systemic* capillaries, this would be harmful in the lungs. A layer of tissue fluid in the alveoli (Chapter 13) would greatly slow down diffusion of gases into and out of the blood. As diffusion through liquid is much slower than through a gas it is therefore essential that the pressure exerted by the right ventricle is *low*, otherwise tissue fluid would form in the lungs.**
2. **Another advantage of high systemic blood pressure is that the kidney requires a high capillary pressure for rapid filtration. The pressure in the kidney capillaries where filtration occurs is nearly double that in the other systemic capillaries.**
3. **During intense exercise the heart increases its output about four times (up to seven times in marathon runners), but blood flow through the muscles may increase up to 20 times. The rest is due to diversion of blood from other parts of the body. It is essential that this massive diversion does not result in a fall in blood pressure in the brain (something like this happens when somebody turns on a tap in the kitchen when you are in the shower). The reason why this does not happen is because the systemic blood pressure is sufficiently high to provide a 'reserve'.**

Veins

These carry blood *toward* the heart. They have much thinner walls and wider cavities than arteries (Fig. 12.7). All veins *except* the pulmonary veins carry deoxygenated blood.

Veins have two features that help the flow of blood:

1. Their wide cavities reduces the resistance to flow (the wider the cavity, the slower the flow).
2. By the time the blood has reached the veins, pressure has fallen, and return of blood to the heart is helped by **valves**. During exercise, veins are squeezed by contraction of

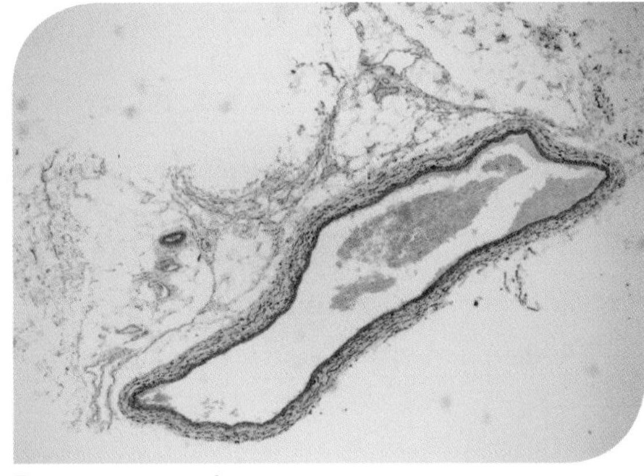

Transverse section of a vein

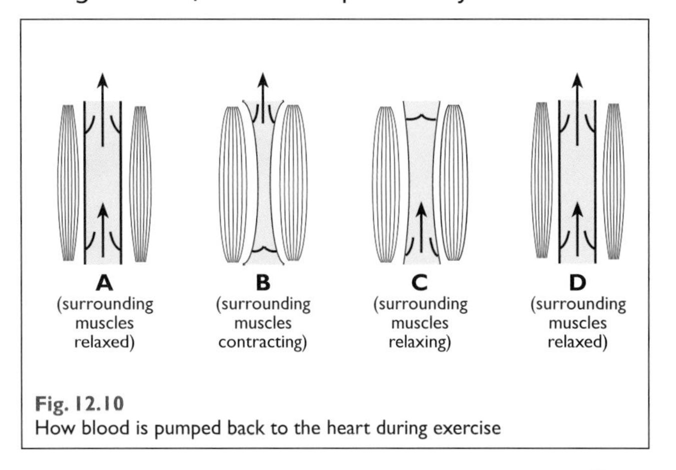

Fig. 12.10
How blood is pumped back to the heart during exercise

ISBN 9780170191340

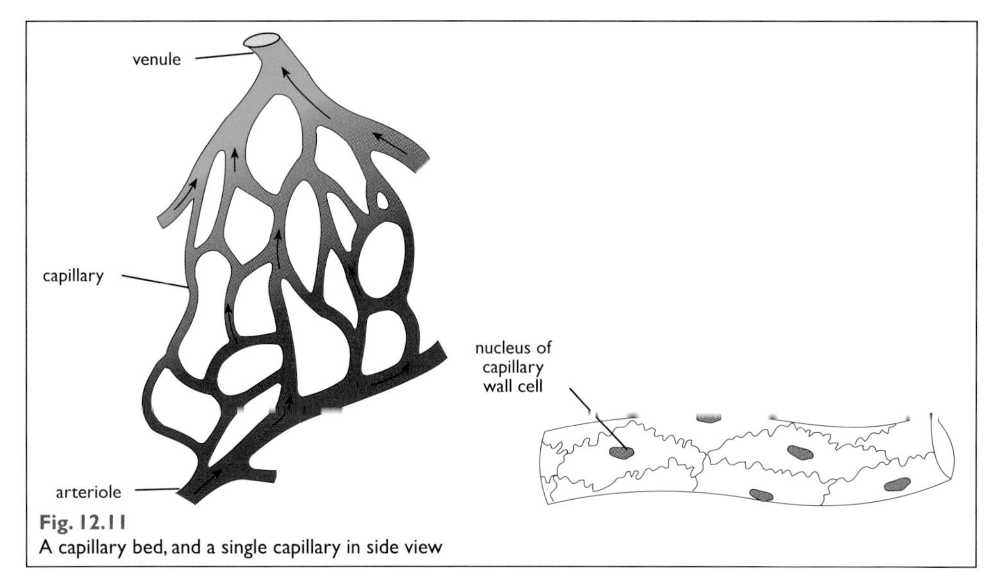

Fig. 12.11
A capillary bed, and a single capillary in side view

neighbouring limb muscles (Fig. 12.10). Active limb muscles therefore pump blood back to the heart, and this is one reason why blood flow is sluggish if you stand still for too long. Blood is also pumped back to the heart by the action of breathing. When you breathe out the pressure in the chest increases, squeezing the venae cavae. The valves ensure that the blood is pushed toward the heart.

Capillaries

It is only in these microscopic vessels that the composition of the blood changes. Capillaries have a number of features that enable substances to enter and leave the blood:

- Their walls are extremely *thin* (about 0.2 μm, or 1/5 the width of a typical bacterium!), consisting of a single layer of very flat cells (Fig.12.11).
- Because they are extremely *small* and *numerous*, they have a *large total surface area* — over 6,000 m² in an average adult — larger than the area of a hockey pitch.
- They have a *huge total cross-section area*, so the blood flows very *slowly* (about 0.5 mm s^{-1}), giving plenty of time for substances to diffuse into and out of the blood.

Most capillaries are about or slightly less than the diameter of a red blood cell, and about half a millimetre long. Because there are so many, no cell is more than a fraction of a millimetre from the nearest capillary. As the red cells squeeze through the capillaries, their shape becomes distorted into a bell shape (Fig. 12.12). As a result most of the haemoglobin is displaced to the periphery of the cells, reducing the distance that oxygen has to diffuse to the surrounding cells.

How substances enter and leave capillaries

Substances can enter and leave the blood in two ways (Fig. 12.13):

1. By *diffusion*, for example oxygen and CO_2. These have molecules small enough to pass through the partially permeable membranes of the capillary wall cells.
2. Through tiny pores between capillary wall cells. These are wide enough for glucose, amino acids and salts to pass through, but too small for the plasma proteins. Capillary walls therefore act as *filters*.

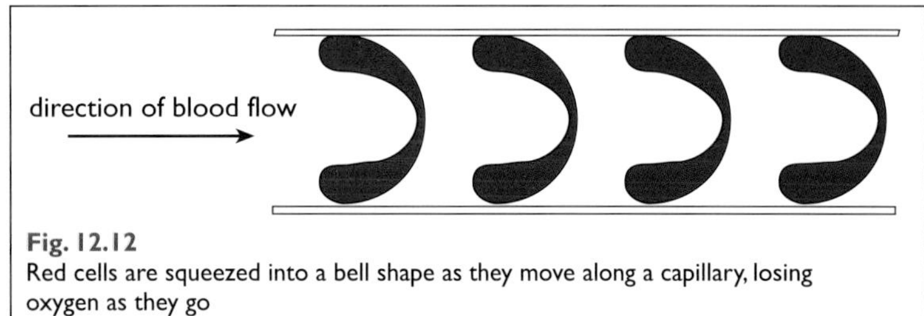

Fig. 12.12
Red cells are squeezed into a bell shape as they move along a capillary, losing oxygen as they go

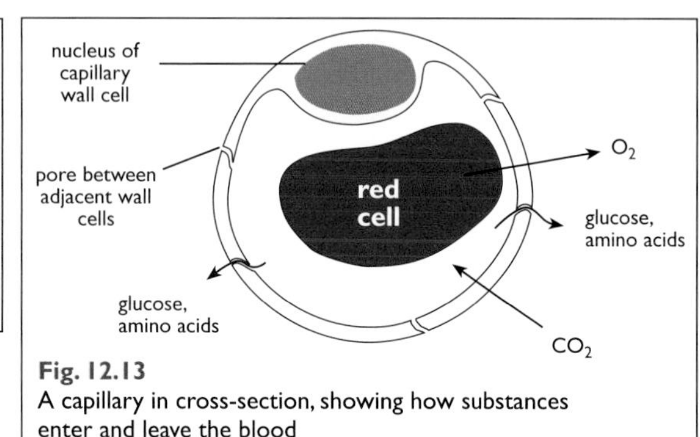

Fig. 12.13
A capillary in cross-section, showing how substances enter and leave the blood

ISBN 9780170191340

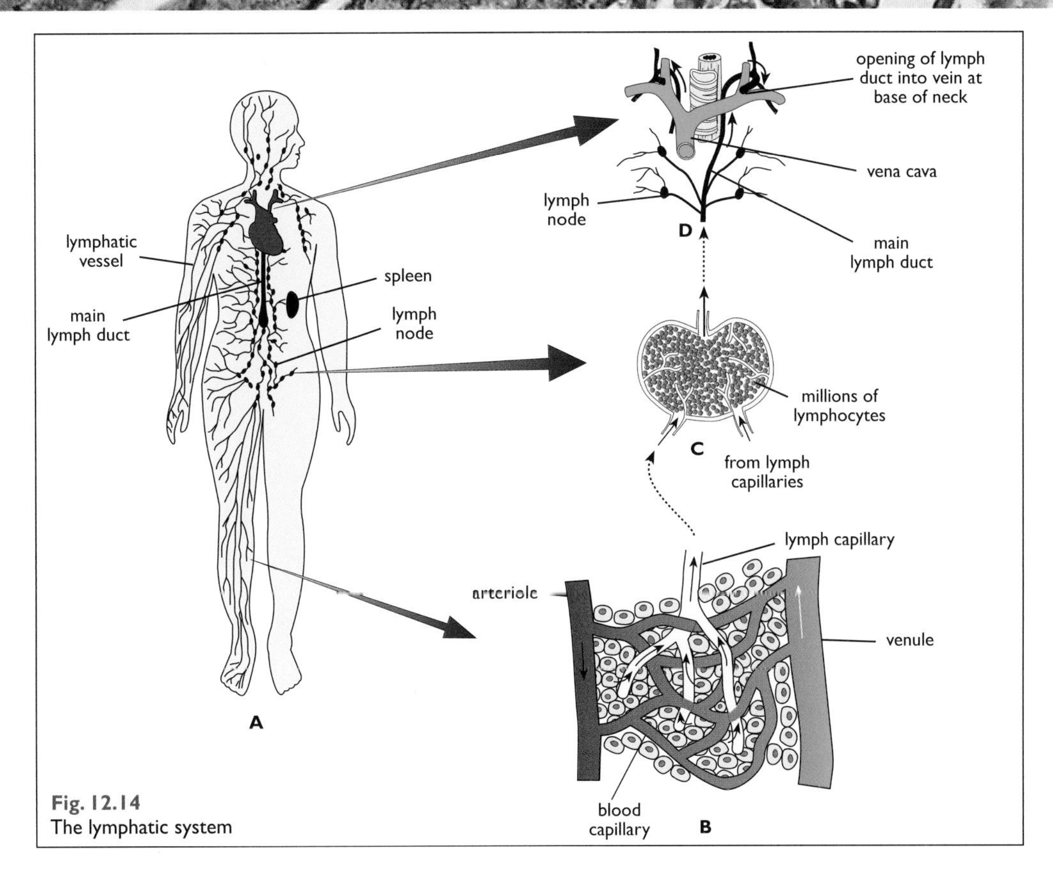

Fig. 12.14
The lymphatic system

The lymphatic system

Like the blood system, the lymphatic system is like a tree extending throughout the body (Fig. 12.14). Minute pores in capillary walls make them very leaky, allowing all substances (except most of the plasma proteins) to pass through. As a result, colourless **tissue fluid** seeps out from the arterial end of capillaries (where the pressure is higher) into the surrounding tissues. Tissue fluid is like plasma but with less protein. The cells of the body are therefore bathed in an 'internal environment' containing dissolved glucose, amino acids, salts, CO_2 and oxygen.

Since tissue fluid contains less protein than plasma, it has a slightly higher water concentration. There is a tendency therefore for water to re-enter the capillaries by osmosis. At the arterial end of a capillary the pressure of blood is more than enough to prevent this. Near the venous end, however, the blood pressure has fallen and is not enough to prevent some tissue fluid from re-entering the blood. But, slightly less fluid re-enters the capillary than leaves it. Tissue fluid would therefore accumulate in the tissues if it were not for the **lymphatic system**, a kind of 'overflow' system that returns surplus tissue fluid to the blood (Fig. 12.5).

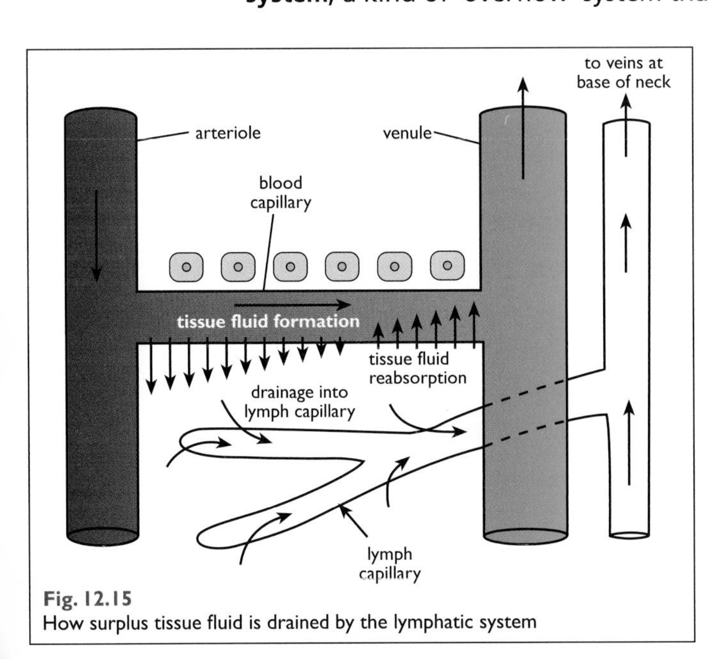

Fig. 12.15
How surplus tissue fluid is drained by the lymphatic system

The lymphatic system differs from the blood system in that its finest branches or *lymphatic capillaries* are *blind-ending*. They eventually join to form two main lymphatic vessels, which open into veins near the base of the neck. The liquid in the lymphatic vessels is called **lymph**. About 2–3 dm^3 (litres) of lymph returns to the blood each day.

Lymph flow is sluggish and is helped by contractions of smooth muscle in the walls of the larger lymph vessels, one-way flow being ensured by *valves*. Flow is greatly accelerated by the massaging action of skeletal muscles during exercise.

At intervals along lymphatic vessels are **lymph nodes**. These are part of the defence system of the body and are packed with white blood cells. During

ISBN 9780170191340

infection, the number of white blood cells increases, causing the nodes to become enlarged and tender. This condition is often called 'swollen glands'.

Degeneration of the circulatory system

Unless you have the misfortune to inherit the wrong genes (choose your parents carefully!), your blood system may continue to function for many decades. As we get older, our cardiovascular system (heart and blood vessels) undergoes gradual degeneration, but these changes can be greatly accelerated by lifestyle choices such as bad diet, smoking or lack of exercise.

Atherosclerosis

Atherosclerosis accounts for more than half of all premature deaths in industrial countries. It literally means 'hardening of the arteries' but it is actually a narrowing of the arteries, which reduces blood flow. It results from deposition of fatty materials in the arterial walls, forming a mound or *plaque* which bulges into the arterial cavity. The plaque is covered by a cap of fibrous tissue, and if this ruptures it can lead to the formation of a blood clot. This may block the artery, killing all cells 'downstream'. When this happens in the coronary arteries it results in a heart attack or *cardiac infarction*. If it happens in the brain it causes a 'stroke'.

Even if a plaque does not rupture, it reduces the blood supply to tissues downstream. This is called *ischaemia* and in the heart it can cause pain (*angina*).

These changes begin in childhood or early adulthood — long before any symptoms appear. The causes include too much saturated fat and cholesterol in the diet, and also cigarette smoking.

Summary of key points in this chapter

- The essential components of the blood system are the *blood*, *heart*, *arteries*, *veins* and *capillaries*.
- Dissolved substances are carried in the *plasma*; oxygen is carried in the red cells.
- Oxygen is carried in chemical combination with *haemoglobin*.
- Carbon dioxide is carried as *hydrogen carbonate* ions in the plasma.
- White blood cells are of two main kinds: *phagocytes* and *lymphocytes*.
- The heart is a *double* pump, the left side pumping oxygenated blood and the right side pumping deoxygenated blood. Humans and other mammals therefore have a *double circulation*, in which blood passes *twice* through the heart in each cycle.
- Each side of the heart has a receiving chamber or *atrium* and a pumping chamber or *ventricle*.
- The left ventricle pumps blood at much higher pressure than the right, but the two sides pump equal volumes.
- Heart muscle contracts spontaneously; it does not require nervous stimulation.
- The heart is supplied by two sets of nerves, one tending to speed it up, the other to slow it down.
- Though the heart is full of blood, it has its own *coronary* circulation.
- Backflow of blood is prevented by four valves, two on each side of the heart.
- Except for the pulmonary artery, arteries carry blood under high pressure.
- The walls of the larger arteries are rich in elastic tissue, which smooths out the pulse.
- The walls of the arterioles contain a lot of smooth muscle, which regulates the flow of blood.
- Veins carry blood under low pressure and have thin walls.
- During exercise, skeletal muscles help pump blood back to the heart, assisted by the valves in the veins.
- In the capillaries, exchange of substances with the tissues is promoted by their thin walls, huge total surface area, and slow flow of blood.
- Substances are conveyed in a solution between blood and tissues called *tissue fluid*.
- Surplus tissue fluid is drained from the tissues by the *lymphatic system*.
- The blood system degenerates slowly with age, but lifestyle choices can greatly accelerate this process.

ISBN 9780170191340

Test your basics

Copy and complete the following sentences.

a) Blood is carried away from the heart in ___*___ and returns to the heart in ___*___. Between these are the microscopic ___*___, in which substances enter and leave the blood.

b) The function of red blood cells is to carry ___*___ around the body. This is carried in combination with the ___*___-containing protein ___*___ as ___*___.

c) Red cells are unique in the human body in that they lack a ___*___ , enabling them to contain more haemoglobin and hence more ___*___. Their peculiar shape gives them a high ___*___ compared with their ___*___.

d) Like all blood cells, red cells are produced in the bone ___*___. After about four months they are destroyed in the ___*___.

e) White blood cells defend the body against ___*___. ___*___ feed on bacteria by engulfing them and digesting them. ___*___ grow into cells that make defence proteins called ______.

f) The heart is a double pump, the right side pumping ___*___ blood to the ___*___ via the ___*___ arteries and the left side pumping ___*___ blood to the rest of the body via the ___*___. Each side consists of a receiving chamber or ___*___ and a pumping chamber or ___*___. Blood is prevented from flowing backward by four ___*___. On each side, a ___*___ valve prevents blood flowing from the artery to the ventricle, and an ___*___-___*___ valve prevents blood re-entering the ___*___. Although the left side exerts a much higher ___*___ than the right, the two sides pump equal ___*___.

g) Besides being strong, artery walls are very ___*___, which enables them to expand with each surge of pressure. As a result the intermittent flow of blood leaving the heart becomes ___*___ by the time it reaches the capillaries.

h) The walls of arteries, especially the ___*___ ones, have ___*___ muscle, which enables the blood flow to an organ to be varied according to demand.

i) Oxygen and carbon dioxide pass through capillary walls by ___*___. This is extremely rapid because the walls are extremely ___*___, consisting of a ___*___ layer of flattened cells, so the blood is very ___*___ to the cells it serves. Also, because capillaries are so narrow, they have a very large ___*___ compared with their ___*___.

j) Between adjacent capillary wall cells are tiny ___*___, through which ___*___ fluid is forced through by blood pressure. This fluid is similar to plasma minus its ___*___. Although some tissue fluid drains back into the venous ends of the capillaries, this is less than the amount seeping out of the capillaries. The surplus is drained by the ___*___ system, the larger vessels of which open into the veins at the base of the neck.

ISBN 9780170191340

13 Respiration, gas exchange and breathing

HOW CELLS EXPEND ENERGY: ATP

All living cells are constantly using energy, for such purposes as:

- **Biosynthesis**, or making large molecules from small ones, e.g. making proteins from amino acids, and glycogen from glucose.
- **Active transport**, moving substances across a cell membrane from a lower to a higher concentration.
- **Movement**, for example the movement of chromosomes in cell division, the contraction of muscle cells and movement of cilia.

The immediate source of energy for these processes is **ATP** (*adenosine triphosphate*). When this is converted to **ADP** (*adenosine diphosphate*), energy is released:

$$ATP + H_2O \rightarrow ADP + P_i \text{ (inorganic phosphate)} + \textbf{energy}$$

The energy released when ATP is converted to ADP is enough to 'drive' energy-requiring processes in cells, with a little left over. The 'left-over' energy is wasted as heat (Fig. 13.1).

ATP is continuously being used by a cell, so it must be produced continuously, and this process *requires* energy:

$$ADP + P_i \text{ (inorganic phosphate)} + \textbf{energy} \rightarrow ATP + H_2O$$

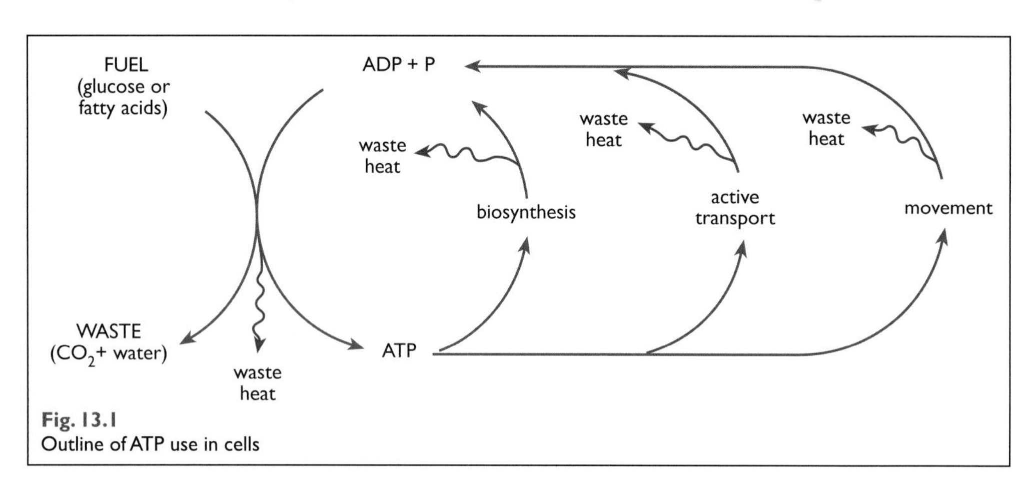

Fig. 13.1
Outline of ATP use in cells

The energy to make ATP comes from the breakdown of organic molecules such as glucose. This occurs in two stages:

1. **Glycolysis**: Glucose is split into two molecules of pyruvic acid, yielding enough energy to make two molecules of ATP. Glycolysis occurs in the general cytoplasm and *does not require oxygen* (Fig. 13.2).
2. **Respiration**: Pyruvic acid is oxidised to CO_2 and water. This yields enough energy to make 30 molecules of ATP for each original glucose used. *Respiration requires oxygen.*

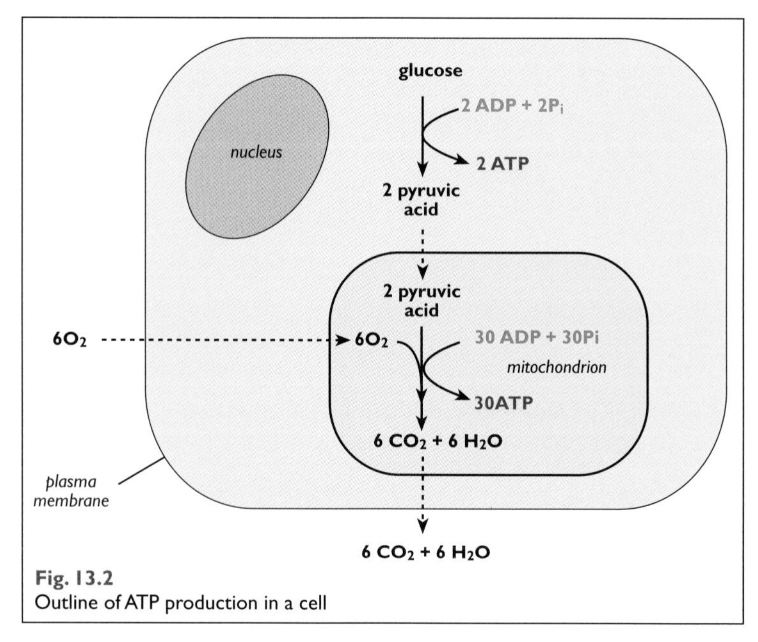

Fig. 13.2
Outline of ATP production in a cell

ISBN 9780170191340

Respiration occurs in small sausage-shaped structures in the cytoplasm called **mitochondria**. A mitochondrion is an example of an **organelle** — a part of a cell specialised for a particular function. Cells that are particularly active can have over a thousand mitochondria.

Since the oxidation of each glucose molecule yields enough energy to make over 30 ATP molecules, an ATP molecule represents the 'small change' in the cell's energy 'currency'.

Explosive exercise

In a 100 metres sprint, your muscles are expending energy far faster than the ability of the heart to deliver oxygen and fuel. Muscle tissue can do this because of two adaptations:

1. It stores its own carbohydrate in the form of **glycogen**. Like starch, glycogen can be broken down into glucose.
2. It can obtain energy from glucose anaerobically (in the absence of oxygen). Muscles do this by a process that is essentially like the *fermentation* that occurs when milk goes sour, glucose being converted to lactic acid (Fig. 13.3).

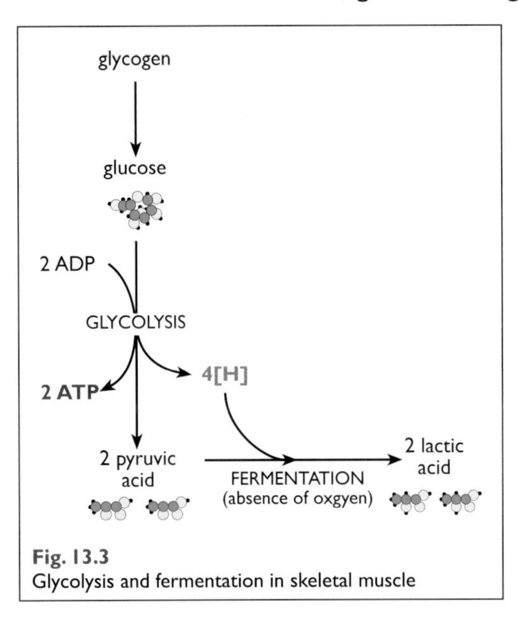

Fig. 13.3
Glycolysis and fermentation in skeletal muscle

Glycolysis is very inefficient in that it only yields two molecules of ATP for every glucose used (about 1/15 as many as are produced in respiration). Despite this, an active muscle can generate ATP extremely rapidly, because it can make glucose from its own glycogen store.

Lactic acid fermentation is only suitable for relatively short periods because lactic acid is slightly toxic and eventually it accumulates and the muscle works less efficiently. This is why you cannot sprint flat out for more than about 200 metres.

The lactic acid produced by muscle is transported to the liver. Here, a small proportion is burnt in respiration, supplying enough ATP to convert the rest of the lactic acid back into glycogen.

In doing without oxygen during a sprint, you are incurring an *oxygen debt*, which is 'repaid' during the recovery. This is why you continue to pant for some time *after* finishing a sprint.

GAS TRANSPORT

Cells need to continuously import oxygen and glucose, and to export CO_2. The movement of oxygen and CO_2 into and out of the cell occurs by **diffusion** (Chapter 9). Glucose molecules are too large to enter by simple diffusion, and are carried into the cell by a process called *facilitated diffusion*.

Over very short distances diffusion is rapid and in small animals the cells are close enough to the body surface for oxygen to reach the mitochondria by diffusion alone. In mammals, oxygen and CO_2 are transported between the cells and the environment by the blood system (Chapter 12).

Oxygen reaches the red cells in chemical combination with **haemoglobin** as *oxyhaemoglobin*:

$$\text{oxyhaemoglobin} \rightarrow \text{haemoglobin + oxygen}$$

CO_2 is transported away from the cells in the plasma as HCO_3^- (*hydrogen carbonate*) ions:

$$CO_2 + H_2O \rightarrow HCO_3^- \text{ (hydrogen carbonate)} + H^+$$

These processes are reversed in the lungs, where haemoglobin picks up oxygen to form oxyhaemoglobin, and hydrogen carbonate ions break down into CO_2 and water. Haemoglobin contains *iron*, which is one reason why it is needed in the diet.

ISBN 9780170191340

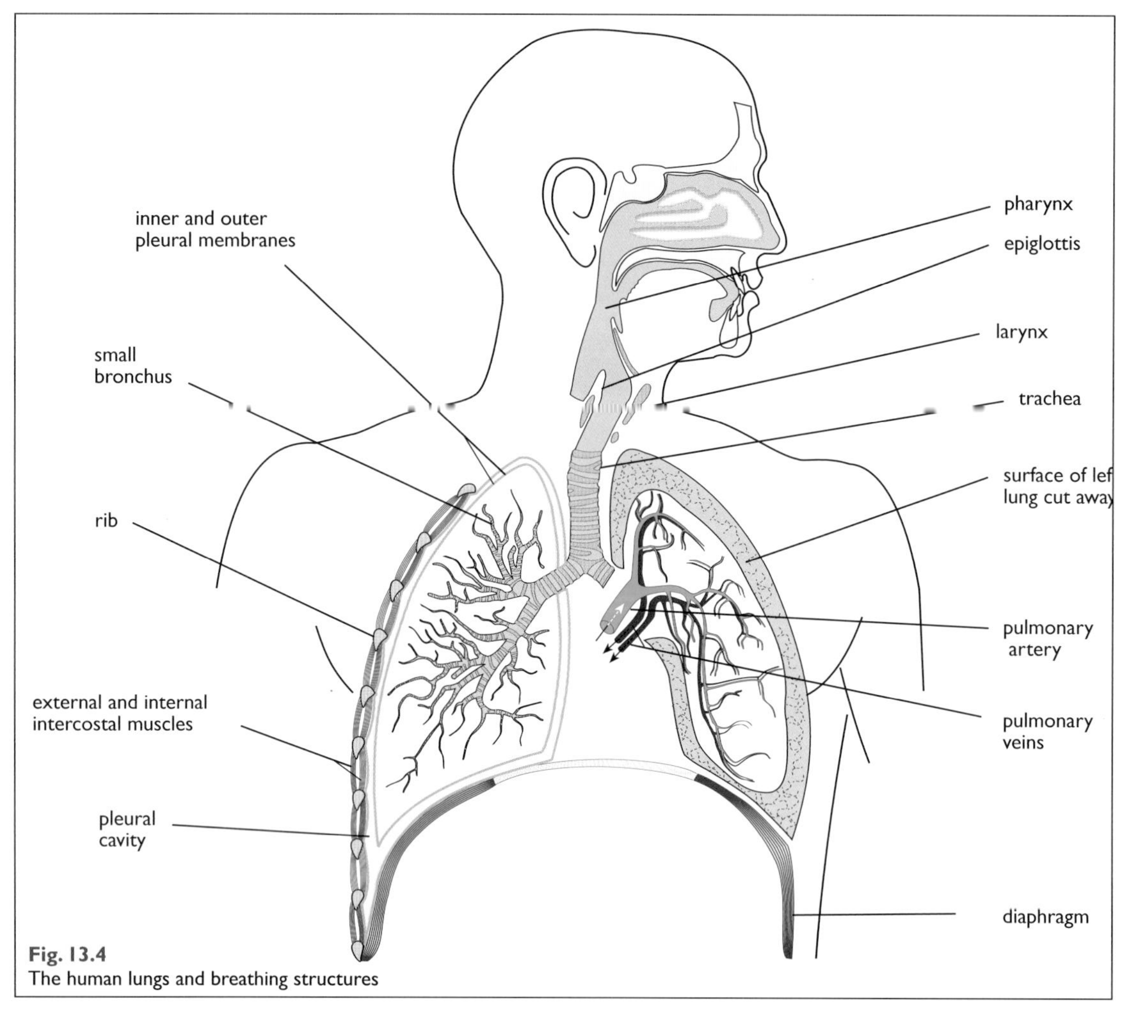

Fig. 13.4
The human lungs and breathing structures

GAS EXCHANGE: THE LUNGS

The two-way diffusion of oxygen and CO_2 across a surface is called **gas exchange**. It occurs at the surfaces of cells and also at the surface of the body. In mammals the part of the body surface where gas exchange occurs is deep in the **lungs**. These are spongy organs on either side of the heart (Fig. 13.4). Each lung lies in a **pleural cavity**, filled with a thin layer of fluid that lubricates the lung surface during breathing.

Air is delivered to the gas exchange surface by a system of tubes called the *bronchial tree*. The 'trunk' of the tree is the **trachea** (windpipe). This branches into two main *bronchi*, one to each lung. Each main **bronchus** branches repeatedly into smaller and smaller bronchi. Like the trachea, the walls of the bronchi are supported by *cartilage*. The smallest bronchi lead into **bronchioles**, which are not supported by cartilage but have smooth ('involuntary') muscle in their walls. The bronchioles lead into clusters of air sacs or **alveoli** (singular: *alveolus*), where gas exchange occurs (Fig. 13.5). There are about 300 million in the average human lung, with an average diameter of about 0.2 mm. Deoxygenated blood reaches the alveolar capillaries from the right side of the heart via branches of the **pulmonary artery**. It leaves via branches of the **pulmonary veins**, which take the blood to the left side of the heart.

Figure 13.6 shows the capillaries in the lung of a frog. The network is so dense that they form an almost continuous sheet of blood.

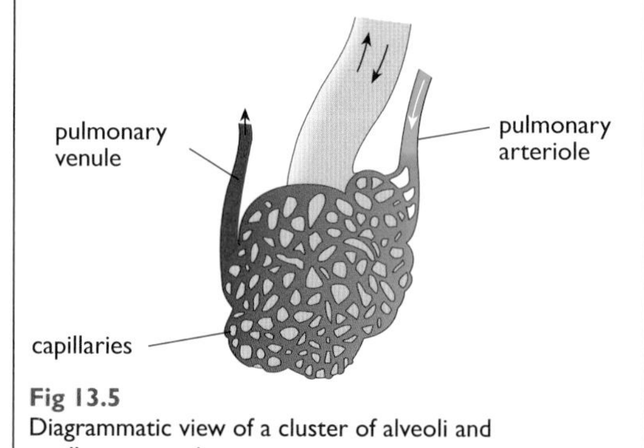

Fig 13.5
Diagrammatic view of a cluster of alveoli and capillary network

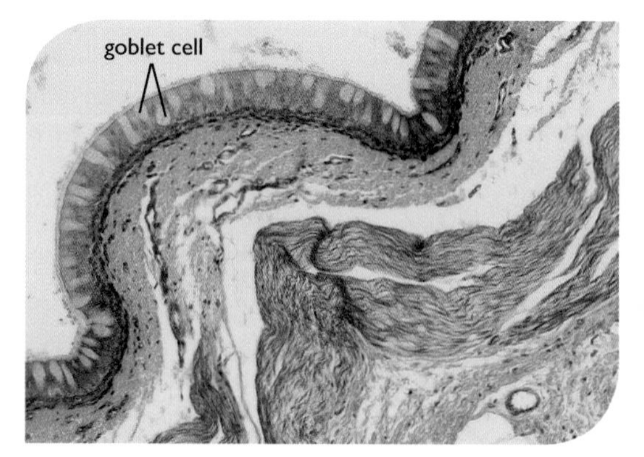

Section through wall of bronchus, showing goblet cells

ISBN 9780170191340

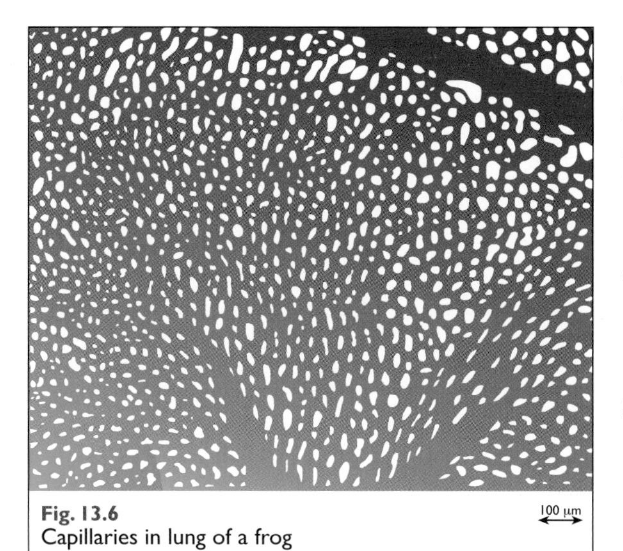
Fig. 13.6
Capillaries in lung of a frog

Each capillary lies *between* alveoli, so that the capillaries are actually *surrounded by air*. The capillaries and alveoli have extremely thin walls; blood and air are separated by only about 0.5 μm (half the width of an average bacterium!). This is important because diffusion is only rapid over short distances.

To sum up, the fine structure of the lungs is adapted for gas exchange in two ways:

1. The distance between the blood and the air is extremely short, so the *concentration gradient* is steep.
2. Blood and air come close together over an extremely *large surface area*. Their total area in a human is estimated to be about 75 m^2. Strictly speaking, it is the area of the capillaries that is the important factor, rather than the area of the alveoli, since the spaces between the capillaries do not take part in gas exchange.

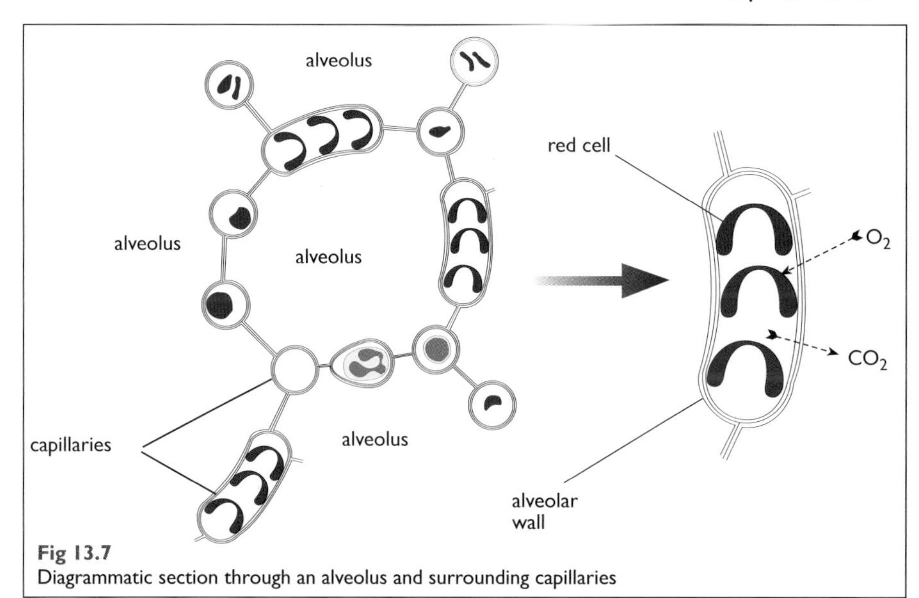

Fig 13.7
Diagrammatic section through an alveolus and surrounding capillaries

Figure 13.8 shows that oxygen reaches the blood (and the reverse for CO_2) in three stages:

1. Bulk flow in the air passages of the bronchial tree.
2. Diffusion through still air in the alveoli.
3. Diffusion through liquid in the alveolar and capillary walls.

During quiet breathing, only about 10% of the alveolar air is changed in each breath. Each inhaled breath therefore simply 'tops up' a much larger volume of stagnant air already present. As a result the warm, humid air in the alveoli is very different from the atmosphere outside (Table 13.1).

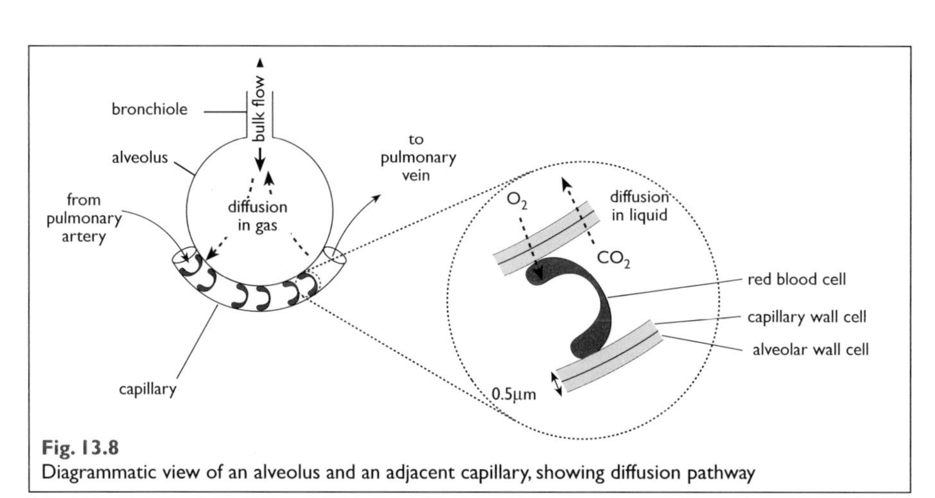

Fig. 13.8
Diagrammatic view of an alveolus and an adjacent capillary, showing diffusion pathway

The data refer to dried air. This is because although the exhaled air is saturated with water vapour (which at 37 °C contains 6.2% water), the water content of *inhaled* air varies considerably. Since water vapour 'dilutes' the other gases, the percentage of these other gases also varies with the humidity of the inhaled air.

Gas	Inhaled air	Exhaled air	Alveolar air
Nitrogen	79.0%	79.5%	80.7%
Oxygen	20.96%	16.4%	13.8%
CO_2	0.04%	4.1%	5.5%

Table 13.1 Composition of dried inhaled, exhaled and alveolar air

Notice that exhaled air contains slightly more oxygen and slightly less CO_2 than alveolar air. The reason is that when we breathe in, the last bit of air to enter the nose does not reach the alveoli, but stays in the bronchial

ISBN 9780170191340

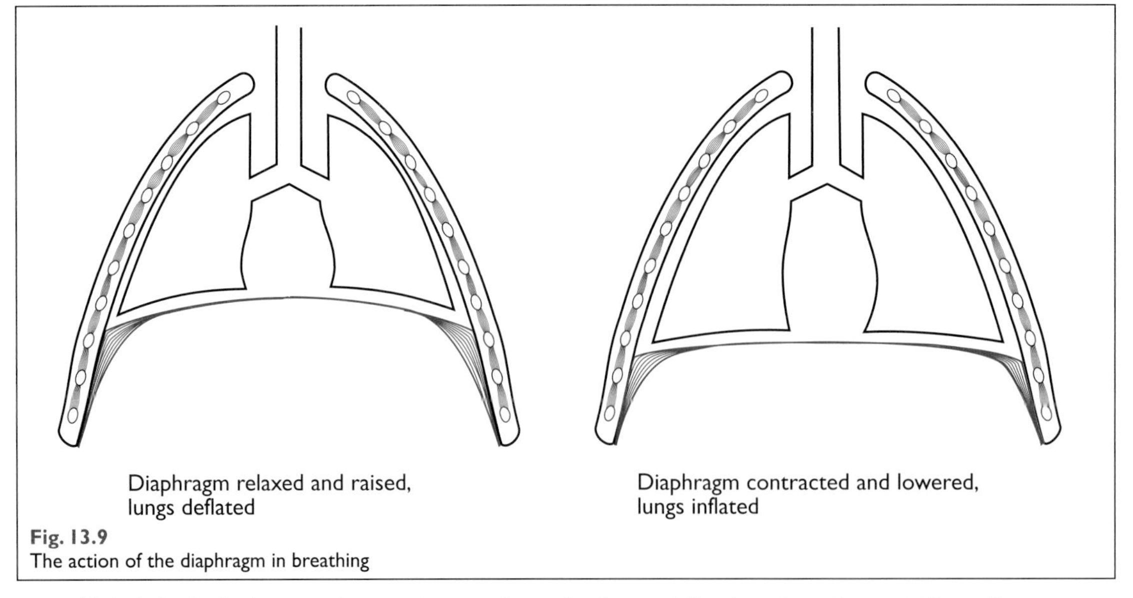

Fig. 13.9
The action of the diaphragm in breathing

tree. This inhaled air remains unchanged, and mixes with alveolar air next time the person breathes out.

Notice also that the percentage of nitrogen in exhaled air is slightly higher than it is in inhaled air. This is because the body uses slightly more oxygen than it produces CO_2. This has the effect of slightly increasing the concentration of nitrogen (but not its amount). This is similar to the effect of evaporating water from sugar solution; the amount of sugar stays the same, but its *concentration* increases.

BREATHING

Alveolar air is continually 'topped up' with fresh air by breathing. The breathing cycle has two phases: **inspiration** (breathing in) and **expiration** (breathing out). Inspiration is active, and at rest, expiration is passive.

There are two main sets of breathing muscles, shown in Fig. 13.9.

1. The **diaphragm**, a dome-shaped muscle between the chest and abdomen.
2. The **intercostal muscles** between the ribs (Fig. 13.10). There are two sets, one internal to the other.

Inspiration

To lower the pressure in the lungs, their volume must be increased:

- The diaphragm contracts, lowering the floor of the chest.
- The external intercostal muscles contract, raising the ribs.

As a result of the increase in lung volume the pressure decreases to below that of the air outside, and air rushes in. The expansion of the lungs stretches elastic tissue in the lungs, and also stretches the muscles of the abdominal wall. In this way some of the energy expended in inspiration is stored (just as an elastic band can store energy when stretched).

Expiration

The diaphragm and external intercostal muscles relax, and the stretched elastic tissue causes the lungs to partially deflate. In this way the energy for expiration comes from 'elastic' energy stored in the previous inspiration.

In heavy breathing (e.g. during exercise), the force of expiration is increased by contraction of the *internal* intercostal muscles, and also by the active contraction of the

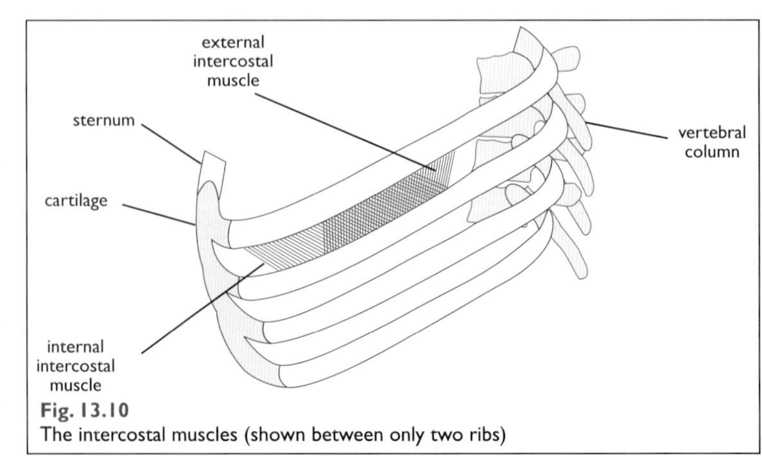

Fig. 13.10
The intercostal muscles (shown between only two ribs)

ISBN 9780170191340

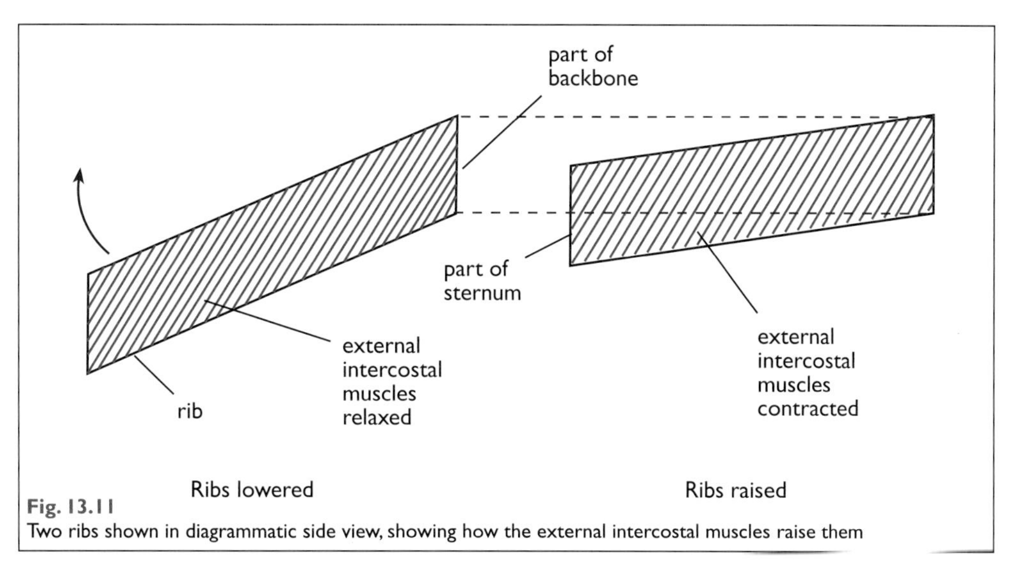

Fig. 13.11
Two ribs shown in diagrammatic side view, showing how the external intercostal muscles raise them

muscles of the abdominal wall, which push the stomach, liver and intestines against the diaphragm, forcing it upward.

Regulating breathing

A resting adult normally takes about 15 breaths per minute, each of about 0.5 dm^3 (cubic decimetres or 'litres'). The total air inhaled each minute ('the minute volume') at rest is therefore about 15 x 0.5 = 7.5 dm^3. During exercise both the rate and depth of breathing increase, delivering more fresh air to the alveoli.

The stimulus for increased breathing in exercise is a rise in CO_2 concentration in the blood. This stimulates the breathing centre in the brain, causing it to send a higher frequency of nerve impulses to the breathing muscles. As a result they work harder, increasing the rate at which CO_2 is expelled from the blood (Fig. 13.12).

If you try to hold your breath, the desire to breathe is due to the increase in CO_2 content of the blood rather than the decrease in oxygen content. The more that CO_2 builds up in the blood, the greater the desire to breathe.

Similarly when we take exercise, the harder we work the higher the level of CO_2 is in the blood and the faster the CO_2 is got rid of. As a result of this self-correcting mechanism, *the composition of exhaled air hardly changes during exercise*. What does change is the *volume* of air inhaled and exhaled.

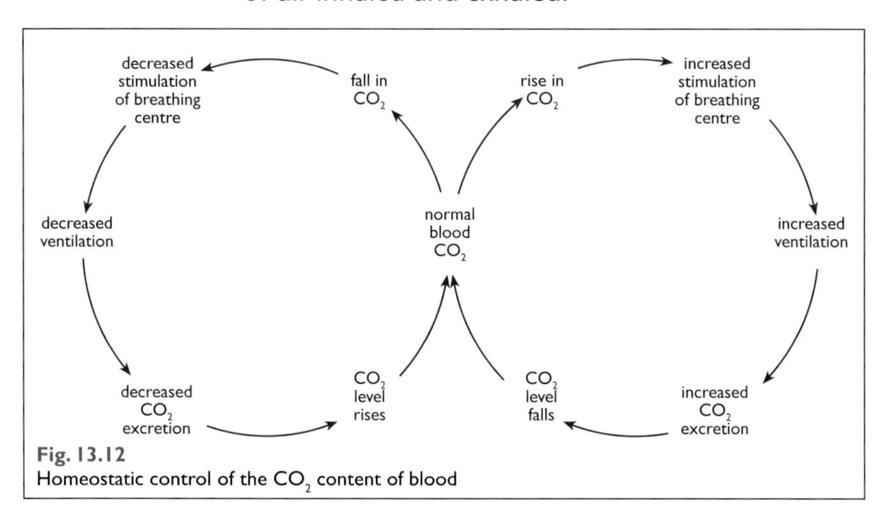

Fig. 13.12
Homeostatic control of the CO_2 content of blood

The regulation in CO_2 content of the blood is an example of **homeostasis**. In homeostasis the greater the disturbance (in this case a rise in CO_2), the greater the tendency is to correct it. This is called **negative feedback**.

Although CO_2 is the main stimulus for breathing, this mechanism indirectly regulates the oxygen content of the blood as well, since a build-up of CO_2 normally accompanies a shortage of oxygen.

Cleaning the air passages

The gas exchange surface is potentially vulnerable to attack by pathogens (disease-causing organisms); it is moist and separated from the blood by an extremely thin layer. It is, however, well-protected by a mechanism that cleans the airways of bacteria and dust (Fig. 13.13).

ISBN 9780170191340

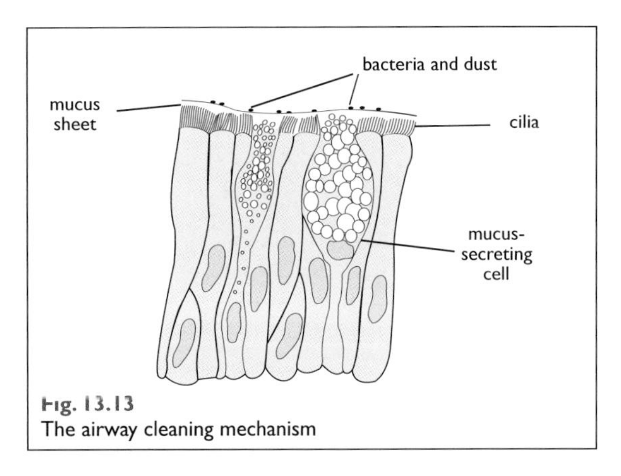

Fig. 13.13
The airway cleaning mechanism

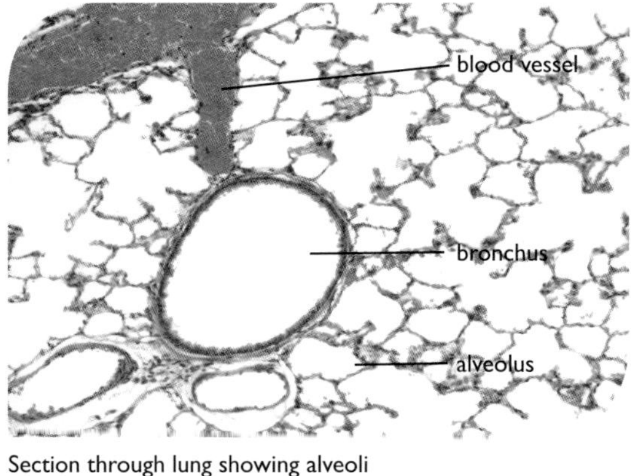

Section through lung showing alveoli

The trachea, bronchi and nasal passages are lined by two kinds of cell:

1. Some cells ('goblet cells', so-called because their shape resembles a wine goblet) secrete **mucus**, a slimy substance that traps bacteria and dust.
2. Other cells have thousands of **cilia**. These are tiny hair-like extensions of the plasma membrane, which beat and propel the mucus and trapped particles toward the pharynx, where it is periodically swallowed.

HOW THE ENVIRONMENT AFFECTS GAS EXCHANGE

Gas exchange can be affected by a number of environmental influences, some natural, some accidental and some self-induced.

Emphysema

This is one of many conditions resulting from smoking. The alveolar walls tend to break down, resulting in a smaller number of larger air sacs (Fig. 13.14). This has three serious effects:

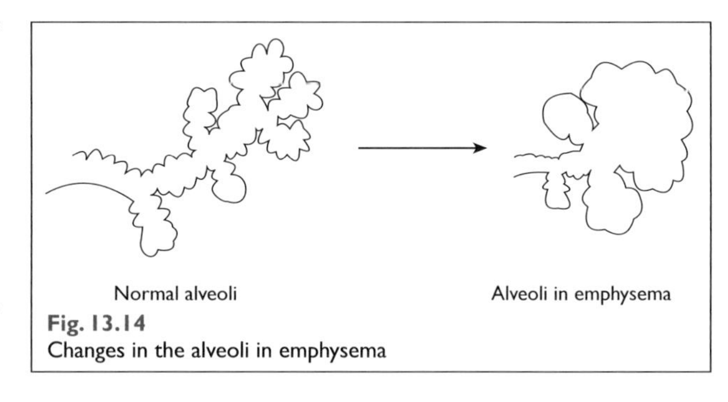

Fig. 13.14
Changes in the alveoli in emphysema

1. It reduces the area of the gas exchange surface.
2. The smaller number of capillaries increases the resistance to blood flow, leading to an enlargement of the right side of the heart. The increased pressure of blood causes tissue fluid to leak out and accumulate in the alveoli. Diffusion is much slower in liquid, so for this and the previous reason, gas exchange is seriously affected.
3. The elastic tissue, which normally helps keep the bronchioles open, is attacked by white cells. As a result the bronchioles become narrower, leading to difficulty in breathing.

The patient may be assisted by extra oxygen, delivered via a narrow tube to the nose. Over 20 or more years of extreme difficulty in getting enough oxygen, death eventually puts an end to a patient's suffering.

Chronic bronchitis

One of the effects of cigarette smoke is that it temporarily paralyses the cilia that help keep the airways free of bacteria. As a result smokers are more prone to **bronchitis**, or inflammation of the bronchi. In heavy smokers the condition may become permanent or *chronic*. The airways become narrower due to the build up of mucus, which leads to increased airway resistance and frequent coughing. Like emphysema, it contributes to premature death.

Carbon monoxide

Among the 200 or so harmful chemicals in cigarette smoke is **carbon monoxide** (CO). This colourless, odourless gas combines with haemoglobin to form *carboxyhaemoglobin* (COHb), which is even brighter red than oxyghaemoglobin:

$$CO + Hb \text{ (dark red)} \rightarrow COHb \text{ (cherry red)}$$

ISBN 9780170191340

Carbon monoxide combines 250 times more strongly with haemoglobin than oxygen does, and it is only released at very low CO concentrations. About 5% of the haemoglobin in a 20 a day smoker is in the form of COHb, and in a chain smoker it can be as high as 25% COHb. As a result a significant proportion of a smoker's haemoglobin is unavailable for oxygen transport.

The treatment for accidental CO poisoning is administration of pure oxygen, which slowly displaces the CO from the blood. The treatment is even more effective if the oxygen is accompanied by 5% CO_2, which greatly stimulates breathing.

Asthma

Asthma is caused by an extreme allergic reaction of the immune system to foreign substances (antigens). As a result the smooth muscle in the bronchi and bronchioles contract, narrowing the airways. The air passages also become further narrowed by build up of mucus resulting from over-activity of the mucous gland cells lining the airways.

Extension: Measuring breathing and metabolic rate

Figure 13.15 shows a simple way of measuring the volume of a breath. A more sophisticated way is to use a spirometer (Fig. 13.16). The apparatus is normally filled with medical grade oxygen rather than air. The lid is pivoted at one end and sealed from the environment by water. When the subject breathes into the apparatus the lid rises up and makes a recording on a revolving drum or **kymograph**. It is important that the breathing effort is normal, therefore the lid is counter-weighted so that it does not exert pressure on the air beneath.

To ensure that the subject does not rebreathe the same air, the mouthpiece contains a one-way valve. One of the two hoses connected to the mouthpiece contains soda lime, which absorbs carbon dioxide.

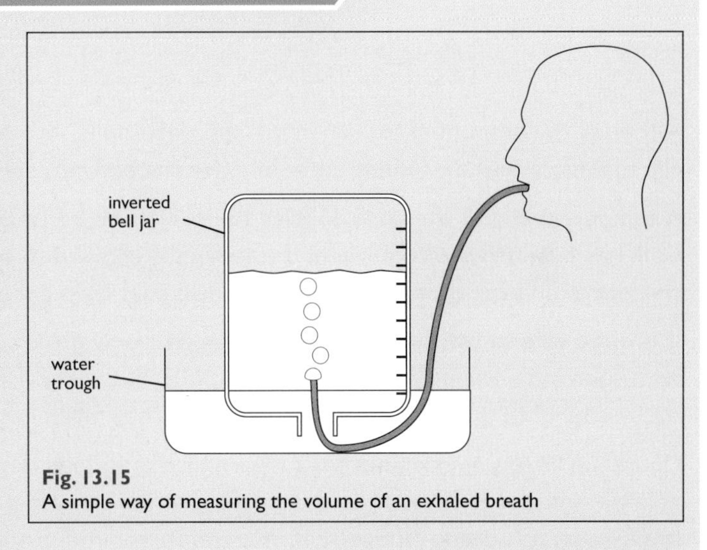

Fig. 13.15
A simple way of measuring the volume of an exhaled breath

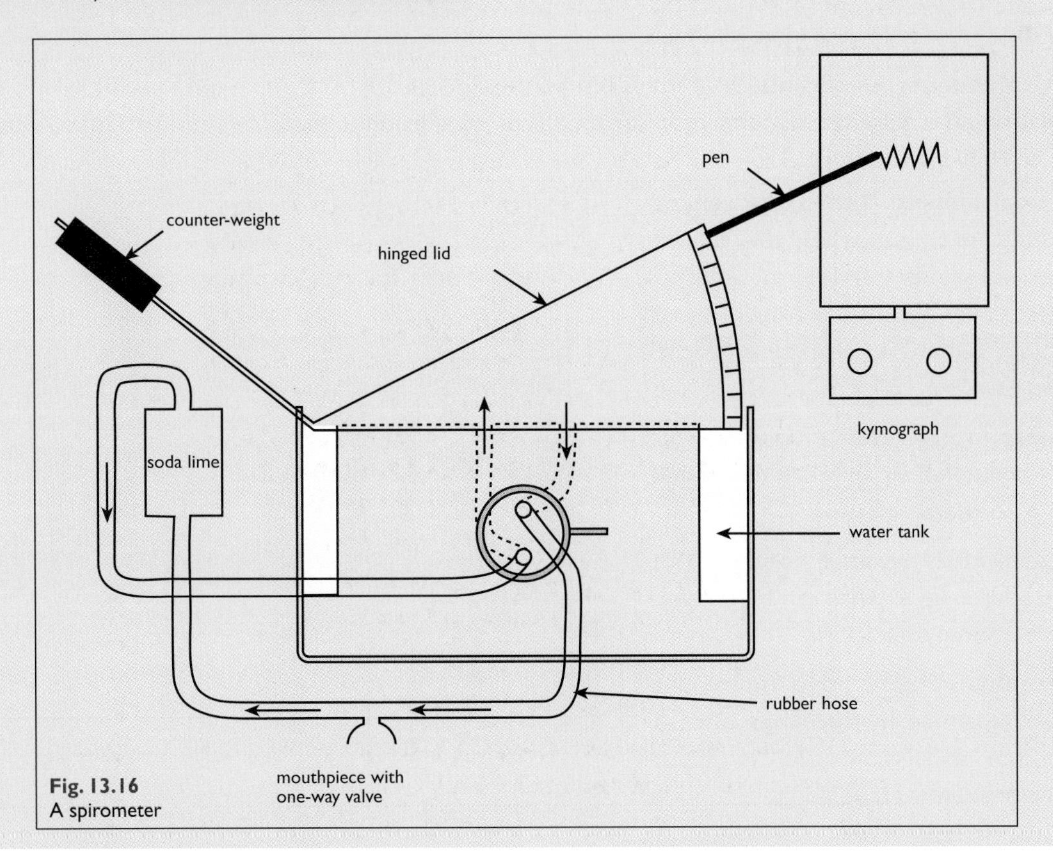

Fig. 13.16
A spirometer

ISBN 9780170191340

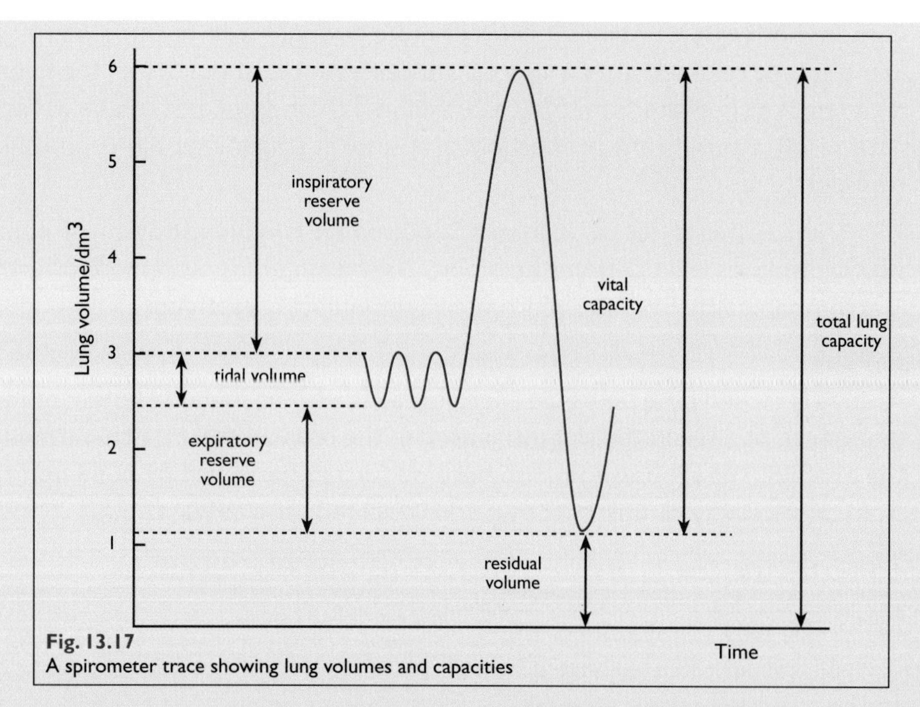

Fig. 13.17
A spirometer trace showing lung volumes and capacities

Although the amount of oxygen in the apparatus falls, its *percentage* remains the same. The subject wears a nose clip to ensure that air cannot enter or leave the body except via the apparatus.

A tap at the end of the hose enables the subject to be isolated from, or connected to, the apparatus. After the tank has been filled with oxygen, the subject is allowed to breathe quietly while isolated from the apparatus. At the end of an expiration the tap is opened so that the next inspiration draws gas from the spirometer.

It is quite safe to use air instead of oxygen, but *only if the soda lime is first taken out of the circuit.* If air is used *with* soda lime, the percentage of oxygen falls but there is no rise in carbon dioxide. There is therefore no desire to breathe until the partial pressure of oxygen has fallen low enough to cause loss of consciousness.

Figure 13.17 is a spirometer trace showing the various lung capacities. The vertical axis represents changes in volume and the horizontal axis represents time. The time scale is worked out from the speed of rotation of the kymograph drum. Notice that changes in volume are *inverted* since a decrease in lung volume pushes the recording pen upward.

The trace shows the following lung volumes:

- The **vital capacity**. This is the largest single breath the subject can take, and is obtained by taking a maximum exhalation after a maximum inhalation (or the other way around). In large, young athletes, vital capacities may be 6000 cm^3 or more.
- The **tidal volume**. This is the volume breathed in or out when the subject is not consciously trying to breathe. At rest, tidal volume averages about 500 cm^3, but it can rise to 2000 cm^3 during exercise.
- The **inspiratory reserve volume**. This is the additional air that can be inhaled after a normal inspiration.
- The **expiratory reserve volume**. This is the additional air that can be exhaled after a normal expiration.
- The **residual volume**. This is the amount of air remaining in the lungs after a maximum exhalation, and is about 1500 cm^3 in an average adult.

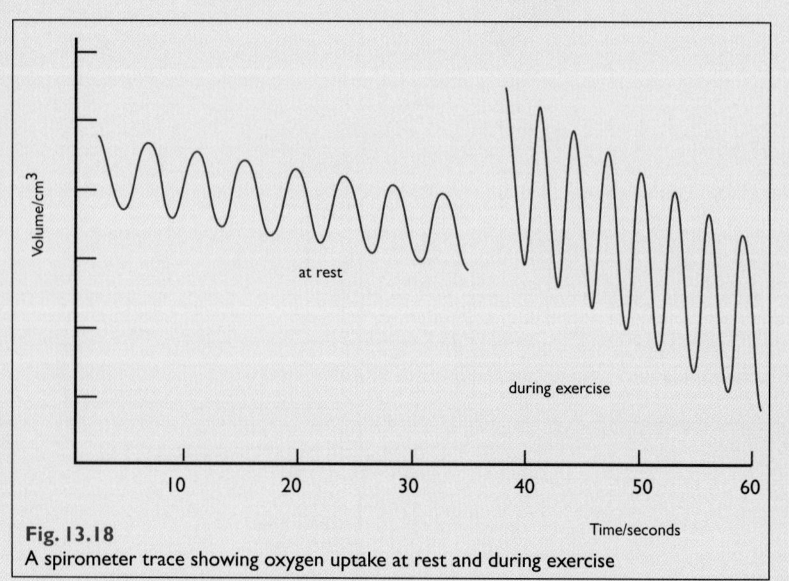

Fig. 13.18
A spirometer trace showing oxygen uptake at rest and during exercise

A spirometer can also be used to measure two other things:

1. The **minute volume**, or amount of air inhaled (or exhaled) in one minute. This is equal to the tidal volume multiplied by the number of breaths per minute.
2. The rate of **oxygen uptake**, indicated by the gradient of the trace (Fig. 13.18).

During exercise the gradient is steeper, showing that the rate of oxygen uptake is greater.

Haemoglobin: An example of *biochemical* adaptation

When we say that an organism is *adapted* to a particular way of life, we mean that it has inherited features that increase its chances of becoming a parent, which means surviving and reproducing. Adaptations can be described as:

- *Anatomical*, for example the cilia and mucous glands that clean the bronchial passages.
- *Physiological*, for example the nervous mechanism that regulates the CO_2 and oxygen content of the blood.
- *Behavioural*, for example slaters move more slowly in humid conditions, so they tend to collect in places where they lose water more slowly.
- *Biochemical*, involving the working of molecules.

An example of a biochemical adaptation is *haemoglobin*, the protein that carries oxygen in the blood. A haemoglobin molecule consists of four polypeptide chains, each of which contains an iron atom and can carry a molecule of oxygen. A haemoglobin molecule can therefore carry up to four molecules of oxygen.

When all the haemoglobin in the blood carries four oxygen molecules, the blood is said to be *saturated* with oxygen. When saturated, blood carries 20% by volume of oxygen, so each cubic decimetre ('litre') of blood carries 200 cm^3 of oxygen.

At sea level, atmospheric pressure is about 100 kPa (kilopascals), of which oxygen contributes about 21 kPa. This is the *partial pressure* of oxygen.

Figure 13.19 is a graph (called an oxyhaemoglobin dissociation curve) showing how the percentage saturation of blood with oxygen varies with the partial pressure of oxygen. Notice three important things about the curve:

1. The partial pressure of oxygen in the alveoli is about 2/3 as much as it is in the atmosphere, yet the blood leaving the lungs is nearly saturated with oxygen.
2. At higher oxygen partial pressures (flat part of the curve), the partial pressure of oxygen in the alveoli has to fall considerably before it has much effect on the oxygen content of the blood leaving the lungs. This 'safety factor' means that, for example, you can hold your breath for some time before it has much effect on the oxygen reaching the tissues.
3. At the lower oxygen concentrations encountered in the tissues (steep part of the curve), a small decrease in oxygen results in a large change in percentage saturation and thus the release of a large amount of oxygen.

The sigmoid ('s'-shaped) shape of the curve also comes in handy when climbing mountains. The dotted lines on the graph show the percentage saturation of oxygen in the blood leaving the lungs at the tops of various mountains: Ruapehu (2779 m), Aorangi Mt Cook (3754 m), Kilimanjaro (5893 m), and Everest (8848 m). People climb to the summit of Ruapehu without being much affected by shortage of oxygen, but at the summit of Aorangi Mt Cook physical performance is noticeably poorer. People who climb Kilimanjaro in Tanzania usually suffer

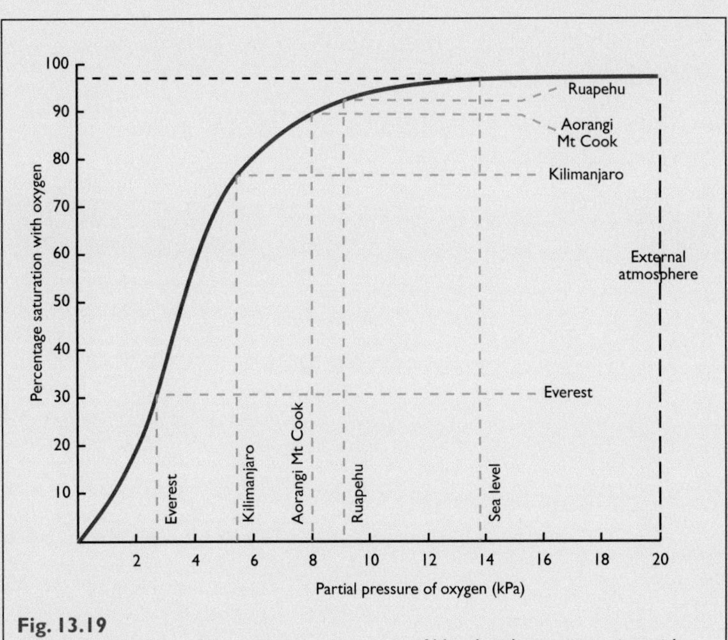

Fig. 13.19
Graph showing how the percentage saturation of blood with oxygen varies with oxygen pressure in the alveoli

ISBN 9780170191340

from altitude sickness, in which the person experiences nausea, vomiting and extreme fatigue.

Over a period of weeks, a person can increase the oxygen-carrying capacity of the blood by increasing the number of red cells from the normal 5 million per mm^3 to nearly 8 million. This is brought about by a hormone called *erythropoietin* (EPO), which is made in the kidneys. Some endurance athletes cheat by dosing themselves with EPO, though they could get the same result by training at altitude.

Haemoglobin also shows adaptation in the foetus. Before birth, a baby makes haemoglobin that is 'greedier' for oxygen than adult haemoglobin. Figure 13.20 shows the oxyhaemoglobin dissociation curve for foetal haemoglobin, compared with that of the mother. At any given partial pressure of oxygen, foetal haemoglobin is more saturated with oxygen than adult haemoglobin. The result is that in the placenta, foetal haemoglobin takes oxygen from the mother's blood.

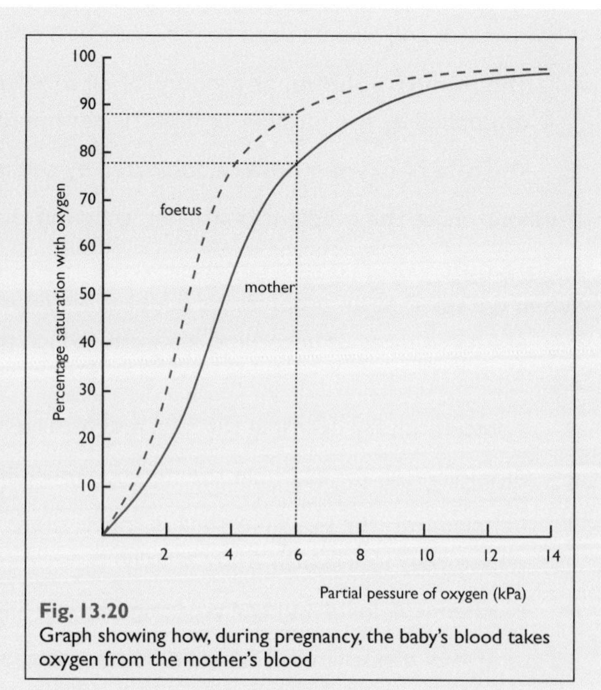

Fig. 13.20
Graph showing how, during pregnancy, the baby's blood takes oxygen from the mother's blood

Summary of key points in this chapter

- Cells are constantly expending energy on *active transport*, *biosynthesis* (making big molecules) and *movement*.
- The immediate source of energy in cells is the breakdown of ATP (adenosine triphosphate).
- ATP is constantly being synthesised using the energy released in the breakdown of organic matter.
- Most ATP is produced in *respiration*, which occurs in the *mitochondria*.
- Oxygen is carried in the red blood cells, combined with *haemoglobin*.
- Carbon dioxide is carried mainly in the plasma as *hydrogen carbonate* ions.
- *Gas exchange* in humans occurs in the millions of minute *alveoli*, which are richly supplied with capillaries.
- Gas exchange is promoted by the very short distance between blood and air, and the huge area of the gas exchange surface.
- The air supply to the alveoli is renewed by *breathing*, in which air flow is tidal (two-way).
- *Inspiration* is active and is due to the activity of the diaphragm and external *intercostal* muscles. At rest, *expiration* is passive.
- The stimulus for increased breathing during exercise is *increased* CO_2 concentration in the blood.
- Gas exchange is adversely affected by conditions such as *emphysema*, chronic bronchitis, asthma and carbon monoxide.

ISBN 9780170191340

Test your basics

Copy and complete the following sentences.

a) Cells are continuously expending energy for processes such as ___*___ (making large molecules such a proteins from small ones such as amino acids), movement, and ___*___ transport. The *immediate* source of energy for these processes is a chemical called ___*___.

b) Cells make most of their ___*___ by the oxidation of organic matter in ___*___. This is quite different from ___*___, which in mammals is the active pumping of air into and out of the lungs.

c) In the lungs, gas exchange occurs in microscopic air-filled cavities called ___*___. Air is carried to and from the ___*___ by a system of tubes called the ___*___ ___*___. The 'trunk' of this 'tree' is the windpipe or ___*___, the walls of which are supported by bands of ___*___. The larger branches of the tree are called ___*___ and are also supported by ___*___. The smallest branches are the microscopic ___*___, and have no ___*___ in the walls.

d) Between adjacent alveoli is a dense network of ___*___, which are supplied by branches of the ___*___ artery and drained by branches of the ___*___ veins.

e) The alveoli and capillaries have a huge ___*___ area and their walls are extremely thin, so the distance between air and blood is extremely ___*___. The blood in the capillaries has a higher concentration of ___*___ and a lower concentration of ___*___ than the air in the alveoli. As a result ___*___ diffuses from air into the blood and ___*___ diffuses in the opposite direction, a process called ___*___ ___*___.

f) Air is pumped into and out of the lungs by two sets of muscles; the ___*___ muscles between the ribs, and the ___*___, which separates chest and abdomen. The lungs are expanded by the raising of the ribs and lowering of the floor of the chest. The expansion of the chest ___*___ the pressure of air in the lungs, causing air to enter.

g) In inspiration (breathing ___*___), the ___*___ ___*___ muscles contract, raising the ribs, and the ___*___ contracts, lowering the ___*___ of the chest. In quiet breathing, expiration is passive, the lungs contracting under their own elasticity. In forced breathing, the ___*___ ___*___ muscles contract, actively lowering the rib cage.

h) The gas exchange surface is protected from pathogens by a cleaning mechanism. Many of the cells lining the air passages secrete a slimy ___*___, which traps dust and ___*___. The layer of ___*___ is slowly propelled up toward the ___*___ by the beating of millions of microscopic ___*___. A similar mechanism operates in the nasal passages, but here the mucus sheet is propelled back toward the ___*___, where it is periodically swallowed.

i) While breathing can be voluntarily interrupted for a short period, the production of ATP in cells is continuous. This is obtained from sugar in two stages:

1. Glycolysis, which occurs in the general cytoplasm and in which the sugar is converted to ___*___ acid with the production of a small amount of ATP.
2. Respiration, which occurs in the ___*___ and in which ___*___ acid is oxidised to ___*___ and ___*___ and the production of a large amount of ATP.

j) Skeletal muscle can store carbohydrate as ___*___, a polymer of ___*___. In vigorous exercise, glycogen is rapidly broken down into ___*___, which is then converted to ___*___ acid and ATP. This process is ___*___, meaning that it occurs without oxygen. During the recovery, some of the ___*___ acid is oxidised in respiration, producing enough ATP to convert the remainder back into ___*___.

ISBN 9780170191340

14 Excretion: Getting rid of waste

All cells carry out chemical reactions, collectively called **metabolism**. Some metabolic reactions result in the production of waste products. In very small animals these leave the body passively by diffusion, but in larger organisms they are actively got rid of, or **excreted**. Excretion is quite different from *egestion*, or the getting rid of undigested food as faeces.

EXCRETORY PRODUCTS

The two main excretory products made by humans are:

1. **Carbon dioxide**, produced in *respiration* and excreted by the *lungs*.
2. **Urea**. Before amino acids can be used as an energy source in respiration, they must first have their nitrogen-containing groups removed. This process is called **deamination** (Fig. 14.1).

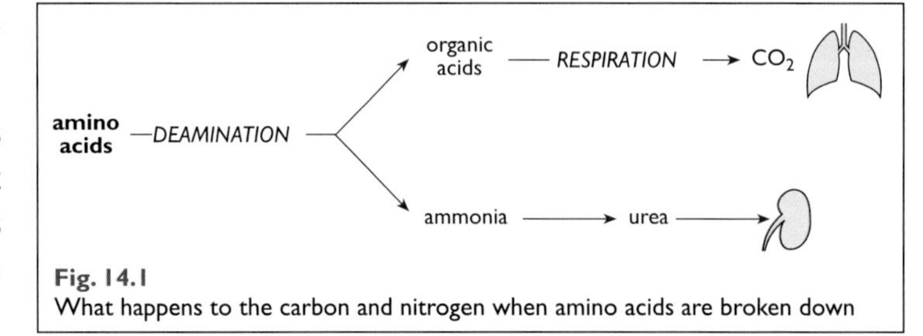

Fig. 14.1
What happens to the carbon and nitrogen when amino acids are broken down

Ammonia is extremely toxic and is immediately converted into *urea*, which is much less toxic. Deamination and urea formation occur in the **liver**.

Urea is excreted by the **kidneys** in solution as **urine**. Because urea contains nitrogen, the kidneys are concerned with *nitrogenous* excretion. Besides excretion of urea, the kidneys have several other functions, including:

- *Osmoregulation*, or the regulation of the water concentration of the blood.
- They play an important role in the regulation of *blood pressure*.
- They help regulate *blood pH*.

THE HUMAN EXCRETORY SYSTEM

Each kidney is supplied with blood by a **renal artery** and drained by a **renal vein**. A **ureter** takes urine from each kidney (Fig. 14.2). This propels urine to the **bladder** by waves of muscular contraction called **peristalsis**. As urine accumulates in the bladder, its walls are stretched, stimulating nerve endings in the bladder wall and producing the desire to urinate. The exit from the bladder is surrounded by a ring of smooth muscle, or **sphincter**. Though most smooth muscle is not under control of the will, we learn to control this particular muscle in early childhood.

You can get an idea of the importance of the kidneys from the fact that they take about 1 dm^3 (litre) of blood per minute, which is 20% of the output of the heart. Since the kidneys account for only 0.5% of the body, each gram of kidney takes about 40 times as much as each gram of the body as a whole.

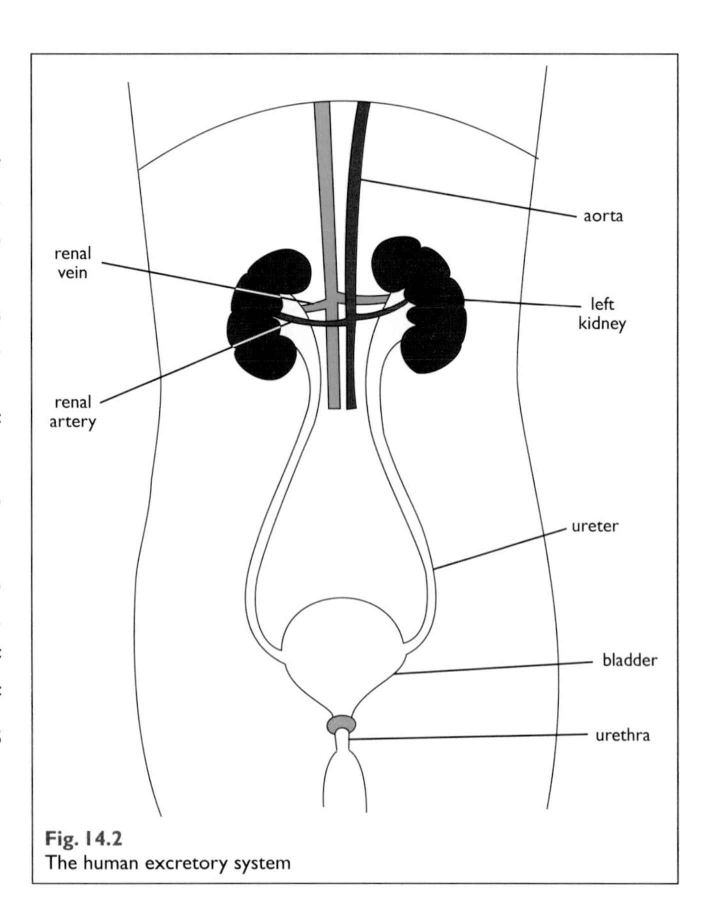

Fig. 14.2
The human excretory system

ISBN 9780170191340

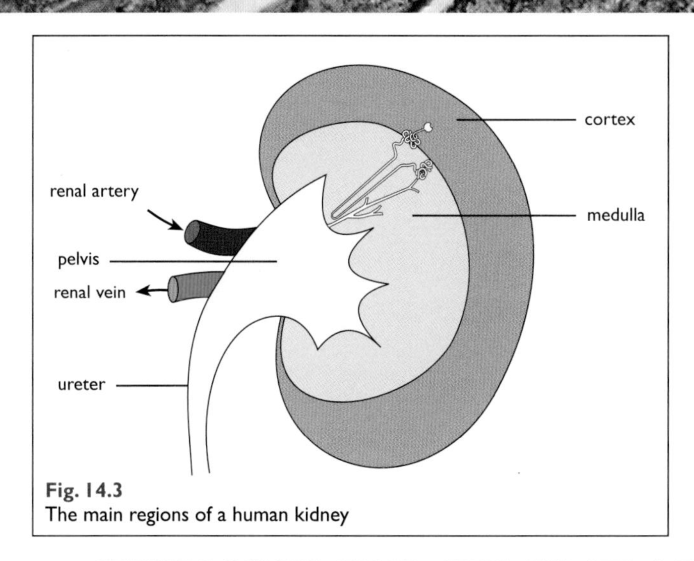

Fig. 14.3
The main regions of a human kidney

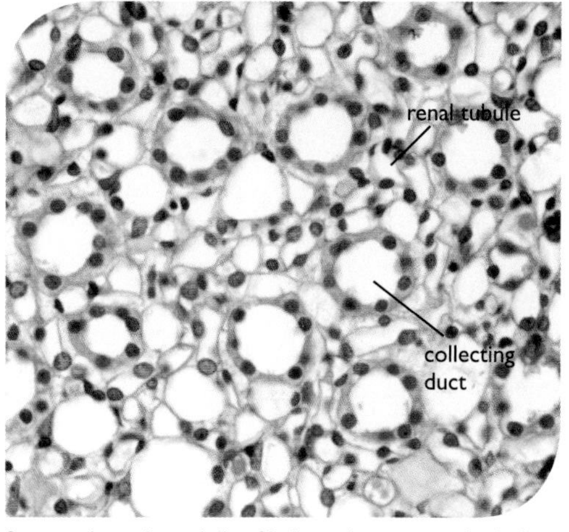

Section through medulla of kidney, showing renal tubules (narrow) and collecting ducts (wider)

STRUCTURE OF THE KIDNEY

If you cut a kidney longitudinally (lengthwise), you can see two layers, an outer red-brown **cortex** and a pink, inner **medulla** (Fig. 14.3).

Each kidney consists of about a million blind-ending tubes called **nephrons**, together with their blood supply (Fig. 14.4).

A nephron and its blood supply consists of two parts with different functions:

1. A blind-ending, cup-shaped **Bowman's capsule** into which dips a cluster of capillaries called a **glomerulus**. Together these *filter* the blood. The resulting **glomerular filtrate** then passes into (2).
2. A long **renal tubule** which modifies the filtrate and converts it into urine.

Like twigs of a tree, the nephrons join to form larger **collecting ducts**. These pass through the medulla and eventually open in a cavity called the *pelvis*, from which the ureter leads.

Filtration

Together with its glomerulus, each Bowman's capsule is a microscopic *filter* (Fig. 14.5). The energy for filtration comes from blood pressure provided by the *heart*. The glomerular filtrate is identical to blood plasma minus most of its protein, as shown in Table 14.1. The figures vary somewhat, depending on the state of the body. For example, when the body is dehydrated the urine is more concentrated, and vice-versa.

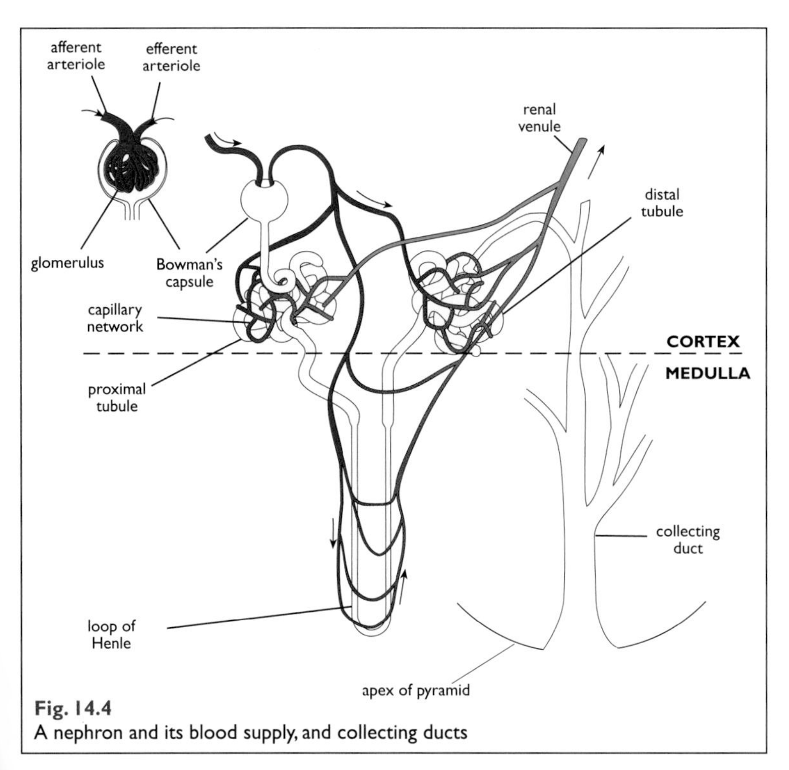

Fig. 14.4
A nephron and its blood supply, and collecting ducts

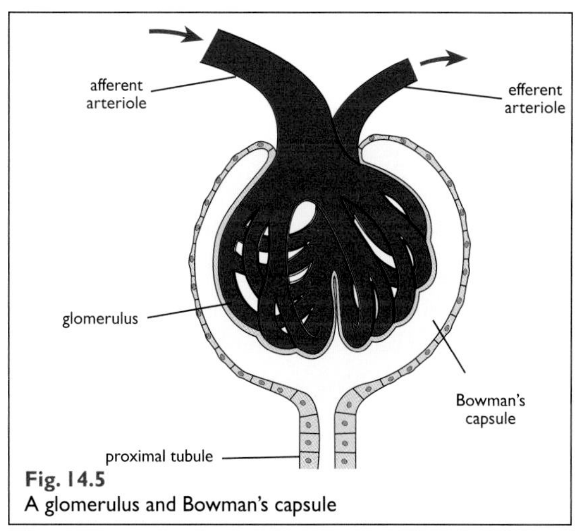

Fig. 14.5
A glomerulus and Bowman's capsule

 ISBN 9780170191340

	Percentage Composition		
	Plasma	Filtrate	Urine
Protein	7–9	0.01	0
Glucose	0.1	0.1	0
Sodium	0.32	0.32	0.35
Urea	0.03	0.03	2

Table 4.1 Composition of urine compared with that of the plasma and glomerular filtrate

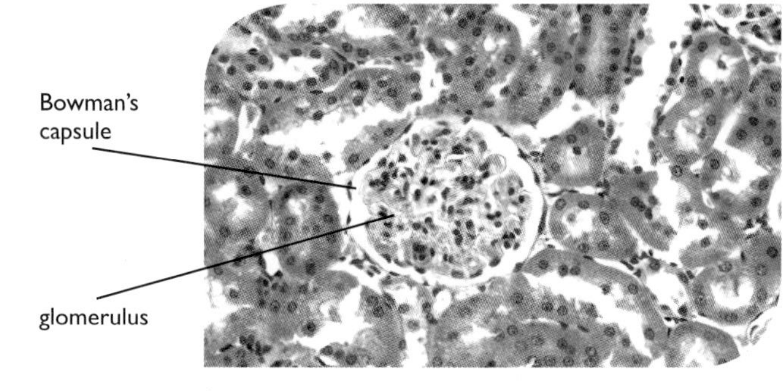

Section through cortex of kidney showing renal tubules, glomerulus with its surrounding Bowman's capsule

The kidney is adapted for filtration in three ways:

1. The *blood pressure* in the glomeruli is about 50% higher than in other capillaries. This is partly because the efferent arteriole (Fig. 14.5) is narrower than the afferent arteriole, causing a 'bottleneck' effect.
2. The capillary wall cells have tiny window-like holes, making them more *permeable* than those of most other capillaries.
3. The glomerular capillaries have a *large surface area* — about a square metre for both kidneys.

Filtrate is formed at about 7.5 dm^3 ('litres') per hour. This is about the same as the entire plasma volume, so it is essential that most of this liquid be recovered. Besides water, the filtrate contains other substances too valuable to lose, such as glucose, amino acids, minerals and vitamins. The second phase of urine formation is concerned with reabsorbing these valuable materials, leaving the urea more concentrated. This second phase of urine-formation depends on *active transport* into and out of the renal tubule, and the energy for this is supplied by the tubule cells.

Extension: The filter in more detail

In Fig. 14.5 the inner layer of Bowman's capsule is shown as a simple layer, but the reality is very different. The cells of the inner layer of Bowman's capsule are among the weirdest in the body. They are called *podocytes* ('foot cells') and each looks like a starfish, with long arms that surround a capillary (Fig. 14.6).

The capillary wall cells rest on an ultra-thin *basement membrane* consisting of a meshwork of protein fibres. The spaces between these fibres are large enough to allow most dissolved substances to pass through, except most of the plasma proteins. The basement membrane is thus the actual filter, rather than the capillary wall cells.

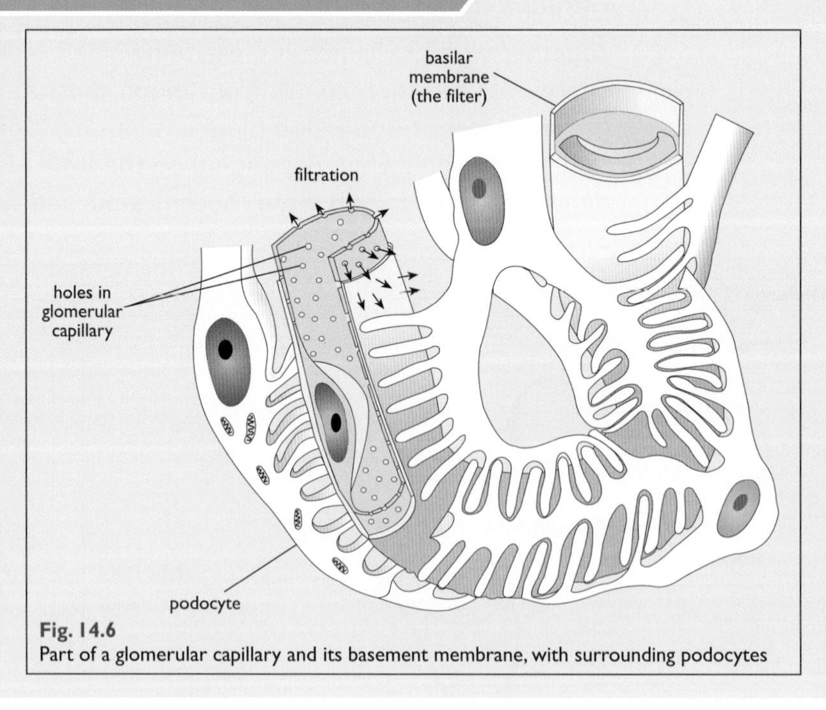

Fig. 14.6
Part of a glomerular capillary and its basement membrane, with surrounding podocytes

Modification of the filtrate: Active transport by the tubule

As the filtrate passes along the **proximal tubule**, various substances are actively transported into the blood (Fig. 14.7). The tubule cells are adapted for this in two ways (Fig. 14.8):

1. The plasma membrane facing the tubule lumen (cavity) has thousands of *microvilli*, which greatly increase the surface area. The total surface area of the proximal tubules of a kidney is about 5m^2.

ISBN 9780170191340

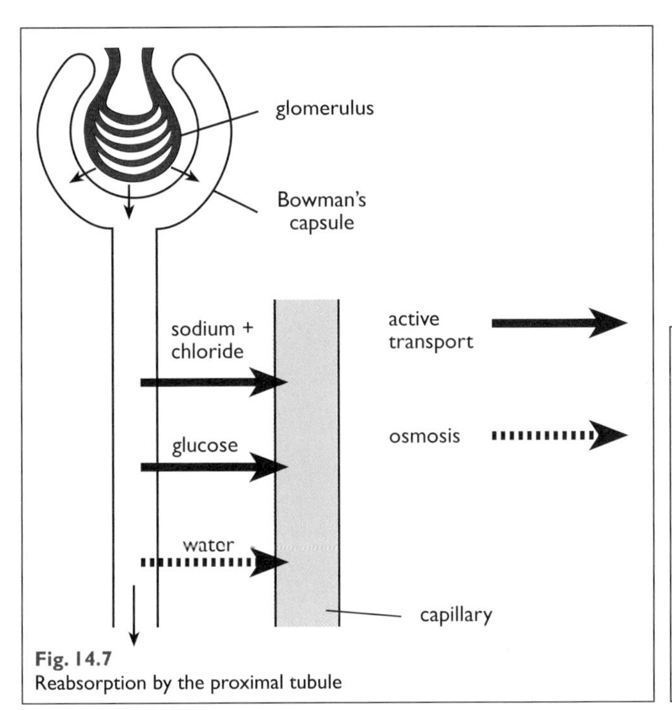

Fig. 14.7
Reabsorption by the proximal tubule

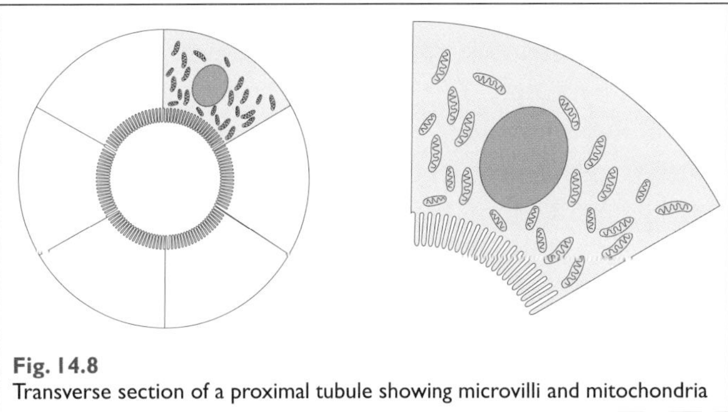
Fig. 14.8
Transverse section of a proximal tubule showing microvilli and mitochondria

2. They have abundant *mitochondria*, which provide the necessary ATP.

Reabsorption of glucose

Normally, all the glucose has been absorbed by the time the filtrate reaches the end of the proximal tubule. In diabetics, however, the blood glucose level is abnormally high and not all of the glucose can be reabsorbed, so some appears in the urine.

In the old days glucose in the urine had to be detected by taste, so this form of diabetes was called diabetes *mellitus* (mellitus means 'honeyed').

Reabsorption of sodium and water

Approximately 4/5 of the sodium chloride is reabsorbed in the proximal tubule. As a result the tubular fluid becomes more dilute than the blood, and water follows by osmosis. Thus 80% of the filtered water is reabsorbed into the blood by the proximal tubule.

This reabsorption of water by the proximal tubules happens regardless of the whether you are under- or over-hydrated.

Reabsorption by the distal tubules

The liquid leaving the proximal tubule has the same water concentration as the blood. As the tubular fluid reaches the **distal tubules**, more salt is reabsorbed, so that the liquid becomes more dilute than the blood. This very dilute fluid is delivered to the collecting ducts, where it may be further modified, depending on the state of hydration of the body.

Osmogulation by the kidney

It is everyday experience that if we drink more water we produce more urine, and if we are short of water we produce less. By varying the rate of urine production, the kidney plays an important part in **osmoregulation**, or the regulation of the water concentration of the blood.

While the proximal tubules reabsorb a constant proportion of the water in the filtrate, a *variable* amount is reabsorbed from the collecting ducts. The amount reabsorbed depends on the state of hydration of the body; if the body is dehydrated more water is reabsorbed, so smaller amounts of more concentrated urine are produced, and vice-versa.

The reabsorption of water from the collecting ducts depends on the **loop of Henle**. By a process involving active transport, the loop creates a high solute concentration (and therefore low water concentration) in the liquid between the collecting ducts. If the collecting ducts are permeable to water, water leaves the ducts by osmosis, concentrating the urine. On

ISBN 9780170191340

the other hand, if the collecting duct walls are impermeable to water, no more water is reabsorbed and the urine has the same concentration as the liquid leaving the distal tubules.

The various activities involved in urine formation are summarised in Fig. 14.9.

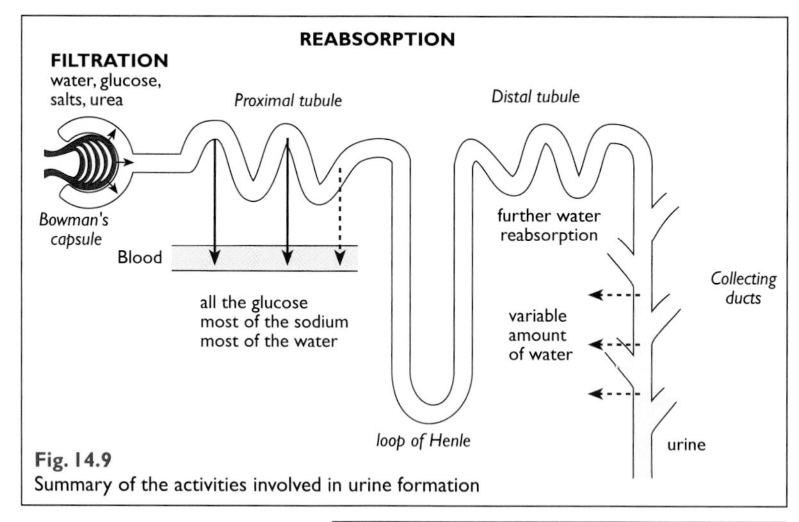

Fig. 14.9
Summary of the activities involved in urine formation

Role of the brain in osmoregulation

The kidneys do not 'know' if the body is over- or under-hydrated. This is the responsibility of the **hypothalamus** in the brain, which contains **osmoreceptors**. These are nerve cells (neurons) that monitor the state of hydration of the blood. Extending from the osmoreceptors are nerve fibres (axons) that end in the **posterior pituitary** gland just below (Fig. 14.10). The osmoreceptors produce **antidiuretic hormone (ADH)**, which is transported down the axons to the posterior pituitary, ready for secretion.

If the body water content falls, nerve impulses from the osmoreceptors travel down the axons to the posterior pituitary, causing them to increase their secretion of ADH. This increases the permeability of the collecting ducts to water, concentrating the urine. When the body is over-hydrated, ADH secretion decreases and urine output rises.

If the nerve fibres connecting the hypothalamus and posterior pituitary are damaged (as can happen after head injury), the person can no longer secrete ADH into the blood. The result is **diabetes insipidus**, in which the kidneys produce up to 20 dm^3 ('litres') of urine a day. The person is perpetually thirsty and has to drink huge volumes of water.

The hypothalamus also responds to dehydration by producing the sensation of *thirst*, thus promoting increased water intake. When we drink a large quantity of water, the blood is diluted and secretion of ADH decreases. There is no immediate effect on the kidney, since the existing ADH has to be destroyed by the liver before urine output increases.

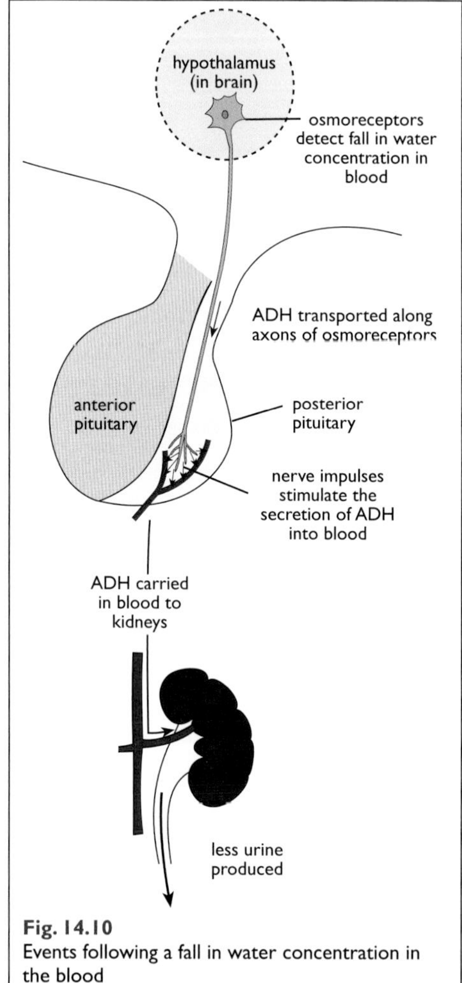

Fig. 14.10
Events following a fall in water concentration in the blood

Osmsoregulation is an example of **homeostasis**, or the regulation of internal conditions in the body. It also illustrates the concept of *negative feedback*, in which the results of a disturbance act to correct the disturbance. The processes involved in are shown in Fig. 14.11.

Effect of diet on urine

Besides the state of hydration, the composition of urine is also affected by our protein intake. A high protein diet leads to increased urea production and hence a higher concentration of urea in the urine.

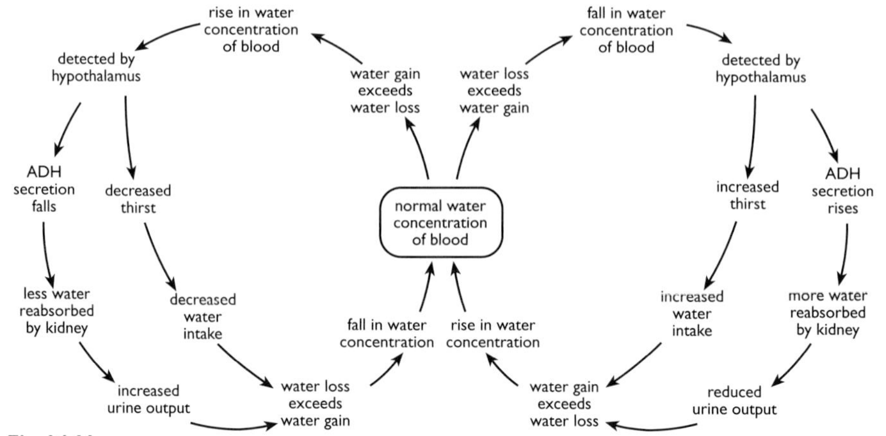

rise in water concentration of blood
detected by hypothalamus
ADH secretion falls
decreased thirst
less water reabsorbed by kidney
decreased water intake
increased urine output
water loss exceeds water gain
fall in water concentration
normal water concentration of blood
water gain exceeds water loss
water loss exceeds water gain
rise in water concentration
fall in water concentration of blood
detected by hypothalamus
increased thirst
ADH secretion rises
increased water intake
more water reabsorbed by kidney
reduced urine output
water gain exceeds water loss

Fig. 14.11
Summary of the processes involved in osmoregulation

ISBN 9780170191340

Summary of key points in this chapter

- The main excretory products of humans and other mammals are CO_2 and urea.
- Urea is produced in the deamination of amino acids in the liver.
- Urine formation involves two distinct processes:
  1. *Filtration*, energised by pressure developed by the heart, producing a liquid resembling plasma minus its proteins.
  2. Active reabsorption of useful materials back into the blood, for which the energy is provided by the tubule cells.
- In the *proximal tubules* all the glucose and 4/5 of the salt and water are reabsorbed.
- Depending on state of hydration of the body, a variable proportion of the remaining water is reabsorbed in the *collecting ducts*.
- Reabsorption of water from the collecting ducts is under the control of the hormone ADH, secreted by the *posterior pituitary* gland.

Test your basics

Copy and complete the following sentences.

a) Excretion is the getting rid of __*__ substances produced in __*__ .

b) The two main excretory products are CO_2 produced in __*__, and __*__ produced by __*__ of __*__ __*__.

c) The organs of nitrogenous excretion are the __*__. These also play an important part in __*__, or the regulation of the water content of the blood.

d) Blood is delivered to the kidney by the __*__ __*__ and leaves by the __*__ __*__. Urine leaves each kidney by a __*__ and is stored temporarily in the __*__.

e) Each kidney consists of about a million units or __*__. Each __*__ begins as a blind-ending, cup-shaped __*__ __*__, leading to a long renal __*__. This has three parts; the __*__ tubule, the loop of __*__, and the __*__ tubule. The __*__ and __*__ tubules are in the cortex and the loop of __*__ is in the medulla. The nephrons join to form __*__ ducts, which pass through the __*__ and open into a cavity called the pelvis, which is drained by the __*__.

f) Dipping into __*__ capsule is a cluster of capillaries called a __*__. The walls of the __*__ capillaries are permeable to all plasma constituents except most of the __*__, and as a result liquid filters through into __*__ capsule under pressure. This liquid is called the __*__ __*__ and is identical to plasma minus its __*__.

g) As the liquid passes along the __*__ tubule, all the __*__, together with most of the salts, are absorbed back into the blood by __*__ transport, and water follows by __*__.

h) As a result of the activity of the loop of __*__, the liquid surrounding the collecting ducts has a higher salt concentration (and therefore a lower __*__ concentration) than the blood. When the body is underhydrated (short of water), the walls of the collecting ducts become more __*__ to water. Water therefore passes by osmosis from the collecting ducts into the blood, resulting in a __*__ quantity of more __*__ urine. If the body becomes over-hydrated the collecting ducts become less __*__ to water, so less water leaves and a __*__ quantity of more __*__ is produced.

i) The permeability of the collecting ducts to water is under the control of __*__ hormone (__*__) secreted by the __*__ gland.

j) The water content of the blood is monitored by the __*__ in the brain, which connects with the __*__ by nerve fibres. When the body is dehydrated, the __*__ stimulates the __*__ to secrete more __*__, causing the kidneys to produce more concentrated urine. Upon drinking a large volume of water, the __*__ secretes less __*__, causing the kidneys to produce more __*__ urine.

k) The regulation of body water content is called __*__ and is an example of __*__, or the regulation of stable conditions in the body.

ISBN 9780170191340

15 Support and movement: The skeleto-muscular system

Most animals above a certain size have some form of **skeleton**. Skeletons can be of two basic kinds:

1. **Hydrostatic skeletons**. These occur in many soft-bodied animals such as earthworms, and depend on the fact that fluid can change its shape but not its volume.
2. *Jointed skeletons*. These are formed from some kind of stiff material, which acts as a system of levers. They are of two basic kinds:

   A. **Exoskeletons,** found in insects, crustaceans and other arthropods. The skeleton is made mainly of chitin and is produced by the skin, thus forming the outer layer of the body. Because the skeleton is non-living it cannot grow, so it has to be cast off periodically as the animal increases in size.

   B. **Endoskeletons**, found in *mammals* and other vertebrates (backboned animals). The skeleton of vertebrates consists of *bone* and *cartilage*, described below.

Skeletons have a variety of functions:

- They *transmit forces* from muscles to loads. In hydrostatic skeletons, this is the main function. In jointed skeletons the skeletal parts transmit forces by acting as *levers*.
- *Support*. They help to maintain the shape of the body.
- *Protection*. In humans and other animals with hard skeletons, certain parts protect vital organs. The skull protects the brain, the backbone protects the spinal cord, and the ribs protect the heart and lungs.
- Bone acts as a store of *calcium*.
- The marrow inside certain bones *produces blood cells*.
- The tiny bones in the middle ear *transmit sound* to the inner ear.

PARTS OF THE HUMAN SKELETON

A very young child has over 300 bones, but many of these fuse together during development with the result that the adult has an average of 206 bones. The human skeleton is shown in Fig. 15.1. The skeleton is conventionally divided into two regions:

1. The *axial skeleton*, consisting of skull, vertebral column ('backbone'), ribs and sternum.
2. The *appendicular skeleton*, consisting of limbs and the girdles to which they are anchored.

Tissues of the skeleton

The human skeleton consists of two tissues, *bone* and *cartilage*.

Bone

Bone is a living tissue, containing a rich internal blood supply. The bone cells or **osteocytes** are embedded in a *matrix* of **calcium phosphate** and the protein **collagen**, which they secrete. The bone cells secrete the matrix and are connected to each other by fine channels through which oxygen and nutrients reach them by diffusion.

Collagen is like rope; it is very resistant to *tension* but cannot resist *compression*. Calcium phosphate is very resistant to compression but is very brittle. The two combined together are very resistant to both compression and tension and make bone very strong.

Fig. 15.2 shows a longitudinal (lengthwise) section through a long bone such as a femur. The shaft consists of *compact bone*, in which the bone cells are arranged in concentric cylinders around a central **Haversian canal** containing a capillary (Fig. 15.3).

ISBN 9780170191340

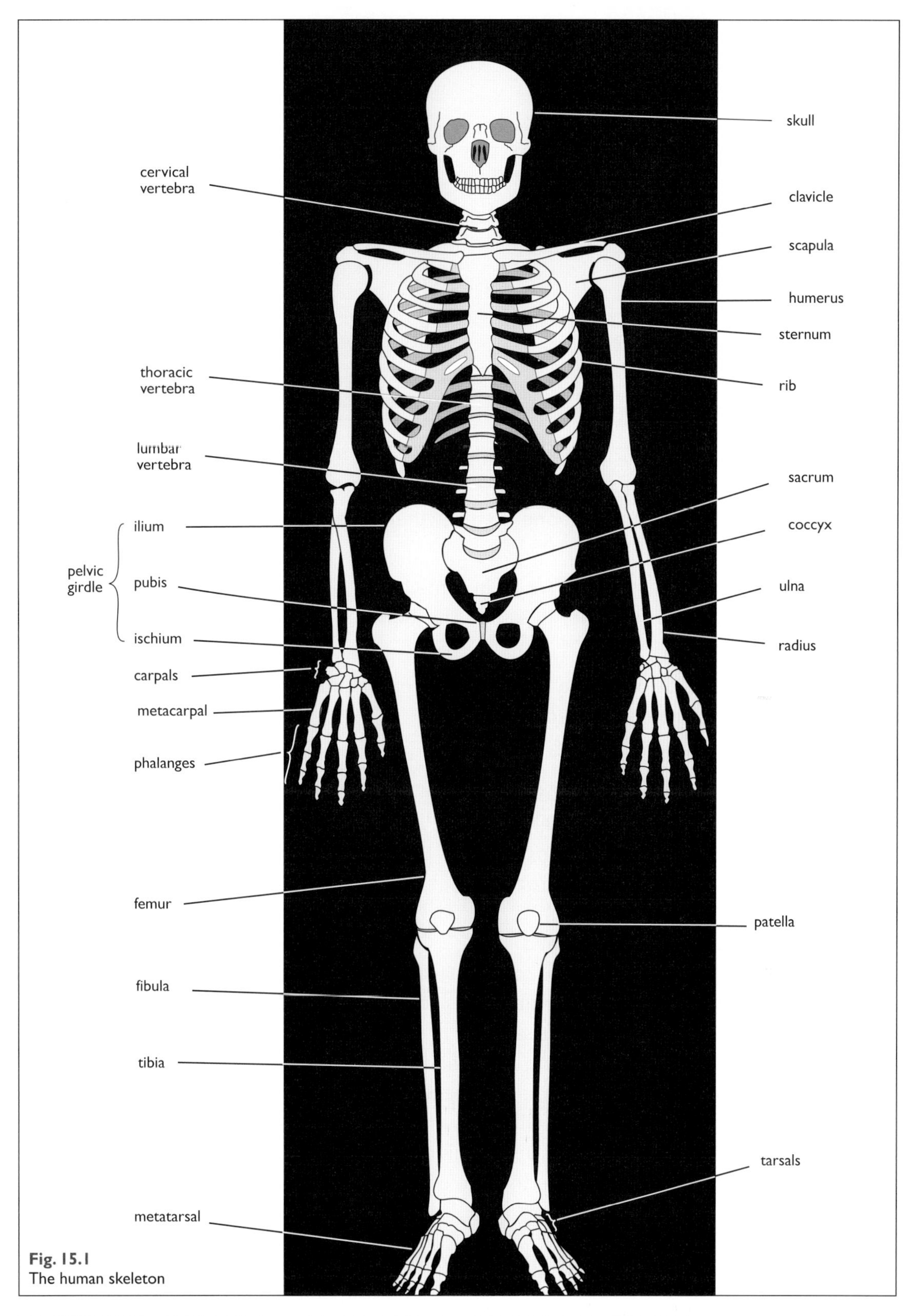

Fig. 15.1
The human skeleton

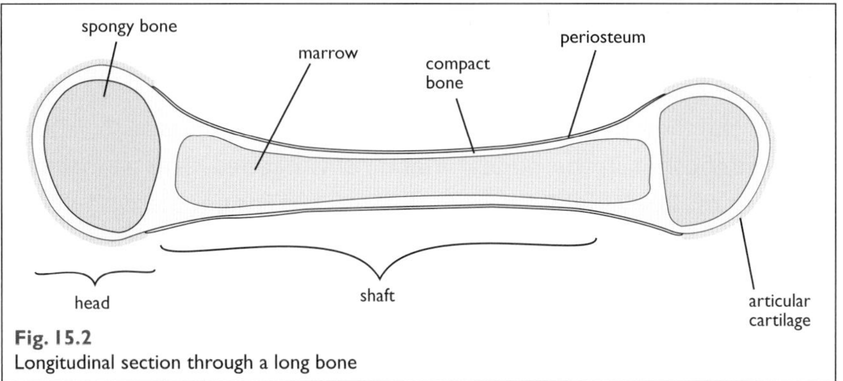

Fig. 15.2
Longitudinal section through a long bone

ISBN 9780170191340

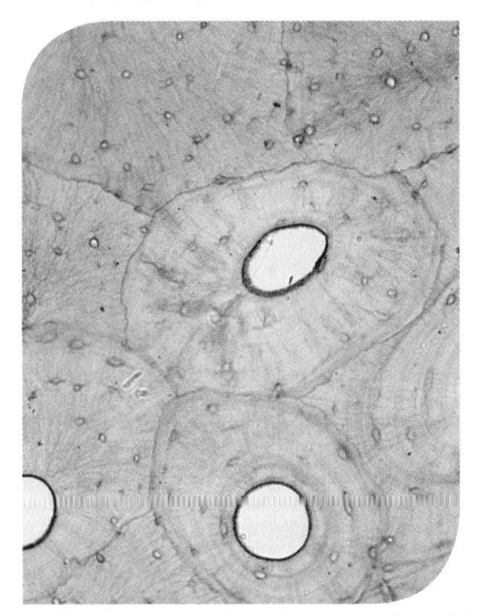
Transverse section through compact bone (the cells and blood vessels are no longer present)

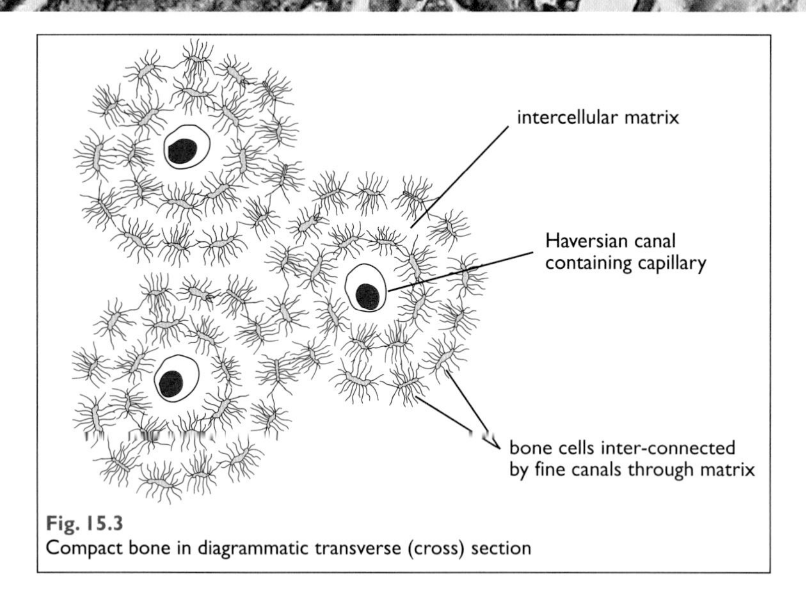

Fig. 15.3
Compact bone in diagrammatic transverse (cross) section

Long bones such as the femur and humerus need to withstand *bending*. The most economical way to do this is for them to be *hollow*, which saves weight. For a given cross-sectional area of bone, a hollow bone is stiffer than a solid one. The scaffolding used on building sites illustrates the same principle.

Inside the shaft is the *marrow*, which in some bones is where blood cells are produced. The ends consist of *spongy bone*, in which the bone and blood spaces form a network running in all directions.

Covering the outside of a bone is a tough, fibrous membrane, the *periosteum*, to which *ligaments* and *tendons* are anchored.

How bones grow

In early development, the skeleton of a human foetus consists entirely of cartilage. As the child develops, most of this is replaced by bone in the process of *ossification*. In a long bone, such as a humerus, parts remain as unossified as *epiphyseal cartilages* (Fig. 15.4). At one side of an epiphyseal cartilage, cartilage cells multiply and add to the cartilage, and at the other side the cartilage is being replaced by bone. Throughout the growing period, growth and replacement of cartilage keep pace with each other, so the epiphyseal cartilages remain the same thickness. Eventually, growth of cartilage ceases and the epiphyseal cartilages are completely replaced by bone. This occurs at a characteristic age for each bone, so it is possible for an expert to judge the age of a young person from the skeleton.

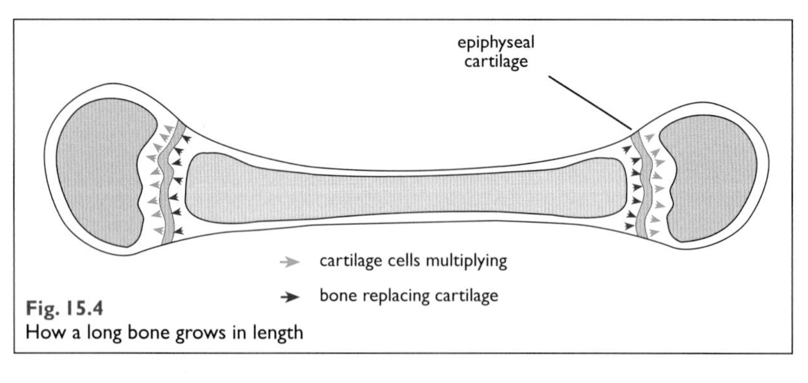

Fig. 15.4
How a long bone grows in length

Bones, diet and parathyroids

For strong bones, the diet must contain enough calcium, especially during growth and pregnancy. Nursing mothers also need more calcium since milk is rich in the mineral. In men, and in women who are not 'eating for two', deficiency of calcium is rare, since it occurs in so many foods. The best sources of calcium are milk and cheese. Other sources include green vegetables and fish.

The supply of calcium in the diet is not the only important factor, however. To absorb calcium from the gut and to deposit it in the bones, **vitamin D** is needed. This is produced in the skin under the influence of ultraviolet light from the sun, but for people who do not get enough sunlight, it must be present in the diet. Good dietary sources are fish (especially fish oils), and dairy products.

ISBN 9780170191340

If the diet is deficient in vitamin D, the bones become weak and easily broken. In children, the deficiency disease is **rickets**, in which the bones are mis-shaped. In adults the bones become more liable to fracture — a condition called **osteomalacia**.

Though the skeleton accounts for 99% of the body's calcium, calcium is even more important in other ways. For example it is vital in communication between nerves and muscles, and in muscle contraction and blood clotting. If calcium were to be suddenly removed from the blood, you would be dead within seconds.

The bones are therefore an important store of calcium; if the diet is temporarily deficient, calcium is removed from the bones to maintain the blood concentration.

The level of calcium in the blood is controlled by **parathyroid hormone (PTH)** secreted by the *parathyroid glands*. These are four tiny patches of tissue embedded in the thyroid gland in the neck. PTH stimulates the release of calcium from the bones into the blood. If the level of calcium in the blood decreases, PTH output increases and calcium is released from the bones.

Like other tissues, bone is in a constant state of *turnover*, meaning that it is simultaneously being added and removed. In a healthy adult about 10% of the bone is renewed each year. Normally, new bone is being added at about the same rate as which it is removed, but in women after menopause, addition may fail to keep up with removal. This condition is called **osteoporosis**, in which the bones become weaker.

As a living tissue, bone responds to environmental change. When bones are stressed, extra bone tissue is deposited and the bones become stronger, so weight training can greatly reduce osteoporosis. If a bone breaks and is set in a slightly 'incorrect' way, new bone is deposited and the bone is re-modelled to suit the new pattern of stresses.

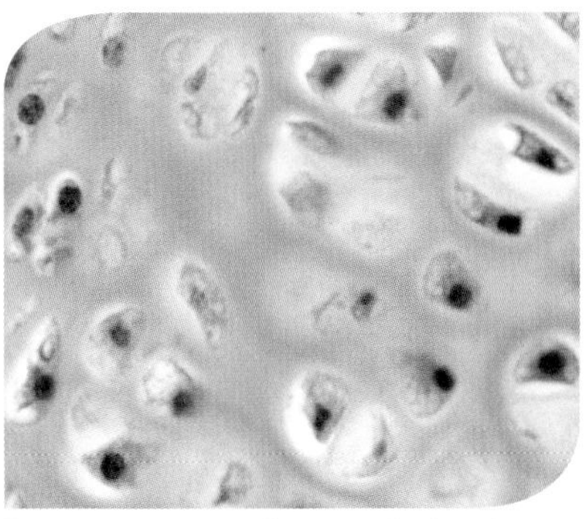
Section through cartilage in wall of trachea

Cartilage

Like bone, **cartilage** consists of cells embedded in a matrix they secrete, but is much more flexible. The most common form of cartilage has a glassy appearance and covers joint surfaces. It also supports the walls of the trachea and bronchi, and forms the ends of the ribs where they join the sternum (Fig. 15.5). Cartilage forms the embryonic skeleton, though most is later replaced by bone.

Unlike bone, cartilage has no *internal* blood supply, so the cartilage cells within depend on nutrients diffusing from the capillary network on the surface of the cartilage. Probably for this reason, cartilage cannot repair itself.

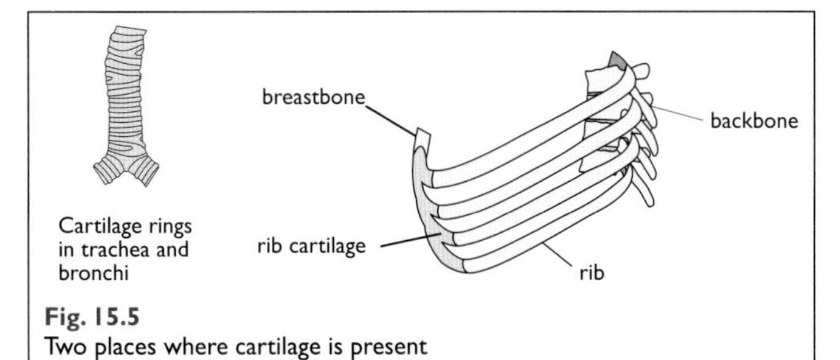

Fig. 15.5
Two places where cartilage is present

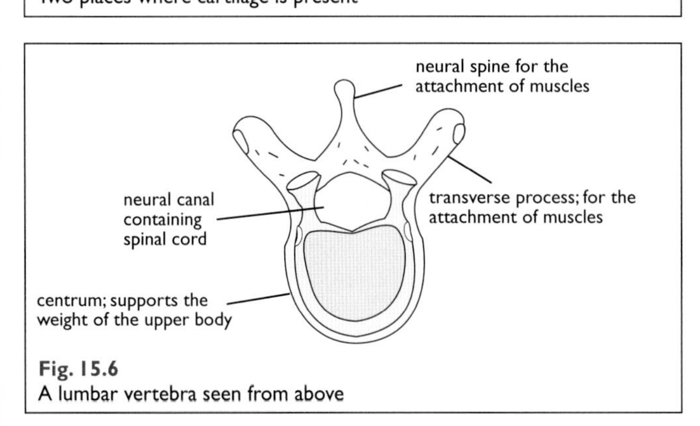

Fig. 15.6
A lumbar vertebra seen from above

The vertebral column

The 'backbone' is not a single bone but consists of many individual bones called **vertebrae**, and is more correctly called the *vertebral column*. A lumbar (lower back) vertebra is shown in Fig. 15.6.

The backbone is divided into five regions:

1. *Cervical* (neck) region, with seven vertebrae.
2. *Thoracic* (chest) region, with 12 vertebrae.
3. *Lumbar* (lower back) region, with five vertebrae.
4. *Sacrum*, consisting of five fused (joined) vertebrae. The first articulates (forms a joint) with the hip girdle and so transmits the weight of the body to it.
5. *Coccyx*, consisting of four fused vertebrae.

ISBN 9780170191340

A lumbar vertebra has the following parts: The *centrum* consists of spongy bone and resists compression due to the weight of the body above: the *neural arch* surrounds and protects the spinal cord; the *neural spine* and *transverse processes* provide extra surface for the attachment of muscles.

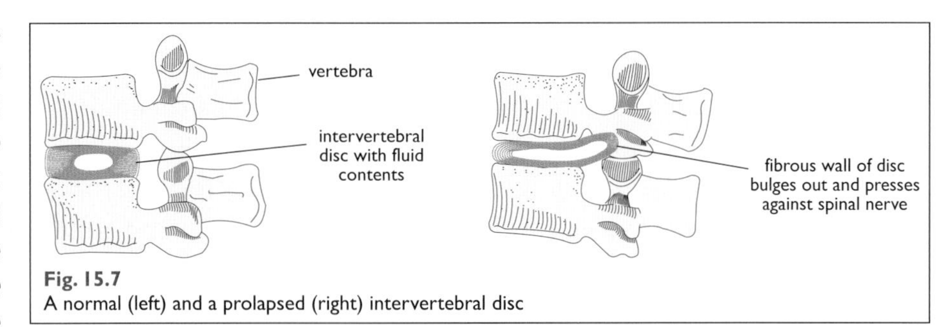

Fig. 15.7
A normal (left) and a prolapsed (right) intervertebral disc

Between the vertebrae are pads that act as shock absorbers. Each is like an extremely tough, thick-walled balloon, with a fluid centre surrounded by a thick wall of a special kind of tough cartilage. When bearing the body weight the fluid is compressed and the fibrous wall is under tension, just like the wall of a balloon.

Provided that the pressure on the discs is evenly distributed, they can withstand considerable compression. This is why you should not bend your back when lifting a heavy weight. When doing so you should bend at the knee and hip joints, keeping the back straight and the discs evenly loaded (Fig. 15.7). Failure to do this can result in a *prolapsed* or 'slipped' disc. The disc does not 'slip', but the wall bulges out and may press against a spinal nerve, causing considerable pain.

Joints

A joint is formed where two or more bones meet. Joints can be divided into three classes according to the kind of tissue separating the bones:

1. In *fibrous joints* the bones are held together by fibrous tissue, for example the bones of the skull. In this type little or no movement is possible.
2. In *cartilage joints* the bones are held together by cartilage, for example the joints between the ribs and breastbone, and between the vertebrae.
3. *Synovial joints* allow free movement between the ends of bones (Fig. 15.8).

In a synovial joint the joint surfaces are covered by a smooth layer of cartilage. Movement is restricted by **ligaments**, which consist of dense bundles of collagen fibres. Ligaments are immensely resistant to tension and sometimes it is a bone that breaks rather than the ligaments that tie them together. Ligaments have a poor blood supply and so appear white. This is why they heal more slowly than bone. An injury to a ligament is a *sprain*.

The ligaments form a capsule containing a lubricating **synovial fluid,** secreted by the **synovial membrane** which forms the lining of the joint capsule. Normally the amount of synovial fluid is very small (about a teaspoonful in the knee joint), but if the synovial membrane becomes inflamed it produces more fluid and the joint becomes swollen.

Synovial joints can be subdivided according to the kind of movement that occurs. A common type is a **hinge joint**, in which movement is possible in one plane only. Examples are the joints of the fingers, elbow and knee. **Ball and socket joints**, for example the hip and shoulder joints, allow movement in more than one plane.

Muscles and joints

Though they are often treated separately, the skeleton and muscles work so closely together that they are better treated as a single *skeleto-muscular* system. Bones act as anchorage points for muscles, and transmit their forces to loads.

The muscles that move bones are called **skeletal muscles,** and are anchored by **tendons** to the tough periosteum surrounding the bone. A skeletal muscle consists of thousands of *muscle fibres*.

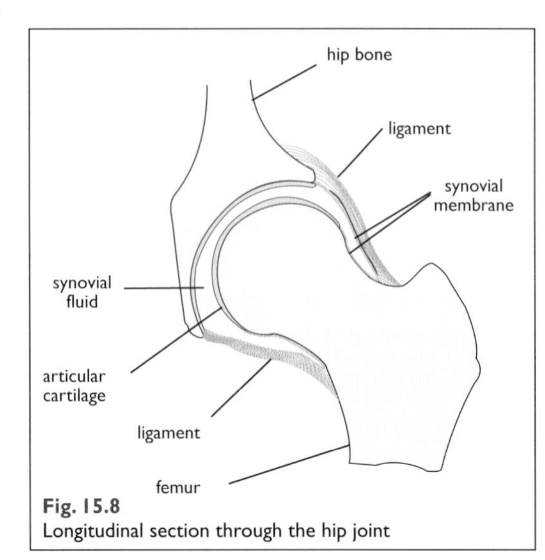

Fig. 15.8
Longitudinal section through the hip joint

ISBN 9780170191340

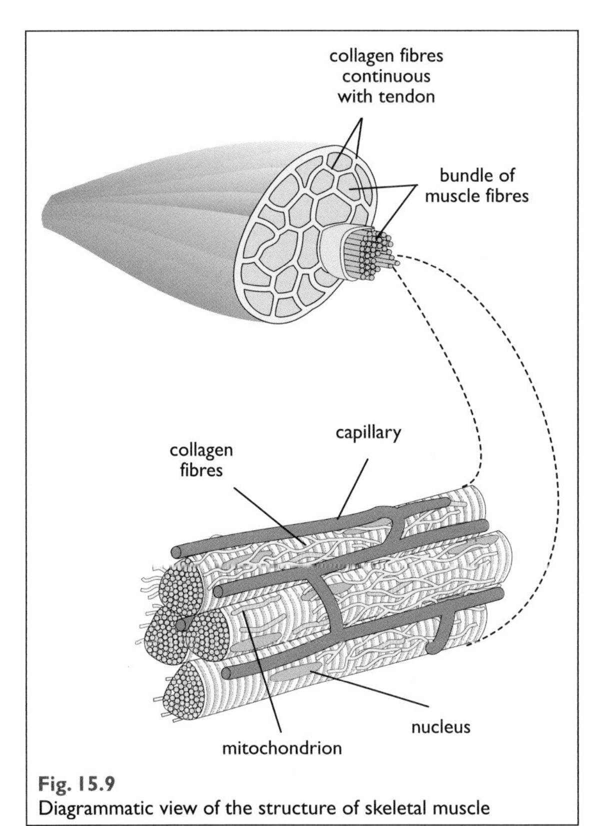

Fig. 15.9
Diagrammatic view of the structure of skeletal muscle

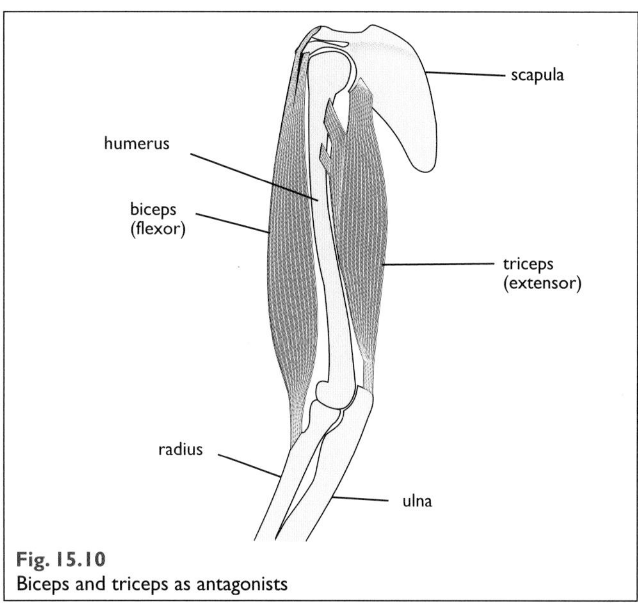

Fig. 15.10
Biceps and triceps as antagonists

Longitudinal section through skeletal muscle, showing characteristic striations. In each fibre there are many nuclei

These are not cells, but are formed by the joining together of individual cells early in embryonic development, so each fibre contains many nuclei.

Skeletal muscle tissue is also called *striated muscle* because under the microscope its fibres have fine lines or striations running across it (Fig. 15.9). This is actually a more appropriate name, because some of the muscles of the tongue, which are striated, have no connection with the skeleton.

Like ligaments, tendons are extremely strong and are composed of thick bundles of collagen fibres. The tendon fibres fan out and extend deep between the muscle fibres. As a result muscle and tendon fibres grip each other over a huge surface area. The tearing of collagen fibres in a muscle or tendon is a *strain*.

Muscles convert chemical energy in sugar into mechanical energy in the form of movement. When a muscle fibre contracts, protein filaments slide past each other, causing the fibre to shorten and thicken. Skeletal muscles can work in two different ways:

1. Some contract rapidly to produce *movement*. These muscles are liable to fatigue.
2. Some maintain steady contraction over long periods to *prevent* movement, for example muscles in the back and abdomen that help maintain posture. Without these we would collapse in a heap.

Muscles can only *pull*; they cannot *push*. This is true of all muscles, including cardiac and smooth muscle. When a muscle has shortened, it must be pulled back to its original length by an external force before it can shorten again. This external force is usually due to contraction of another muscle, which is said to act **antagonistically**. In the elbow joint the biceps flexes (bends) the joint and the triceps extends (straightens) it (15.10). Biceps and triceps are thus an *antagonistic pair*.

Skeletal muscles produce slow, powerful forces, but by means of levers these can be converted into fast-moving (but correspondingly weaker) forces acting on the load. Figure 15.11 illustrates this idea using the elbow joint: The load is 30/5 = six times further from the fulcrum (joint) than the insertion of the biceps. When the biceps contracts, the load therefore moves six times as fast as the muscle shortens. The other side of the coin is that the muscle must exert six times as much force as the load just to hold the load steady. Thus a lever

ISBN 9780170191340

enables a load to be moved much faster than the muscle can shorten. The longer the lever, the faster the load can be moved, but the weaker the force transmitted to it.

The feet of a cheetah in top gear can move at 28 ms^{-1}, which is far faster than its muscles are shortening. This is made possible by having long limbs (levers). On the other hand moles have stumpy legs that act as short levers, thus transmitting greater force to the soil during burrowing.

Figure 15.11 shows a simple case, in which the muscle and load are pulling at right angles to the lever. When a muscle or load acts at a different angle, it gets a bit more complicated. In cases like this, what matters is the distance between the fulcrum and the *line of action* of a force.

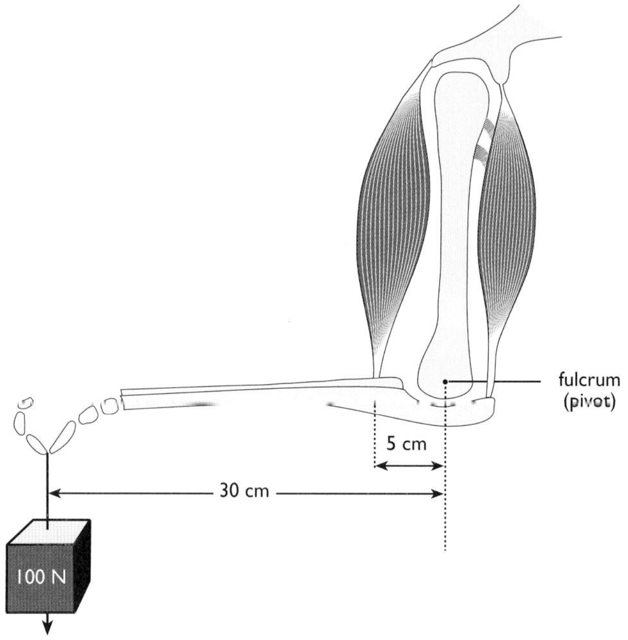

Fig. 15.11
The human forearm as a lever

Effects of training

Like bone, muscles respond to use and disuse. Training for strength and power results in the muscle fibres getting thicker, but their number does not increase (this remains the same as the number we are born with). In endurance events, response to training consists in improving the blood supply by strengthening the heart.

Summary of key points in this chapter

- Many soft-bodied animals have a *hydrostatic* (fluid) skeleton.
- Vertebrate animals have internal skeletons or *endoskeletons*, in contrast to insects and other arthropods which have *exoskeletons* (external skeletons).
- Skeletons transmit forces from muscles to loads, and help maintain body shape. In hard, jointed skeletons they also protect vital organs.
- The human skeleton consists of two tissues: *bone* and *cartilage*.
- Bone contains cells (*osteocytes*) embedded in a *matrix* they secrete.
- The bone matrix consists of the fibrous protein *collagen* and the mineral *calcium phosphate*.
- Compact bone consists of concentric cylinders of bone, nourished by a capillary in a central *Haversian canal*.
- *Vitamin D* is essential for absorbing calcium from the gut and depositing it in the bones.
- The level of calcium in the blood is controlled by *parathyroid hormone*.
- Cartilage is softer and more flexible than bone, and cannot repair itself as it has no internal blood supply.
- In synovial joints, the joint surfaces are covered with cartilage and lubricated by *synovial fluid*.
- In synovial joints the bones are held together by *ligaments*, which consist largely of collagen fibres.
- Since muscles can only pull, they have to work in *antagonistic pairs*, which have opposite effects. *Flexor* muscles bend joints, and *extensors* straighten them.
- The forces exerted by muscles are modified by bones acting as *levers*; strong, slow forces produced by muscles can be converted into weaker but faster-moving forces acting on the load.

ISBN 9780170191340

Test your basics

Copy and complete the following sentences.

a) The human skeleton consists of two kinds of tissue: __*__, which is hard and has a rich internal __*__ supply, and __*__, which is flexible and has no internal __*__ supply.

b) Bone consists of bone cells or __*__, surrounded by a __*__, which consists of the protein __*__ and the mineral __*__ __*__. In the shaft of a long bone such as a femur (thigh bone), the cells are arranged in concentric cylinders around a central canal containing a __*__. A bone is surrounded by a tough membrane called the __*__, to which ligaments and __*__ are attached.

c) Besides its mechanical functions (protection, support and transmitting muscular forces), bone is an important store of __*__.

d) The absorption of calcium from the gut and its deposition in the bones is promoted by vitamin __*__. Lack of this vitamin in young children causes __*__, in which the bones are mis-shapen, and in older people it leads to __*__, in which the bones become weaker.

e) In young adults, bone is deposited and removed at the same rate, but in older people, bone formation tends to lag behind bone reabsorption, so the bones become weaker, a condition called __*__.

f) Long bones grow at sites called __*__ cartilages. At one side of the cartilage, cartilage cells multiply, and on the other side, cartilage is replaced by bone. The process by which cartilage is replaced by bone is called __*__.

g) The 'backbone' or __*__ __*__ consists of 33 bones or __*__, divisible into five regions: the neck or __*__ region, the chest or __*__ region, the lower back or __*__ region, the __*__ and the short 'tail' region or __*__. Between the __*__ in the neck, chest and back there are discs of tough __*__.

h) In __*__ joints, the bones move relative to each other. The joint surfaces are covered with a layer of __*__, and the bones are held together by __*__ which consist of the tough fibrous protein __*__. These joints are lubricated by __*__ fluid.

i) Joints are moved by __*__ muscles, which are anchored to the bones by __*__. These are like ligaments in that they consist of bundles of __*__.

j) Since a muscle can only __*__, once it has contracted and shortened, it cannot get back to its original length by itself, but must be extended by another muscle called an __*__. Muscles thus work in pairs; a muscle that bends a joint is called a __*__, and a muscle that straightens it is called an __*__.

k) One of the functions of long bones is that they act as levers. A lever enables the slow movement of muscular contraction to be converted into a __*__ (but __*__) force acting on the load.

ISBN 9780170191340

16 Sensitivity

One of the key characteristics of living organisms is their ability to respond to changes or **stimuli** (singular: *stimulus*). Stimuli may be potentially beneficial — for example the presence of food in the case of animals, or a change in light direction in the case of plants — or they may be potentially harmful.

In all animals more complicated than sponges, stimuli are detected by special cells (or in some cases parts of cells) called **receptors**. Each receptor is specialised for detecting a particular kind of stimulus. Thus **photoreceptors** detect *light*, *chemoreceptors* detect *chemicals*, and *mechanoreceptors* detect mechanical stimuli such as *pressure, tension* or *vibration*.

Receptors can only *detect* stimuli. The *response* is the job of **effectors**, such as muscle fibres or gland cells. Receptors communicate with effectors by means of nerve cells or **neurons**, which collectively make up the *nervous system*.

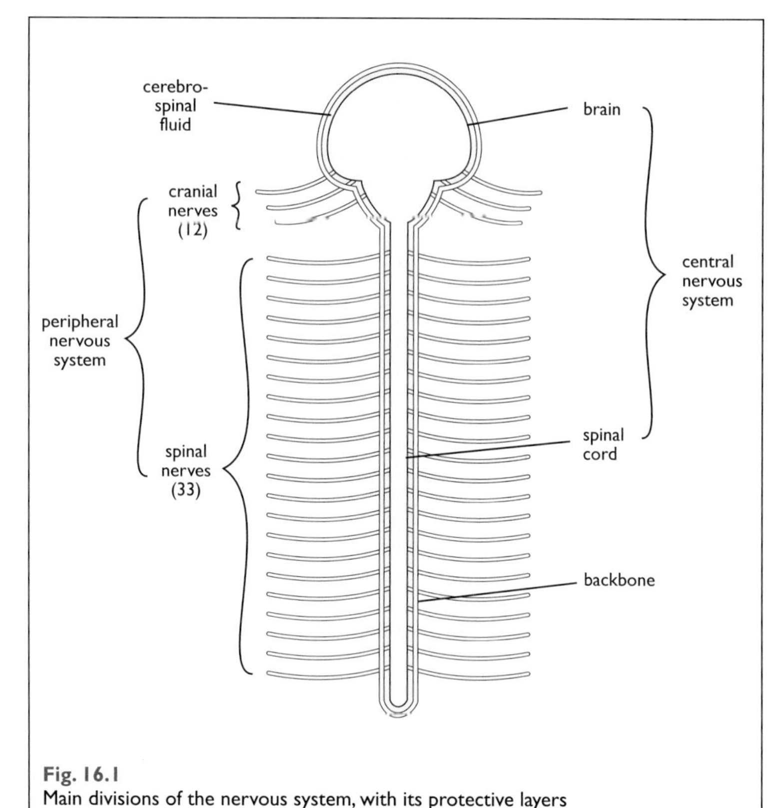

Fig. 16.1
Main divisions of the nervous system, with its protective layers

In general, receptors do not communicate directly with effectors. For an appropriate response, information from various sources must first be processed. This is the function of the **central nervous system** (CNS), which consists of the brain and spinal cord (Fig. 16.1).

Neurons carry information around the body in the form of electrical signals called **impulses**. These travel at high speed (up to 110 m s^{-1}, somewhat faster than the speed of a jumbo jet coming in to land) along nerve fibres or **axons**. An axon is a long, thread-like outgrowth of a neuron and the most rapidly-conducting axons are covered by a fatty, insulating **myelin sheath**. Neurons collect information via branching threads called **dendrites**.

Neurons that carry impulses from the CNS to effectors are called **motor** or *efferent* neurons (Fig. 16.2). Neurons that carry impulses from receptors to the CNS are called **sensory** or *afferent* neurons. Neurons connecting sensory and motor neurons are called **interneurons**.

In a nerve pathway, adjacent neurons do not actually come into contact with each other but are separated by extremely narrow gaps called **synapses**. When an impulse reaches a synapse, it triggers the release of a chemical (e.g. acetylcholine) which diffuses across the gap, producing an electrical change in the next neuron.

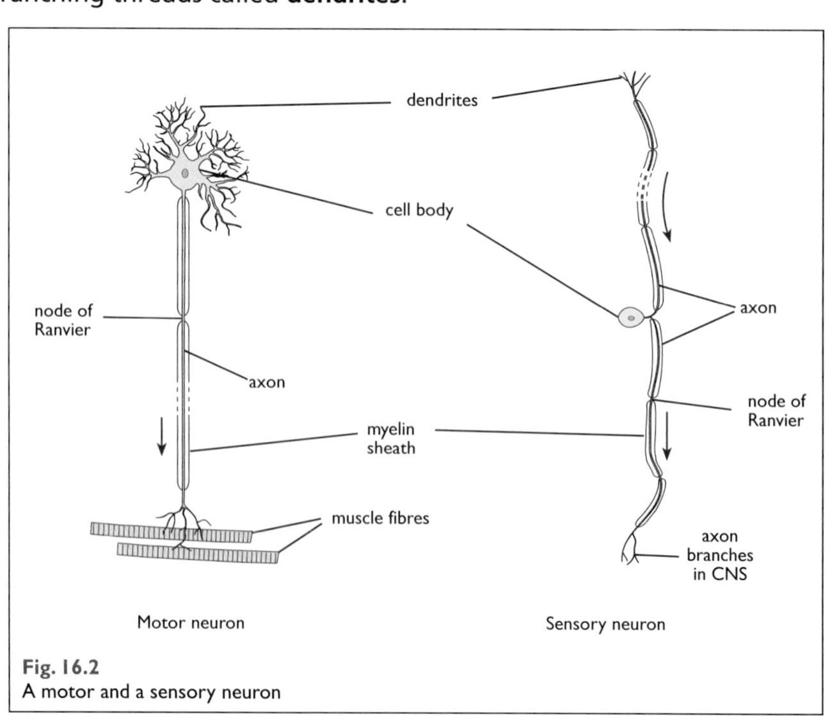

Fig. 16.2
A motor and a sensory neuron

ISBN 9780170191340

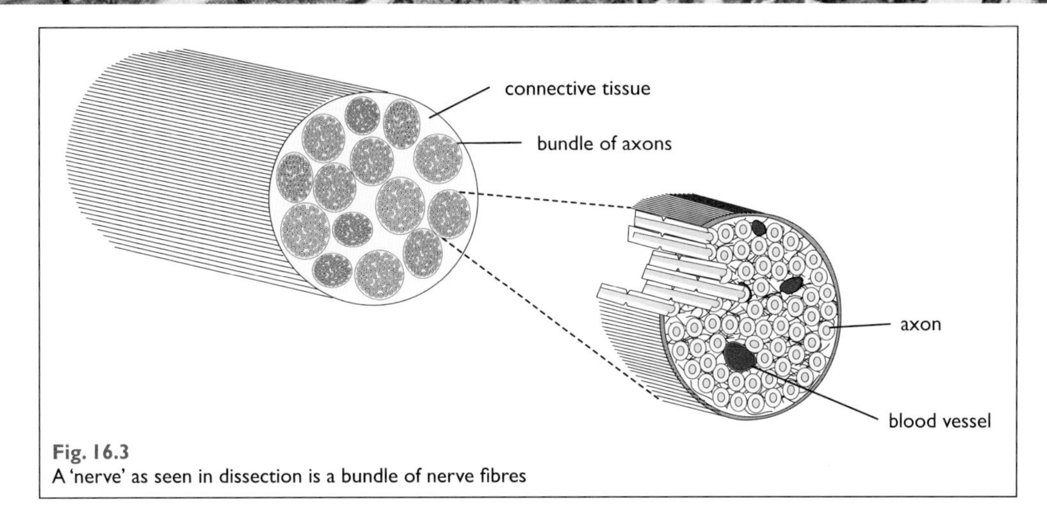

Fig. 16.3
A 'nerve' as seen in dissection is a bundle of nerve fibres

The white stringy nerves you might see in a dissection (i.e. with the naked eye) are actually bundles of thousands of individual axons, held together by connective tissue fibres (Fig. 16.3).

Simple responses: Reflex actions

A **reflex action** is an *automatic* response to a stimulus — in other words, it does not have to be thought about. There are two kinds: *Simple reflexes* do not have to be learned, for example the secretion of tears when grit gets into your eye. *Conditioned reflexes* have to be learned in the first instance but become automatic. An example is when a motorist swerves to avoid a pedestrian.

Reflex arcs

A reflex arc is the simplest arrangement of neurons that can be involved in a reflex action. It is therefore a kind of 'wiring diagram'. The simplest reflex arcs only involve the spinal cord, as shown in Fig. 16.4. The central 'H'-shaped area is the **grey matter** and consists mainly of nerve cell bodies. The surrounding **white matter** consists of axons and owes its appearance to the white myelin sheaths. Notice that all the sensory fibres pass up the dorsal roots and all the motor fibres pass down the ventral roots. The cell bodies of the sensory neurons lie in the dorsal root, causing a swelling called a **ganglion**.

When a receptor is stimulated, impulses travel up sensory axons to interneurons in the spinal cord. These link up with motor neurons, which carry impulses to effectors, causing them to respond.

Reflexes are more complicated than the simple case above would suggest. For example:

- Each axon of a sensory neuron branches in the spinal cord, connecting with many interneurons. These in turn send branches to an even greater number of motor neurons.

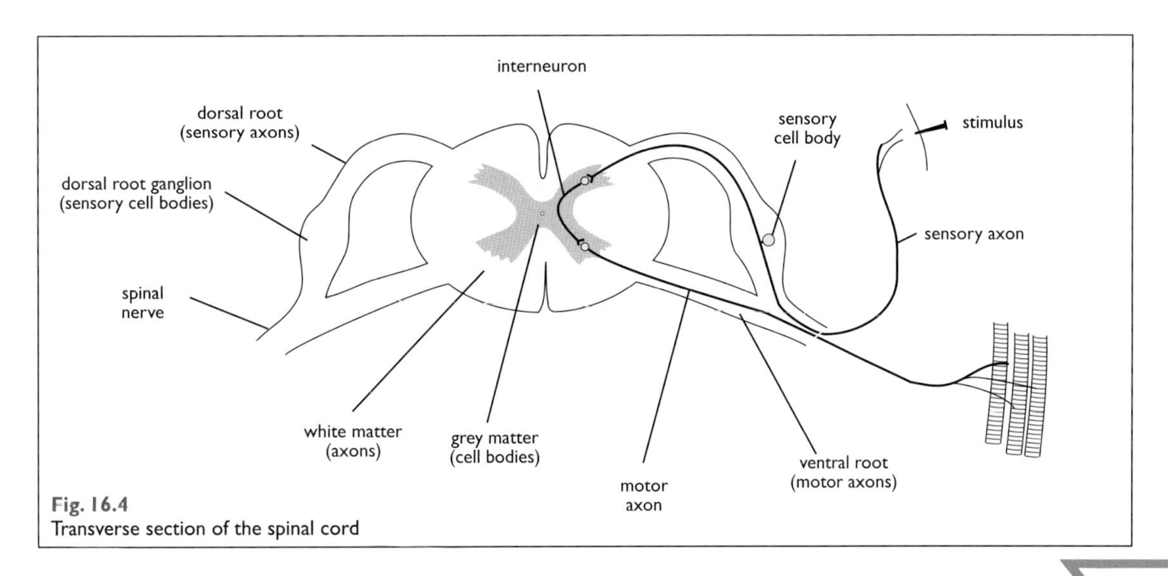

Fig. 16.4
Transverse section of the spinal cord

ISBN 9780170191340

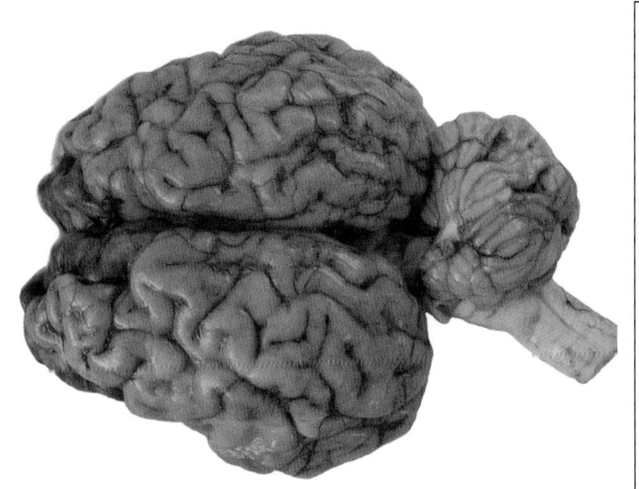

The human brain

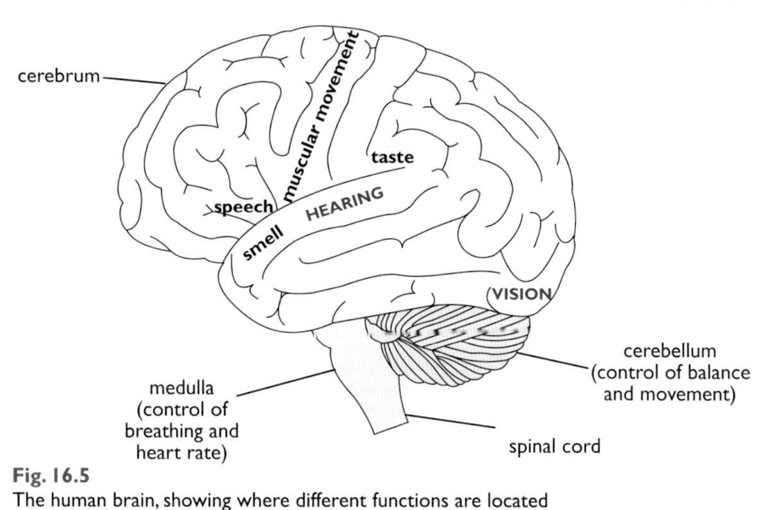

Fig. 16.5
The human brain, showing where different functions are located

- Sensory neurons also send branches to the conscious parts of the brain (this is why we are aware of the stimulus). Though none of us likes pain, it helps us to learn to avoid similar situations in the future.
- Each motor neuron receives connections from many interneurons.

The most complex part of the central nervous system is the *brain*, which is not only relatively enormous, but is divided into many different parts with specialised functions (Fig. 16.5).

One of the most important things about nerve impulses is that in a given axon they all have essentially the *same strength* and travel at the *same speed*. The only thing that varies is the *frequency* of impulses. When we experience severe pain it is because the impulses are travelling along sensory axons at high frequencies (several hundred per second). Similarly, if a muscle is contracting with maximum strength, impulses travelling along the motor nerves are at high frequency.

If impulses coming from photoreceptors are the same as impulses from chemoreceptors, how do we distinguish different stimuli? The answer is that the *sensation* we experience depends on the part of the brain that the impulses reach. To illustrate, imagine you could cut the optic nerve to the brain and the auditory nerve to the ear, and join them up the 'wrong way', so that the information from the eye went along the auditory nerve to the brain. A light stimulus would give you the sensation of hearing!

This 'thought experiment' illustrates an important point: The difference between a 'stimulus' (which is a change in the environment) and a 'sensation' (which is complex phenomenon in the brain). Table 16.1 gives some examples of receptors and their associated sensations.

Stimulus	Sensation
Light	Vision
Chemicals in mouth	Taste
Chemicals in nose	Smell
Vibration in the air	Hearing

Table 16.1 Some stimuli and their associated sensations

Notice that there is no such thing as a 'pain receptor', because pain is a sensation rather than a stimulus. The receptors that detect very strong, potentially dangerous stimuli are called *nociceptors* (from a word meaning 'hurt'), and of course the sensation is — pain!

You can demonstrate the distinction between stimulus and sensation in a simple experiment. Close your eye and apply gentle pressure to the inner side of the eyelids. You see a circle of light. The photoreceptors have been stimulated by an unusual stimulus (mechanical pressure), but the brain does not know the difference!

ISBN 9780170191340

THE EYE AND VISION

The eye is an excellent example of an *organ*: it contains a variety of tissues that cooperate to perform a function that none of the individual tissues can. The actual photoreceptors only account for a tiny proportion of the organ; the other tissues cooperate to produce an image on the photoreceptive tissue and to transmit the information to the brain.

Structure of the human eye

The eye (Fig. 16.6) is protected in a bony socket in the skull called the *orbit*. The wall of the eyeball consists of three layers; an outer *fibrous layer*, a middle *vascular layer* (containing blood vessels), and an inner, nervous layer, the *retina*.

The outer layer of the eye is tough and fibrous, held in a near-spherical shape by the slight internal pressure. The transparent front of the eye is the **cornea**, whose forward bulge enables it to act as a *convex lens*. The rest of the outer layer is the white **sclera**, a tough, fibrous layer in which are inserted three pairs of *extrinsic eye muscles* that can rotate the eyeball.

The anterior part of the sclera is covered by a thin membrane called the **conjunctiva** which also lines the inner surface of the eyelids. The conjunctiva does *not* cover the cornea — which is just as well because it contains blood vessels — not the best for seeing through! The conjunctiva is continuously being washed by tears, secreted by *tear glands*. Tears contain the enzyme **lysozyme**, which kills many bacteria.

The middle layer of the eye consists of the *choroid*, the *ciliary body* (or ciliary muscle) and the *iris*. The **choroid** is rich in blood vessels that nourish the outer layer of the retina. The ciliary body and iris contain the *intrinsic eye muscles* (muscles inside the eye), enabling it to adjust for vision in different conditions. The **ciliary body** is a ring of smooth muscle that adjusts the refracting power of the eye for viewing objects at different distances. The **iris** is a ring of muscle surrounding the **pupil**. By varying the size of the pupil, the iris regulates the intensity of light reaching the retina and also increases the depth of focus when viewing near objects.

The elastic **lens** is anchored to the ciliary body by the fibres of the **suspensory ligament**. The shape of the lens can be changed, enabling the refracting power of the eye to be adjusted for viewing objects at different distances. Like the cornea it consists of living cells, and *has no blood supply*; capillaries would obviously seriously interfere with light transmission.

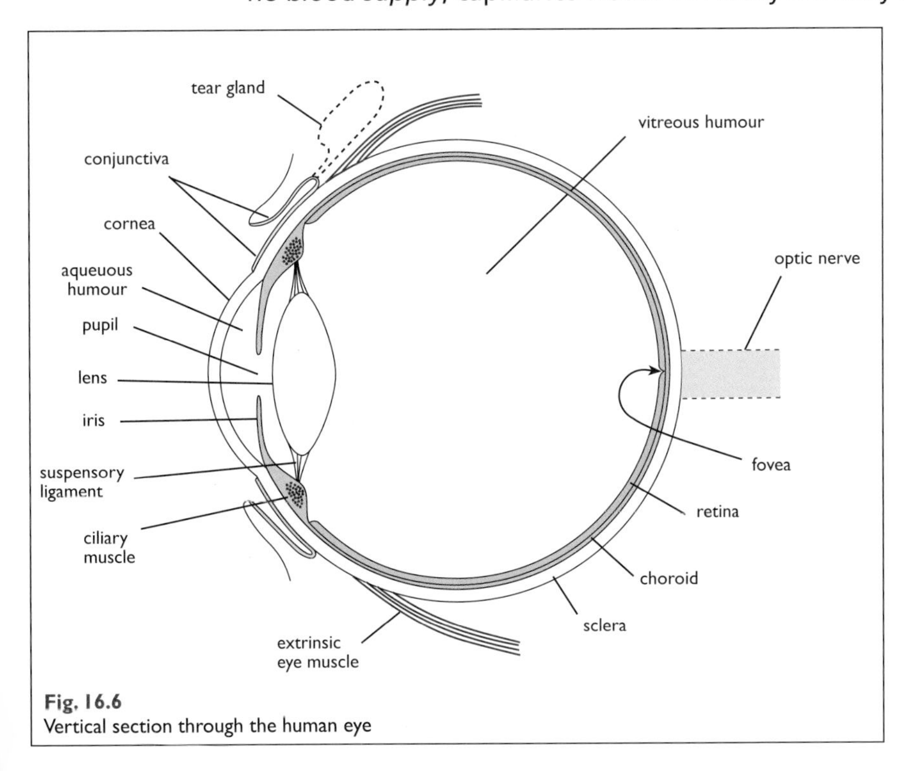

Fig. 16.6
Vertical section through the human eye

The lens and cornea are supplied with glucose and other nutrients by the slowly circulating **aqueous humour**, while the cornea obtains oxygen directly from the atmosphere. The aqueous humour is a watery liquid containing salts, glucose and amino acids. It is continuously being secreted by part of the ciliary body, and drains into a small vein in front of the iris. If drainage is impeded, the pressure in the eye rises, compressing the axons in the optic nerve and also the retinal arteries. If this is prolonged, a form of blindness called **glaucoma** may result.

The **vitreous humour** behind the lens has no nutritive function since it

ISBN 9780170191340

is jelly-like and does not circulate. Its incompressibility, together with that of the aqueous humour, helps maintain the shape of the eyeball.

How the eye forms an image

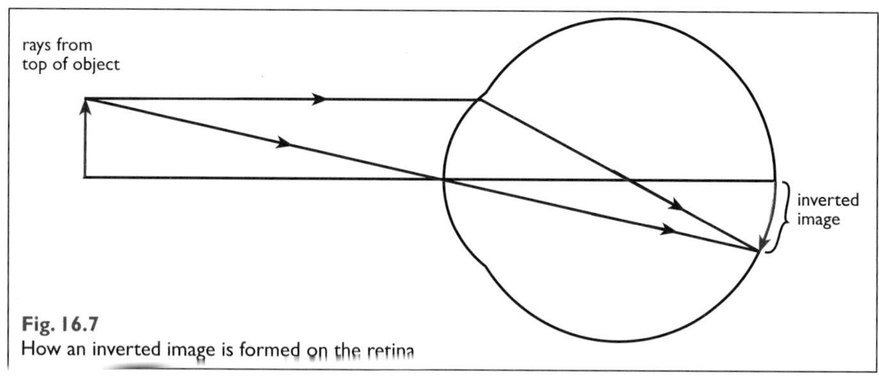

Fig. 16.7
How an inverted image is formed on the retina

The function of the cornea and lens is to form an image on the retina. The image is *inverted* (upside down and back to front), so the brain must 'turn the picture the right way up'. To form an image on the retina, the rays of light entering the eye must be brought to a *focus* on it (Fig. 16.7). This means that all the rays reflected from a given point in the visual field must be bent (refracted) so that they converge to the same point on the retina. Most (about 2/3) of the refraction takes place at the surface of the *cornea*. This is because light is crossing a boundary between two materials of very different density (air and cornea). If you open your eyes underwater, most of this refraction does not occur because water and cornea are similar in density.

Admiring a view and reading a book: Accommodation

You have probably noticed that you cannot see near and distant objects clearly at the same time. Light rays from near objects are diverging (spreading) as they enter the eye more than rays from distant objects — in fact rays from distant objects are almost parallel as they enter the eye (Fig. 16.8). To be brought to a focus, rays from near objects must therefore be bent more than rays from a distant object. To see near objects, the light-bending power of the eye must therefore be *increased*. This is called *accommodation*, and is the function of the *lens* and ciliary body.

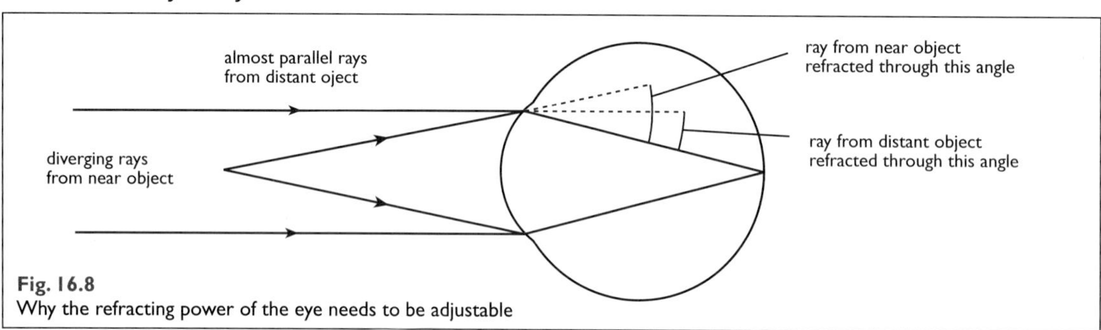

Fig. 16.8
Why the refracting power of the eye needs to be adjustable

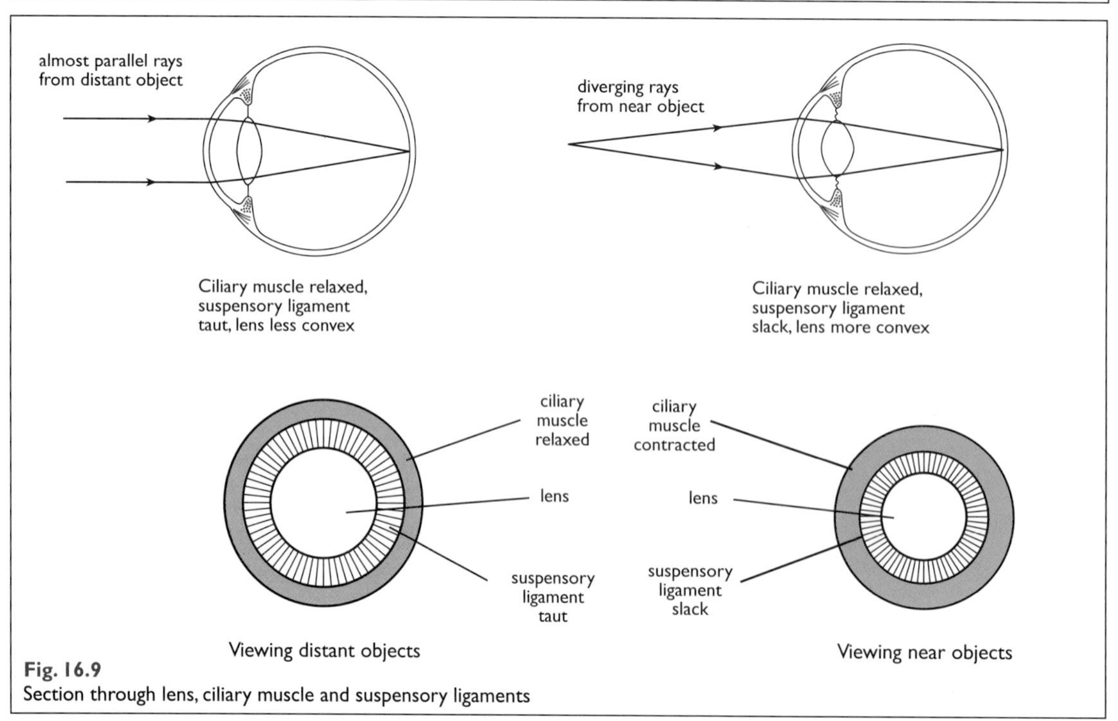

Fig. 16.9
Section through lens, ciliary muscle and suspensory ligaments

ISBN 9780170191340

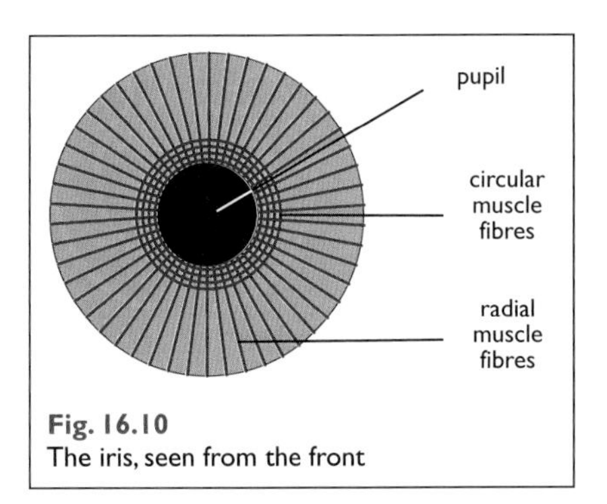

Fig. 16.10
The iris, seen from the front

The lens is elastic, and is normally pulled into a flatter shape by tension in the suspensory ligament (Fig. 16.9). When the circular muscle fibres in the ciliary body contract, its circumference decreases. This slackens the suspensory ligament and allows the lens to become more convex, so the light rays are bent more. When the ciliary muscle relaxes, the elasticity of the choroid pulls the lens back into its original shape.

Control of light entering the eye

If you cover one eye for a few seconds while looking in the mirror with the other, and then uncover your eye, you will see the pupil gets narrower. This *pupil reflex* enables the eye to adjust for different light intensities.

The iris (Fig. 16.10) contains two sets of antagonistic muscle fibres: *circular* fibres that make the pupil smaller, and *radial* fibres that make it larger. In bright light the circular fibres contract and the radial fibres relax, making the pupil smaller. In dim light the opposite happens.

The retina

The retina is the light-sensitive layer of the eye, and it develops in the embryo as an outgrowth of the brain. It contains nerve cells and nerve fibres as well as photoreceptive cells (Fig. 16.11) and blood vessels. The innermost layer of the retina (furthest from the light) consists of a layer of cells containing the pigment **melanin**, which helps prevent internal reflection (like the inside of a camera). This layer extends forward to the back of the iris, giving it its colour; brown-eyed people have more pigment than blue-eyed people.

There are two kinds of photoreceptor in the retina, named according to the shape of the light-sensitive region — approximately 120 million *rods* and 6 million *cones*.

Our dual-purpose retina

Rods are concerned with vision in dim light, but they give poor detail and are not involved in colour vision. This is why we see only shades of grey in dim light. Rods are also completely 'blind' to red light — in very dim light red objects appear black, but green and blue objects are seen as shades of grey.

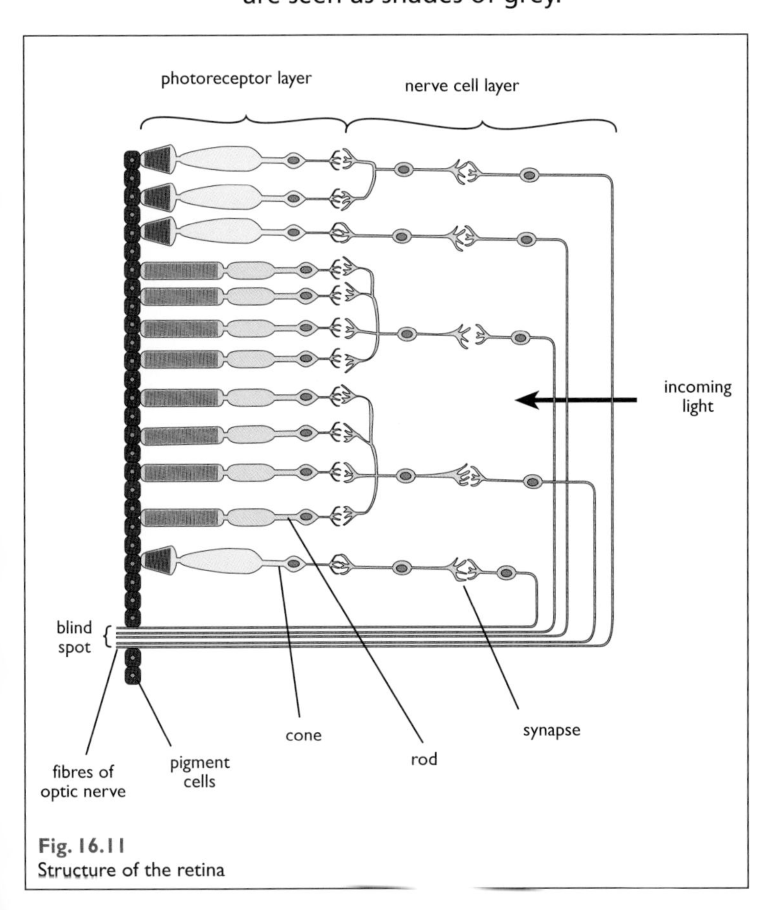

Fig. 16.11
Structure of the retina

Rods contain a pigment called **rhodopsin**, part of which is made from vitamin A. This is why deficiency of vitamin A in the diet leads to *night blindness*. This is rare in developed countries but common in areas where there is poverty.

When a rhodopsin molecule absorbs a photon of light, it becomes bleached (converted to a colourless form). The rod becomes excited and sends an electrical signal to an adjacent nerve cell. In bright light all the rhodopsin is in the bleached form, so in full daylight our rods are inactive. It takes about half an hour for all the pigment to be re-synthesised, which is why it takes that long to fully adjust to very dim light.

Cones are active in *bright light*, and enable *detailed, colour vision*. In humans, apes and monkeys there are three kinds of cone, each containing a pigment maximally sensitive to one of the three primary colours; blue, green and red. The brain perceives colour by comparing the input from the three kinds of cone.

ISBN 9780170191340

People who lack one kind of cone pigment have defective colour vision. Red-green colour blindness is due to defective pigment in the red-sensitive or green-sensitive cones. Most mammals have only two kinds of cone. Dim light is insufficient to activate cones, which is why we cannot see colour in semi-darkness.

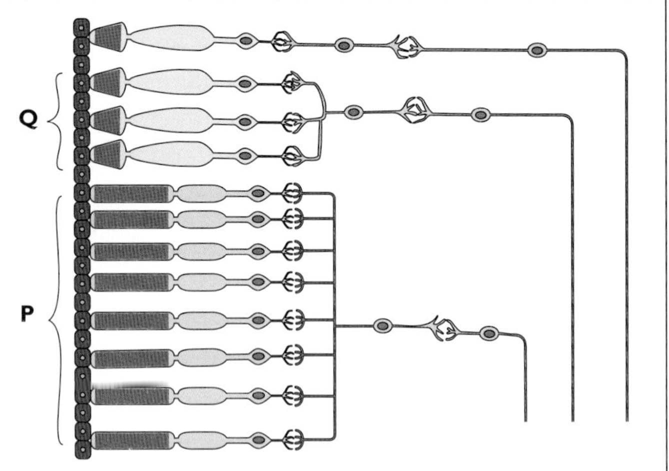

Fig. 16.12
Retinal convergence in rods (pink) and cones (blue). Between them, the rods in group P receive more light than the cones in Group Q, giving greater sensitivity in dim light. However, the brain cannot tell which of the large number of rods was stimulated, so the picture is 'coarse-grained'. Cones are organised into smaller groups, giving 'finer grain' and more detailed vision.

Retinal convergence: Why we cannot see detail in dim light

Although there are over 125 million photoreceptors in a human retina, the optic nerve contains only 1.2 million axons. This means that many photoreceptors must 'share' one optic nerve axon. This *retinal convergence* is much greater in the rods; on average, over 100 rods share a single optic nerve axon, but far fewer cones do (Fig. 16.12).

The advantage of convergence is that the input from many photoreceptors can add together. This **summation** gives greater sensitivity in dim light. The disadvantage is that it gives poorer *acuity*, or perception of detail. This is because the brain cannot distinguish between the input of different photoreceptors sharing the same 'line' to the brain. Each 'population' of photoreceptors connected to a given optic nerve axon thus acts as a single unit. The more photoreceptors in each unit the greater the sensitivity but the poorer the acuity. A 'fine-grained' retina, like a digital camera with a large number of pixels, gives greater detail.

Rods and cones are not randomly distributed over the retina. At the very back of the retina there is a small depression called the **fovea**, where there are cones only, giving detailed vision. Moreover, each cone at the fovea has its own individual 'line' to the brain. The fovea is where an image falls when we look directly at an object, so we see it in detail. Since there are no rods at the fovea we cannot see a faint star by looking directly at it. If you look a bit to one side of the star so that its image falls just outside the fovea where there are rods, you can see it. Rods become more numerous towards the periphery of the retina, and cones become fewer. At the edge of the retina there are no cones, so we see neither colour nor detail at the edge of the field of view.

The human retina is thus dual-purpose; cells in the periphery are used in twilight vision and cells in the central area, especially the fovea, are used for detailed, colour vision, such as when reading or examining an object.

The blind spot

A peculiar feature of the retina is that there is a small area with no photoreceptors at all — the **blind spot**. The reason lies in the fact that the vertebrate retina is *inverted* or 'back to front', with the photoreceptors further from the light than the nerve fibres. To leave the eye, the nerve fibres must pass through the layer of photoreceptors, thus creating the blind spot. You can demonstrate its existence by closing the left eye and looking directly at the cross in Fig. 16.13. When the page is about 20 cm from your eye the spot seems to disappear.

Vision and the environment

The eye is an amazing organ, but it can become defective due to various factors. Eye defects may have genetic or environmental causes, or a combination of the two. Some defects with genetic causes can be remedied by changes in the environment.

ISBN 9780170191340

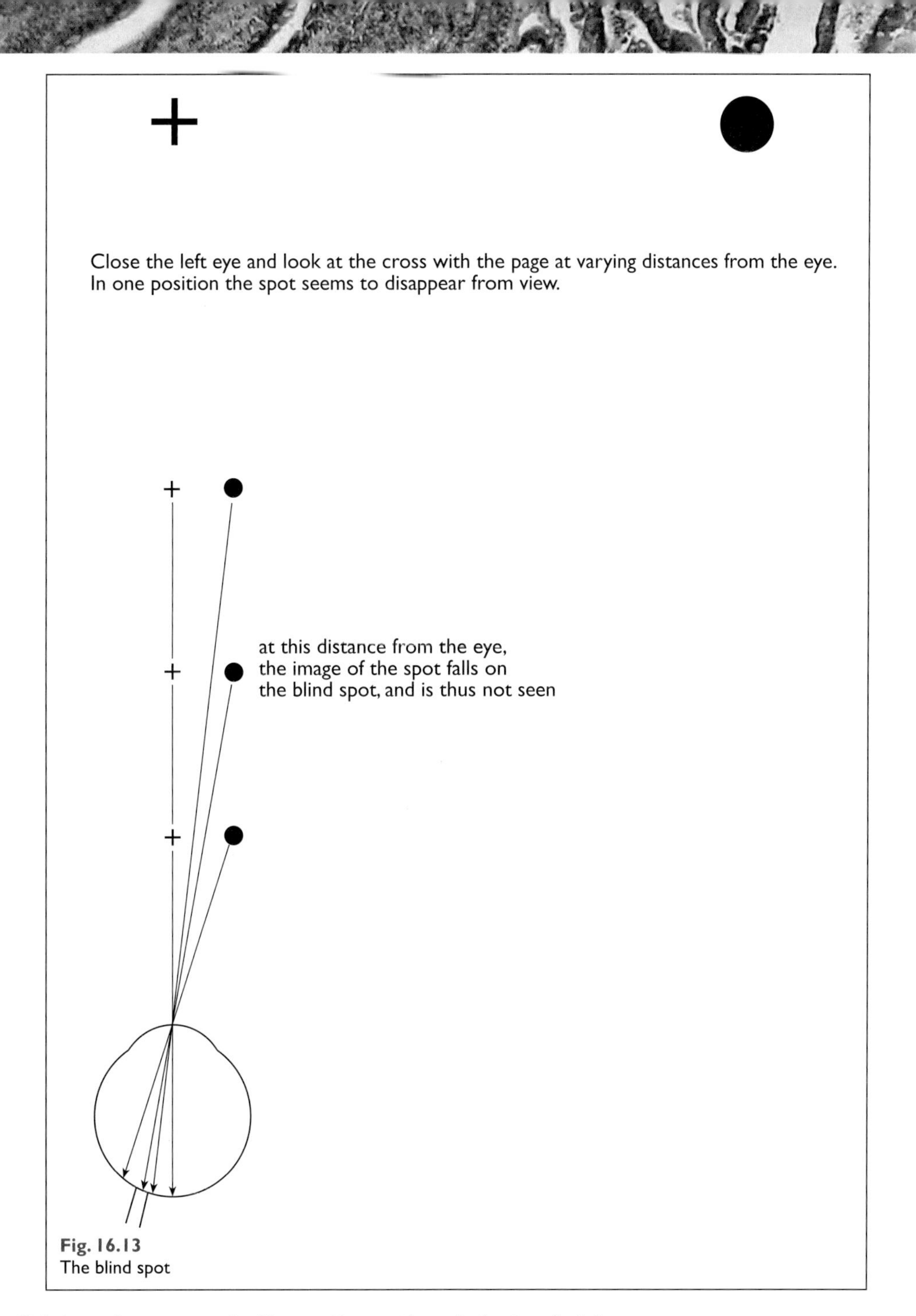

Close the left eye and look at the cross with the page at varying distances from the eye.
In one position the spot seems to disappear from view.
at this distance from the eye, the image of the spot falls on the blind spot, and is thus not seen

Fig. 16.13
The blind spot

Beneficial environmental effects: Correcting defects of vision

In a normal eye, rays from distant objects come to a focus on the retina with the ciliary muscles *relaxed*. There are various reasons why the eye may not focus the light properly. The most common are described below.

Short sight (myopia)

In short sight the image is formed in front ('short') of the retina because the focusing power of the eye is too great (Fig. 16.14). The cause is either the eyeball being too long, or the curvature of the cornea and/or lens being too great. Distant objects cannot be seen clearly unless concave glasses are worn. Another effect of short-sight is that the near point — the closest to the eye at which we can see an object clearly — is less than normal i.e. objects closer than normal can be seen clearly.

Long sight (hypermetropia)

In long sight the focusing power of the eye is insufficient (Fig. 16.15). Either the eyeball is too short, or refractive surfaces are insufficiently curved. Even to see distant objects the converging power of the eye has to be boosted by contraction of the ciliary muscles, so the eye is never at rest unless convex lenses are worn.

ISBN 9780170191340

Loss of accommodating power

As we get older our 'near point' gradually increases. This is because the lens becomes less elastic and is less able to become more convex when viewing near objects. The remedy is to use convex lenses for reading. A person who is short-sighted and has lost some accommodating power may use *bifocal* lenses; the upper part is concave for distant vision, and the lower part convex for near vision.

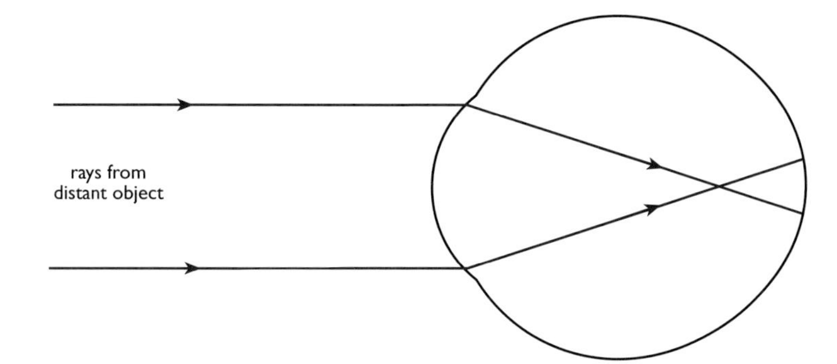

In myopia (short sight) the eyeball is too long for its refracting power. Distant objects cannot be seen clearly but near objects can be seen clearly closer to the eye than in people with normal vision

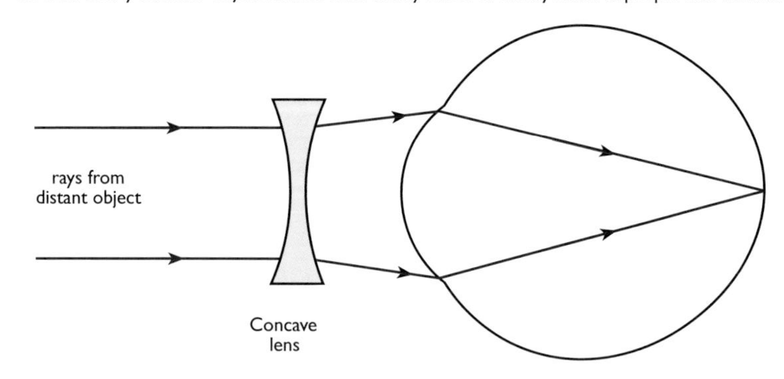

The remedy is to wear concave spectacles, which increase the divergence of the rays entering they eye

Fig. 16.14
Short sight and its correction

Astigmatism

Astigmatism is caused either by the cornea or the lens (or both) being unequally curved in different directions. Instead of being curved like the surface of a soccer ball, the surface is more like that of a rugby ball. The remedy is to use a lens that is distorted in the opposite way.

Harmful environmental effects on vision: Diabetes

The leading cause of blindness in New Zealand is **diabetes,** a condition in which there is failure of the hormone **insulin** to control the blood sugar level. Insulin is produced by tiny clusters of cells called the *Islets of Langerhans* in the pancreas. There are two main forms of the disease:

1. Type I or *insulin-dependent* diabetes, in which the insulin-secreting cells are attacked by the body's own immune system and as a result insulin production falls. The treatment for Type I diabetes is regular injections of insulin.
2. Type II or *non-insulin-dependent* diabetes, in which the pancreas produces insulin but the body loses its sensitivity to the hormone. About 80% of cases of Type II diabetes are linked to obesity, but although obesity is strongly linked to lifestyle, genetic factors are also important.

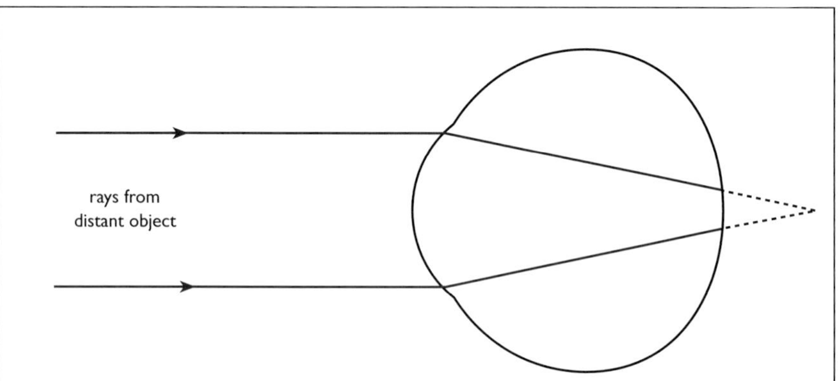

In long sight (hypermetropia) the eyeball is too short for its refracting power, so rays converge towards a point behind the retina. Distant objects can be seen clearly but only by using the ciliary muscles

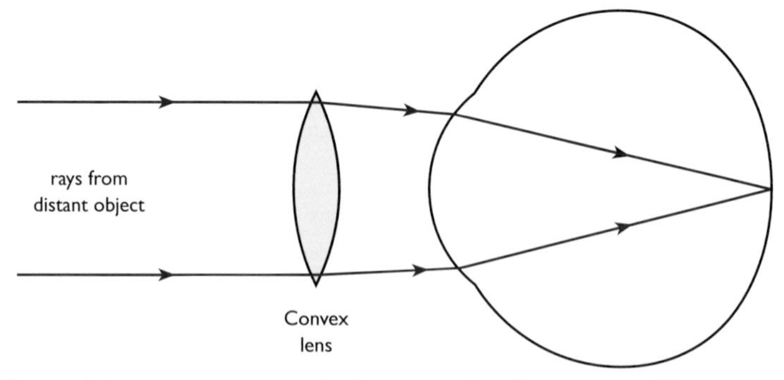

The remedy is to wear convex spectacles that increase the refracting power of the eye

Fig. 16.15
Long sight and its correction

Whatever kind of diabetes is involved, the level of glucose in the blood may rise far above the normal level of about 0.1%. One of the chief effects of this is damage to the capillaries in the retina (and also in other parts of the body). Damage usually begins long before any

ISBN 9780170191340

change in vision is noticed, so it needs to be diagnosed early. If a person has diabetes, it is important to have an eye check at least as often as every two years.

THE EAR: TWO SENSE ORGANS IN ONE

Our ears are actually two sense organs, serving quite different functions — hearing and balance. The functioning of both can be seriously upset by environmental factors; intentional (hearing) and unintentional (balance).

Hearing, and how to avoid losing it

We hear a great deal about heart disease, cancer and other 'killer' diseases, but there is one epidemic that gets little publicity — **noise-induced hearing loss (NIHL)**. It has been called *the silent epidemic* because if present trends continue, three-quarters of the population will be suffering from some degree of permanent hearing loss in decades to come.

Our sense of hearing evolved at a time when the loudest noise our ancestors experienced was a clap of thunder or mammoth rampaging through the undergrowth. Though loud, a thunderclap is very brief, but in the modern world loud, sustained noise is a normal feature of many people's lives.

Our ears are not adapted to cope with noise on this scale. This been clearly demonstrated in survey after survey. A study carried out by the National Foundation for the Deaf published in September 2009 found that 70% of people under 30 have signs of NIHL as a result of listening to loud music (*NZ Herald* 18 Jan 2010). Similar results have been obtained in other Western countries. Unless things change drastically, the majority of the world's population will be hard of hearing in the not too distant future. This does not just mean loss of enjoyment; it can mean loss of ability to do one's job and thus loss of livelihood.

Sound: What is it and how is it measured?

Sound is vibration in the air (or water, if you are a fish). Figure 16.16 shows how a vibrating object (such as a tuning fork) affects the air round it. As it moves in one direction it compresses the air on one side, and when it moves in the opposite direction, it lowers the air pressure on that side. These waves of high and low pressure (compression and rarefaction) travel through the air at a speed of about 340 m s^{-1} at sea level, or just under three seconds to travel a kilometre.

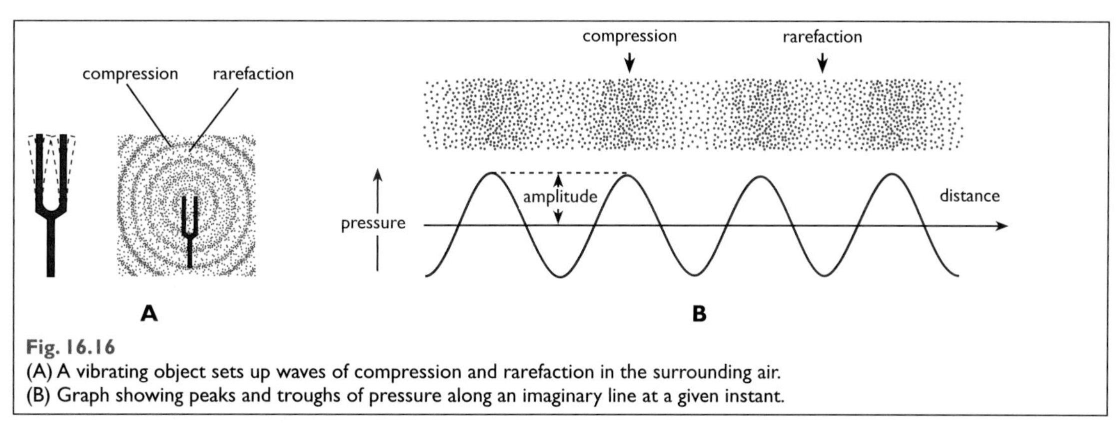

Fig. 16.16
(A) A vibrating object sets up waves of compression and rarefaction in the surrounding air.
(B) Graph showing peaks and troughs of pressure along an imaginary line at a given instant.

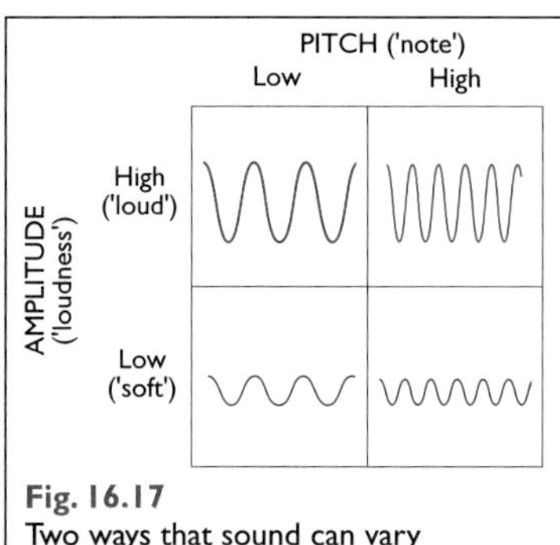

Fig. 16.17
Two ways that sound can vary

As far as hearing is concerned, sound can vary in two important ways (Fig. 16.17):

- Its *frequency* or *pitch*. High frequency sounds we perceive as high notes.
- Its *amplitude* or *loudness*.

Frequency is measured in cycles per second or *Hertz* (*Hz*). Young people can hear sounds from 20 Hz to 20,000 Hz. This declines naturally with age, and elderly people can hear between 50 and 8000 Hz. Some mammals can hear sound of much higher or much lower frequency; dogs can hear sound up to 45,000 Hz and whales can hear frequencies higher than 100,000 Hz.

ISBN 9780170191340

Loudness is measured in **decibels (dB)**. This is a logarithmic scale; a 10 dB increase represents a doubling of what we experience as loudness. What many people do not realise is that a *doubling of loudness represents a 10-fold increase in energy intensity*. Hence an increase from 60 dB to 70 dB doubles the loudness but increases the energy intensity 10 times: 100 dB seems 16 times louder than 60 dB (a 40 dB increase is four doublings of loudness), but is actually 10,000 times the energy intensity.

The necessity for this scale is due to the fact that the human ear can operate over an enormous range of intensities; sound that is so loud as to cause pain has an energy level 10^{12} (a trillion) times greater than a whisper. The audience in a rock concert experiences sound with 100 million times more energy than a whispered voice. Some examples of noise levels are shown in the table below.

Type of noise	Loudness (dB)
Threshold of pain	134
World record for musical noise	132.5
Hearing damage during short-term effect	approx. 120
Jet engine, 100 m distant	110–140
Chain saw at 1 m	110
Pneumatic drill, 1 m distant / disco	approx. 100
Hearing damage from long-term exposure	approx. 85
Traffic noise on major road, 10 m distant	80–90
Telephone dialing tone	80
TV set – typical home level, 1 m distant	approx. 60
Normal talking, 1 m distant	40–60
Whispered voice	30
Calm human breathing	10
Lower hearing limit at 2 kHz	0

Table 16.2 The decibel (dB) scale

NIHL has become epidemic since the advent of iPods and MP3 players, which enable people to have the sound level very loud without other people hearing.

Intensity is not the only important factor; the other is *duration* of exposure. For example, somebody passing a pneumatic drill or jackhammer experiences about 100 dB, but the exposure is brief and no damage results. On the other hand, exposure to 85 dB for more than eight hours a day can result in permanent hearing loss. Car stereos can emit sound well above 85 dB. A useful guide is: *For every 5 decibel increase, reduce the duration of exposure by half.*

If 85 dB is reckoned to be a safe level for prolonged listening to music, what about 88 dB? An increase of 3 dB does not seem much, but it is actually *twice* the energy intensity as 85 dB. The audience at a loud rock concert may be subjected to sound intensities of 115 dB. This is 30 dB or 1000 times the long-term safe level of 85 dB. On 8 June 2007, a noise level of 132.5 dB was achieved by a punk band called The Gallows, performing in a Birmingham studio. This is over 10,000 times the safe level!

NIHL most affects higher frequencies, especially at 4000 Hz. To a person suffering from NIHL, speech seems muffled because he or she cannot hear consonants such as 's' 't' 'f' 'h' and 'sh'. Sufferers who cannot communicate with others tend to become withdrawn and may have difficulty holding down their jobs. The situation is particularly serious because the condition is permanent, and commits the sufferer to life-long use of a hearing aid.

ISBN 9780170191340

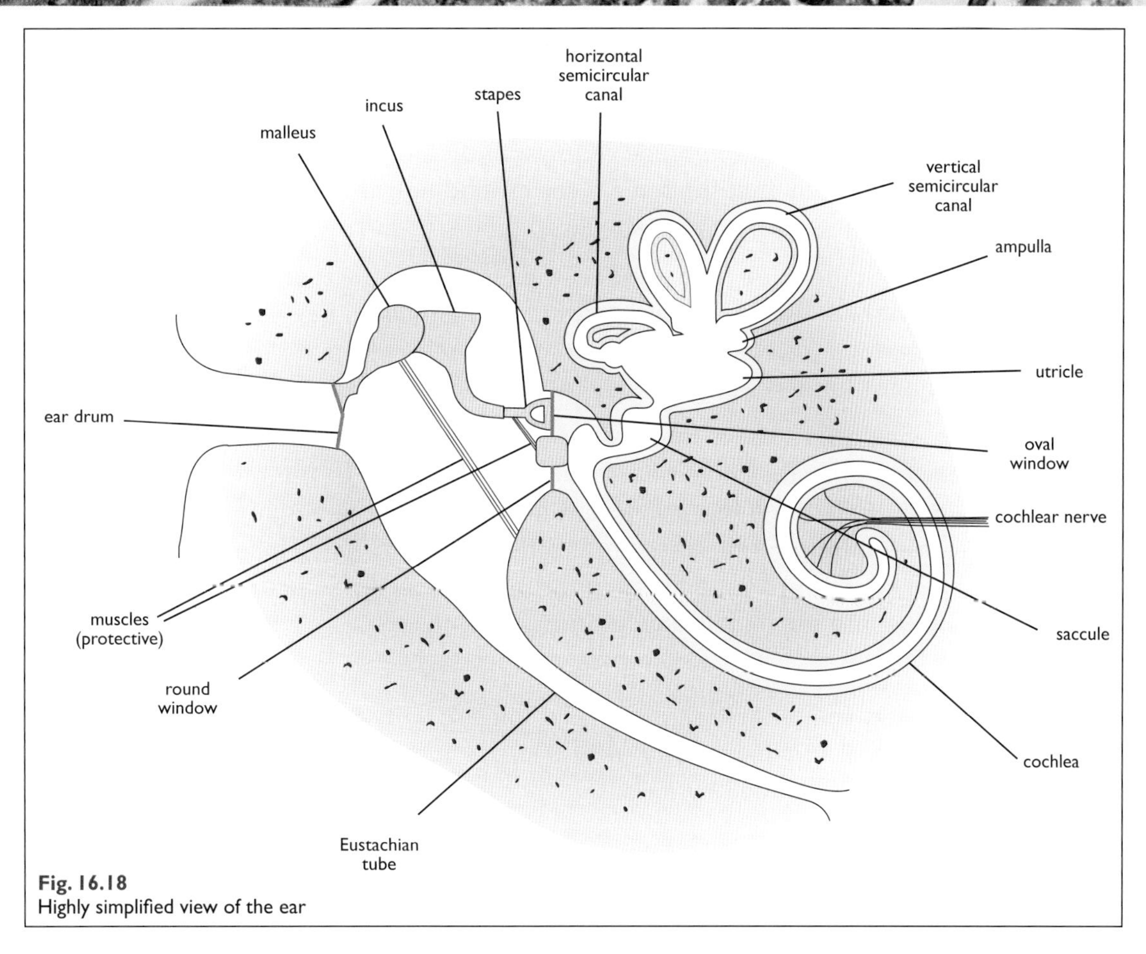

Fig. 16.18
Highly simplified view of the ear

Structure of the hearing apparatus

Why is our hearing so easily damaged by prolonged exposure to loud noise? To understand how, we need to learn a little about how the ear works. Our hearing apparatus consists of three parts: The outer, middle, and inner ears (Fig. 16.18).

- The **external ear** consists of the ear flap and the passage leading down to the **ear drum** or *tympanic membrane*, which separates the outer ear from the middle ear. The ear drum is a sheet of skin which vibrates when sound strikes it.
- The **middle ear** is an air-filled cavity in the skull. It is crossed by three tiny bones ('auditory ossicles'), named according to their shapes:
  1. The *malleus* ('hammer').
  2. The *incus* ('anvil').
  3. The *stapes* ('stirrup'). This, the smallest bone in the body, sits on the **oval window**, a membrane separating the middle ear from the fluid in the inner ear.

  The middle ear is also separated from the inner ear by the **round window**.
- The **inner ear** is a very complex system of fluid-filled cavities, completely embedded in bone. It contains receptors that detect sound vibrations and also receptors concerned with balance.

The middle ear

You have probably noticed that sound is reflected when it hits a solid material such as a cliff face. This is an echo, and happens whenever sound meets a boundary between two materials of different density. It is the reason why sound travels so much better over water; less sound is lost because it bounces off the surface rather than entering the water. The function of the middle ear is to prevent this reflection. Without the middle ear, 99% of the sound energy would bounce back into the air rather than entering the liquid in which the receptor cells are located.

The middle ear works by increasing the *pressure* (force per unit area) of the vibrations. The area of the ear drum is much greater than the area of the oval window, so the force

 ISBN 9780170191340

of the vibrations in concentrated on a much smaller area. This enables the sound to be transmitted from low density air to high density fluid. Note that the ear bones do *not* (as is commonly stated) increase the *amplitude* of the sound; due to a lever effect the amplitude is actually decreased slightly (but the force is correspondingly increased).

Two muscles in the middle ear help (to a very small extent) to protect the inner ear from excessive noise. When there is a loud noise a little muscle contracts and pulls on the stapes, helping to dampen the sound. A muscle attached to the malleus has a similar action. These two muscles thus act in a similar way to the circular muscles of the iris in the eye. They give no protection against sudden bangs, as they cannot contract in time.

The middle ear is connected to the pharynx ('throat') by the **Eustachian tube**. This allows air to pass from the pharynx (the airway behind the tongue) to the middle ear, keeping the pressure equal. Every time you swallow, a muscle at the opening of the tube relaxes, allowing air to pass through if the pressure is different. If you catch a heavy cold the tube can become blocked with mucus and the pressure may become unequal. As a result the eardrum may bulge into the middle ear, reducing hearing.

The inner ear

The inner ear is a very complicated series of fluid-filled spaces containing receptor **hair cells**. It actually consists of an inner cavity with *membraneous* walls, surrounded by an outer cavity with *bony* walls.

The upper part of the inner ear consists of three *semicircular canals*, which are concerned with balance. The lower part consists of the spirally coiled **cochlea**, which is concerned with hearing.

The cochlea

The cochlea (so-called because it resembles a snail shell) consists of three fluid-filled spiral tubes shown at increasing magnifications in Figs. 16.19, 16.20 and 16.21.

The floor of the middle tube is the **basilar membrane** and contains the **Organ of Corti**. This consists of rows of hair cells embedded in a spiral **tectorial membrane**. The hair cells are connected to the brain via the **cochlear nerve**.

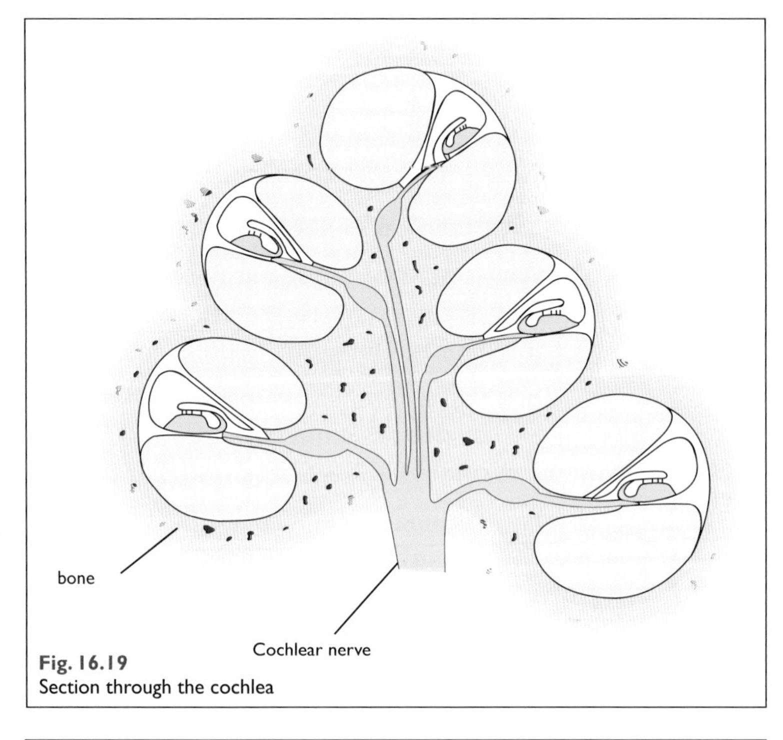

Fig. 16.19
Section through the cochlea

The upper tube communicates with the middle ear by the oval window, and the lower tube communicates with the middle ear by the **round window**. The floor and roof of the middle tube consist of flexible membranes, so when the stapes vibrates, the vibrations are transmitted via the upper tube of the cochlea to the organ of Corti and then to the round window. The round window is necessary because fluid is incompressible; when the oval window bulges inwards the round window bulges outwards, and vice-versa (Fig. 16.22).

Sound reaching the stapes is transmitted to the organ of Corti, causing the floor of the middle tube to vibrate up and down. In doing so the hair cells move relative to the tectorial

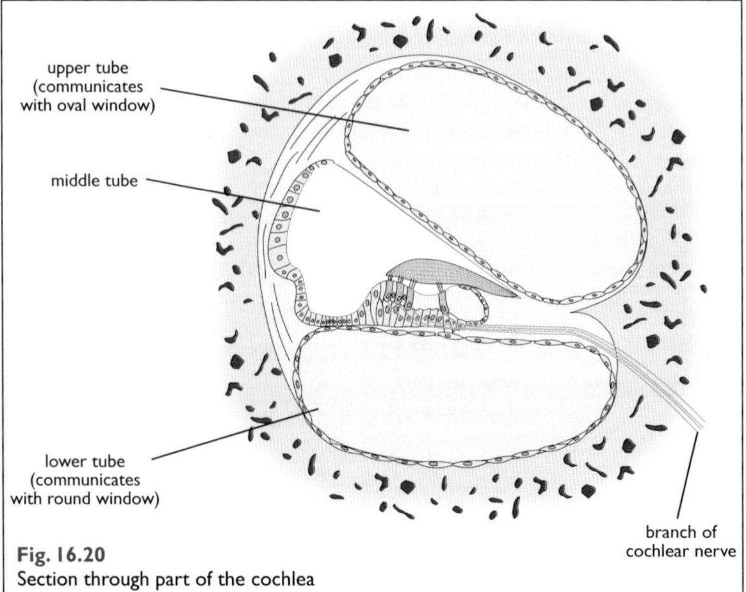

Fig. 16.20
Section through part of the cochlea

ISBN 9780170191340

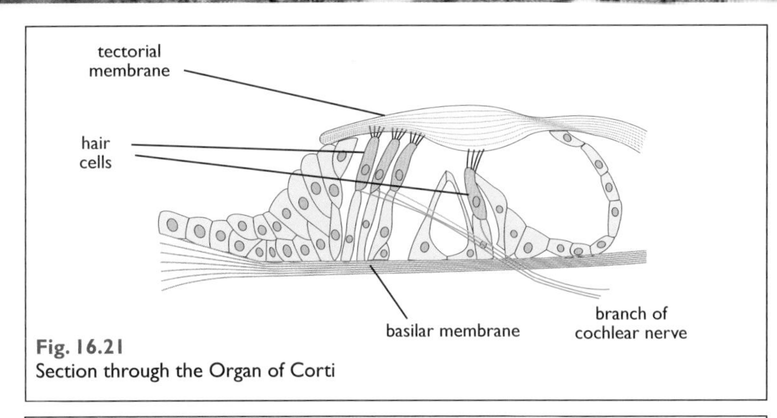

Fig. 16.21
Section through the Organ of Corti

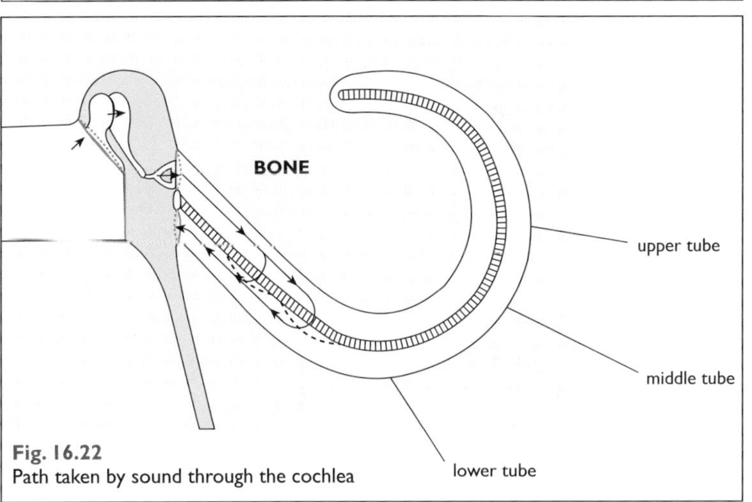

Fig. 16.22
Path taken by sound through the cochlea

membrane, stimulating the hair cells to send impulses to the brain.

The brain is able to appreciate three important things about a sound:

1. *Amplitude.* The louder the sound, the higher the frequency of impulses reaching the brain.
2. *Pitch.* Different parts of the basilar membrane tend to vibrate in response to different pitches. The part nearest the base is most sensitive to high notes, and the part nearest the tip is most sensitive to low notes. Thus the brain can detect pitch from the part of the basilar membrane that is most strongly stimulated.
3. *Direction.* Unless a sound originates directly behind, in front or above, it reaches one ear before the other. Also, the sound is fractionally louder at one ear than the other. By comparing the input from the two ears, the brain can detect its direction.

The hair cells are almost unbelievably sensitive. The faintest sound that can just be heard causes the ear drum and basilar membrane to move through a distance considerably less than the diameter of a hydrogen atom! Hardly surprising, then, that sustained exposure to noise above 85 dB is likely to lead to permanent hearing loss.

Like the rods and cones in the retina, the hair cells cannot repair themselves if damaged. The 15,000 or so hair cells you are born with are all you will ever have, *so look after them*!

Balance and motion sickness

Drivers seldom suffer from motion sickness, but passengers frequently do. The advice usually given is to avoid keeping your eyes on objects in the car but look out of the window and fix your eyes on the scenery. Similarly, you are less likely to get seasick if you fix your eyes on the horizon.

Motion sickness is connected with the experience (known to all of us) or making yourself dizzy by turning round and round, and then when you come to a stop — as well as *feeling* that you are still turning — the surroundings *look* as though they are turning around you. The explanation for these observations depends on the fact that the sense of balance is closely linked to the control of our eye movements.

Our sense of balance actually involves two senses:

1. *Orientation* of the head, detected by sense organs just below the semicircular canals. These sense organs tell us about the *position* of the head in relation to gravity. This is how, for example, we know which way is 'up' when we float, eyes closed, in a swimming pool.

ISBN 9780170191340

2. *Rotation* of the head, detected by the **semicircular canals**. These play a vital part in coordinating movements of the eyes and head, and it is these that are involved in motion sickness.

The semicircular canals and detecting rotation of the head

In each inner ear there are three semicircular canals, two in the vertical plane and one in the horizontal plane. They are thus oriented at right angles to each other (Fig. 16.23).

The two ends of each canal open into a larger chamber, the **utricle**. At one end is a swelling called the **ampulla**, containing a sense organ called a **crista** (Fig. 16.24). Each crista consists of a cluster of hair cells with sensory hairs embedded in a pad of jelly, the **cupola**. The cupola acts as a kind of 'trap door', preventing free flow of liquid through the ampulla.

How do the semicircular canals work? Suppose we consider the two horizontal canals (one in each ear). If you start to turn your head sideways, the liquid in the horizontal canals tends to stay still (due to its inertia). This pushes the cupola to one side, stimulating the hair cells and 'informing' the brain of the head's rotation (Fig. 16.25).

Because the canals are arranged in the three planes of space, *any* rotation of the head must stimulate at least one of the three canals.

Having informed the brain that the head is turning, the brain sends impulses to the eye muscles, causing the eyes to rotate in exactly the opposite direction to the movement of the head. By rotating in their sockets, the eyes are actually staying still relative to the outside world. This enables us to keep looking at an object when our head is turning.

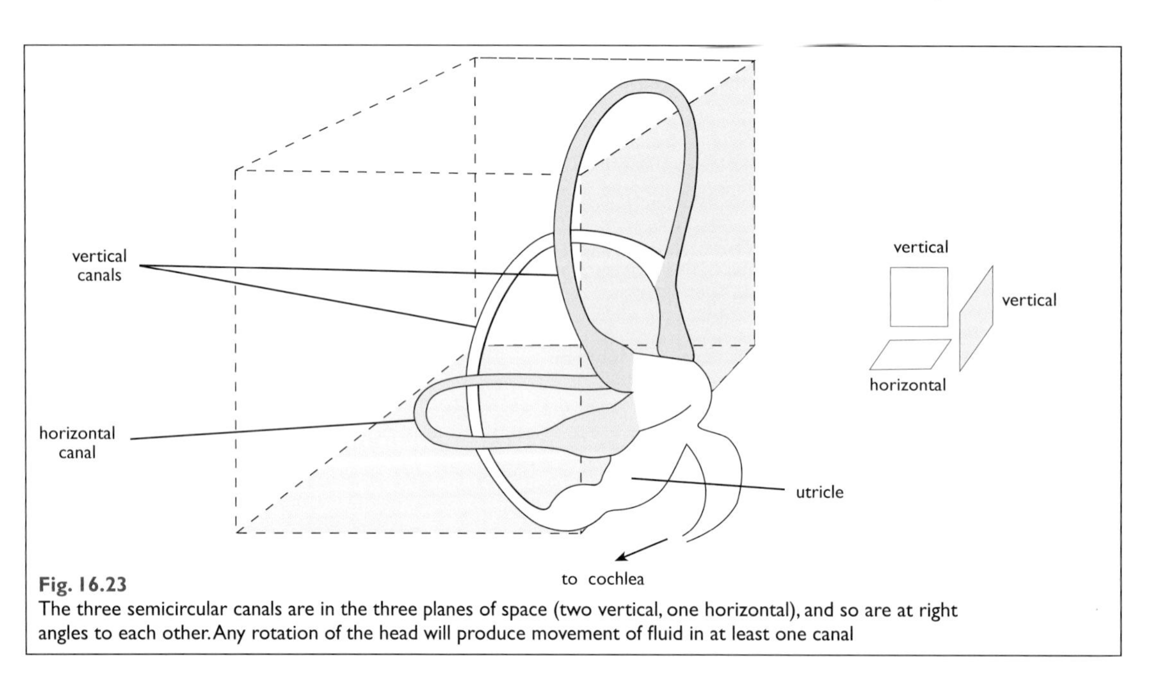

Fig. 16.23
The three semicircular canals are in the three planes of space (two vertical, one horizontal), and so are at right angles to each other. Any rotation of the head will produce movement of fluid in at least one canal

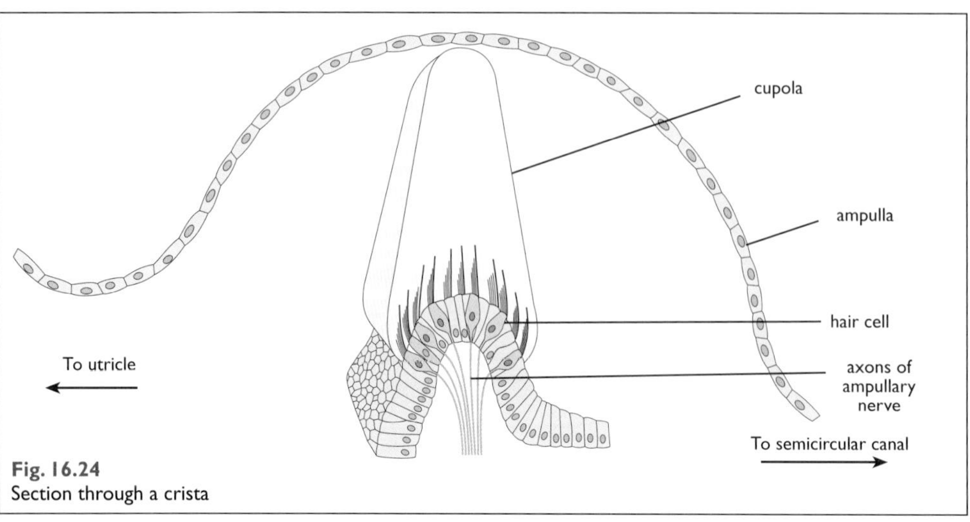

Fig. 16.24
Section through a crista

ISBN 9780170191340

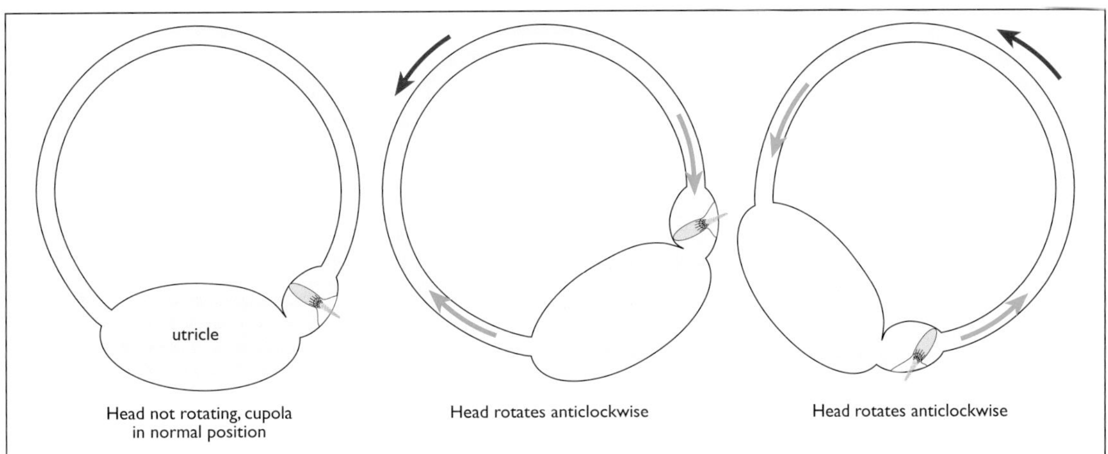

Fig. 16.25
How the brain detects rotation of the head, showing just one semicircular canal. When the head rotates (red arrow), liquid in semicircular canal tends to stay still and thus moves relative to the canal (blue), pushing the cupola and stimulating the hair cells

If you have spun in circles and made yourself dizzy, the brain 'thinks' that you are turning your head and sends impulses to the eye muscles, causing them to move in the opposite direction. In this way the eyes keep moving from side to side, so you seem to see the world going round.

How to avoid motion sickness

When you are in a boat in choppy seas, your semicircular canals are telling your brain that you are moving up and down. If you look at the horizon, your brain sends impulses to the eye muscles, enabling your eyes to make exactly the right movements to keep a visual 'fix' on the horizon. Your eyes and semicircular canals are 'in agreement'.

But suppose you are looking at the inside of the boat. Instead of moving in synchrony with the movement of the boat, you are forcing your eyes to be stationary relative to your head. The brain receives conflicting information; the semicircular canals are telling you that you are moving, but your eyes are telling you that your head is stationary; such mixed messages cause motion sickness. The way to avoid it is to look at an object in the outside world, so the eyes have to move in synchrony with the movements of the head.

ISBN 9780170191340

Summary of key points in this chapter

- Animals detect stimuli by *receptors*, and respond to them by *effectors*.
- Information is transmitted at high speed around the body by specialised cells called *neurons*.
- Neurons transmit information via nerve fibres or *axons*, in the form of electrical *impulses*.
- In a given axon, impulses are individually identical, varying only in *frequency*.
- Information from receptors is processed by the *central nervous system* before being transmitted to effectors.
- The eye consists of the following parts:
  1. The *retina*, which contains the light-sensitive cells or *photoreceptors*.
  2. The *cornea* and *lens*, which bring the light to a focus on the retina.
  3. The ciliary body, which, together with the lens, varies the refractive power of the eye for viewing objects at different distances.
  4. The *iris*, which adjusts the intensity of light falling on the retina.
  5. The *sclera*, the outer, protective layer which anchors the extrinsic eye muscles that rotate the eye.
  6. The *choroid*, which is the middle layer and contains blood vessels that help nourish the outer layers of the retina.
  7. The *aqueous humour*, which circulates slowly, bringing nutrients to the lens and cornea.
  8. The *vitreous humour* which, together with the slight pressure inside the eye, helps keep it in the correct shape.
- The retina has two kinds of photoreceptor that operate under different conditions: Cones work in bright light and give high definition and colour vision, while rods work in dim light but give poor detail and no colour.
- Many rods share one optic nerve axon to the brain, giving high sensitivity in dim light but poor detail. Fewer cones share one optic nerve fibre, giving more detailed vision.
- At the *fovea* there are cones only, and each has its own optic nerve axon to the brain, giving very high definition.
- The retina of humans and other vertebrates is *inverted*, with the photoreceptors furthest from the light. This results in a *blind spot*.
- A major cause of blindness is *diabetes*, one type of which is closely linked to obesity.
- Noise-induced hearing loss (NIHL) is developing into a major epidemic, affecting 70% of young adults. The condition is permanent and can lead to difficulties in maintaining employment.
- Sound intensity is measured on a logarithmic scale because of the huge range of intensity to which human ears can operate.
- The unit of sound intensity is the decibel (dB). An increase of 3 dB represents a doubling of intensity.
- The ear consists of three parts; the external, middle, and internal ear.
- The hearing organ is the *cochlea*, and contains the sound-receptor *hair cells*.
- These hair cells are extremely sensitive to prolonged sound; 85 dB (1/1000 the intensity of a loud rock concert) is considered to be the safe limit for prolonged exposure.
- There are two quite different kinds of sensitivity involved in balance; a sense of *position*, and a sense of *rotation*.
- Rotation of the head is detected by hair cells within the *semicircular canals*.
- Information from the semicircular canals makes it possible for the eyes to maintain a 'fix' on an object while the head is moving.
- When the brain receives conflicting information from the eyes and semicircular canals, motion sickness may result.

ISBN 9780170191340

Test your basics

Copy and complete the following sentences.

a) Environmental changes to which an organism can respond are called ___*___, and are detected by structures called ___*___. Structures that respond to change are called ___*___.

b) Information is transmitted from receptors to effectors in the form of electrical signals called ___*___, by nerve cells or ___*___. These receive information via fine threads or ___*___, and transmit signals via a long cable-like fibres called ___*___.

c) In a given axon, individual impulses are the same, the strength of the signal is indicated by their ___*___.

d) The simplest responses to stimuli are automatic (i.e. do not involve conscious thought) and are called ___*___. Most do not have to be learned, but ___*___ ___*___ do in the first instance, but become automatic.

e) The ___*___ ___*___ ___*___ (CNS) of a mammal consists of the brain and spinal cord. In cross section, the spinal cord has a central 'H'-shaped ___*___ matter, consisting of nerve cell bodies, and an outer ___*___ matter, consisting of nerve fibres. Neurons bringing information into the CNS are called ___*___ neurons, and neurons carrying information away from the CNS are called ___*___ neurons. Connecting these two kinds of neuron are ___*___.

f) The light-sensitive layer in the eye is the ___*___. Beside neurons, the ___*___ contains photoreceptors of two kinds: ___*___, which are sensitive to dim light, and ___*___, which function in bright light and are involved in colour vision. The centre of the retina is used for detailed vision and is called the ___*___. Just to one side of this there are no photoreceptors; this is the ___*___ spot.

g) Light is brought to a focus on the retina by the ___*___, which has fixed shape, and by the ___*___, which can be adjusted for viewing objects at different distances. The shape of the lens is changed by the action of the ___*___ muscle. The intensity of light falling on the retina can be controlled by the action of two sets of muscle fibres in the ___*___.

h) The ear has two functions; hearing and ___*___. Sound reaching the ear causes the ear drum to vibrate, and these vibrations are transmitted to the inner ear via a chain of three tiny bones in an air-filled cavity, the middle ear. The receptors that detect sound are in the part of the inner ear called the ___*___, which is coiled like a snail shell. It consists of three tubes, the middle one contains the actual hearing organ, the Organ of ___*___. This consists of rows of cells with sensory hairs embedded in a ribbon of jelly.

i) The upper of the three tubes connects with the middle ear via the ___*___ window, and the lower tube connects with the middle ear via the ___*___ window. Vibrations are transmitted from the middle ear to the middle tube of the ___*___ by the ___*___ window, and are transmitted back to the middle ear via the ___*___ window. Sound causes the membrane containing the hair cells to vibrate, stimulating the receptors to send impulses along nerve fibres to the brain.

j) The base of the cochlea is most sensitive to high notes and the tip to low notes. This enables the ___*___ of the sound to be distinguished. The direction of the sound is detected by the ___*___ comparing the input from the two ears.

k) The inner ear is also concerned with detecting the rotation of the ___*___, which is the function of the ___*___ canals. Like the cochlea, these are filled with fluid. In each ear there are ___*___ canals, two in the ___*___ plane and one in the ___*___ plane. At one end of each canal is a swelling called an ______. This contains a cluster of cells with long sensory hairs embedded in a pad of ___*___. When the head ___*___, the fluid in the canals tends to stay still and thus ___*___ relative to the canal, pushing the pad of jelly to one side and stimulating the hair cells.

ISBN 9780170191340

17 A new life: Reproduction

Everybody begins life as a fertilised egg or **zygote**, formed by joining together of two cells (Fig. 17.1):

1. An **ovum** or egg, which is one of the largest human cells.
2. A **sperm**, the smallest human cell.

In animals, the gametes are produced in organs called **gonads**.

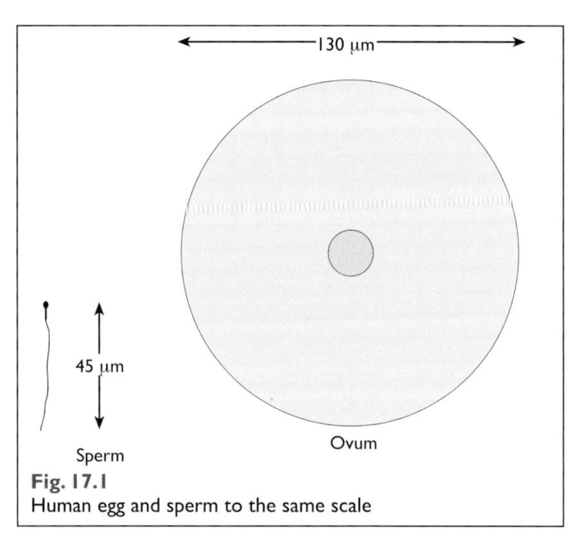

Fig. 17.1
Human egg and sperm to the same scale

THE MALE REPRODUCTIVE SYSTEM

This is shown in side view (Fig. 17.2) and front view (Fig. 17.3). The male gonads or **testes** are about 5 cm long and are situated in a bag-like extension of the body wall called the **scrotum**.

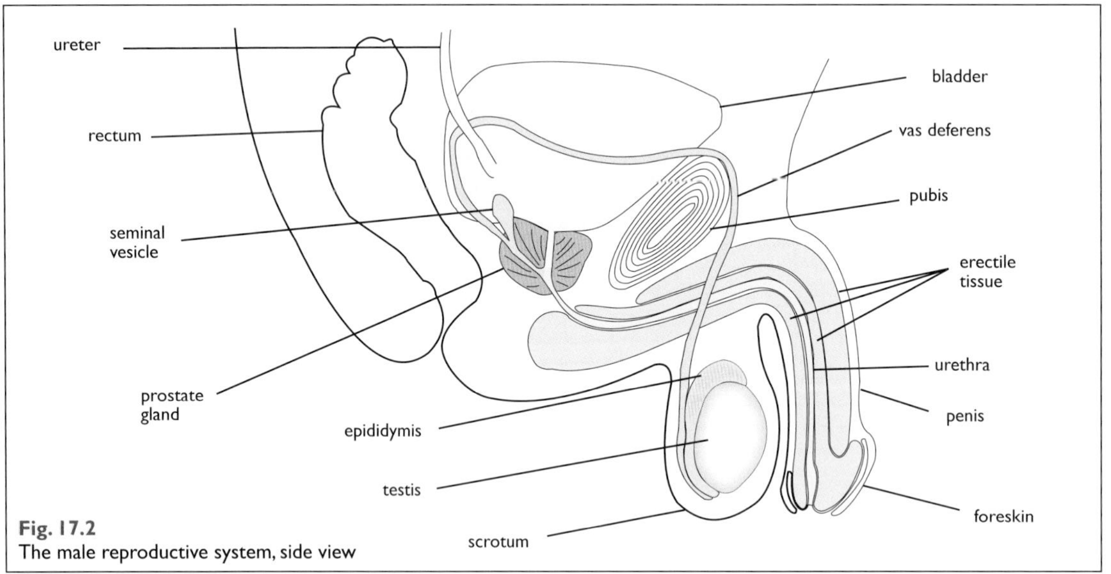

Fig. 17.2
The male reproductive system, side view

They have two functions:

1. Production of male gametes (sperm).
2. Production of male sex hormone **testosterone**.

Each testis contains hundreds of **seminiferous tubules**, the walls of which produce sperm (Fig. 17.4). Between the seminiferous tubules are cells that produce testosterone. The testes are therefore **endocrine** (hormone-producing) glands as well as gamete-producing organs.

After they are produced, sperm have to undergo a period of maturation before they can fertilise an egg. This occurs in a long coiled tube called the **epididymis**, and takes between 10 and 14 days.

During embryonic development the testes are in the abdomen just like the ovaries, but shortly before birth they descend into the

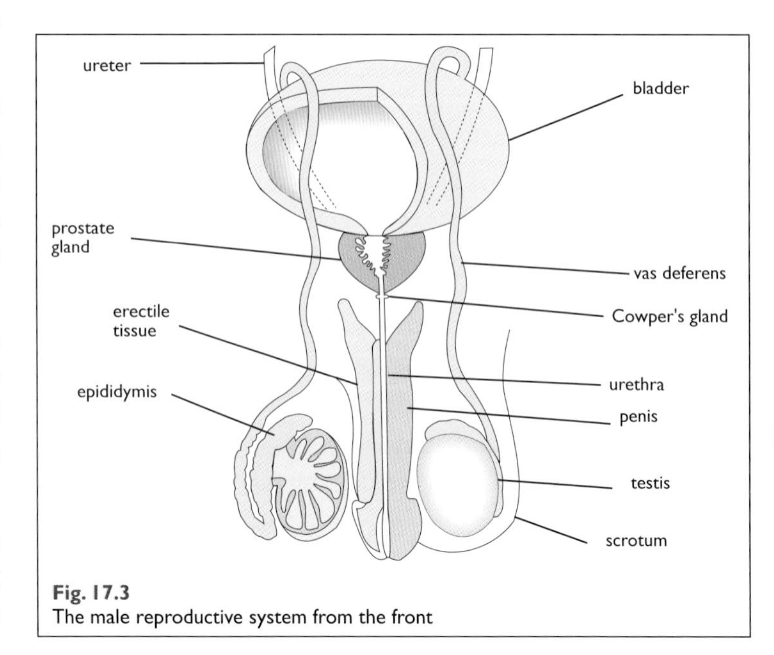

Fig. 17.3
The male reproductive system from the front

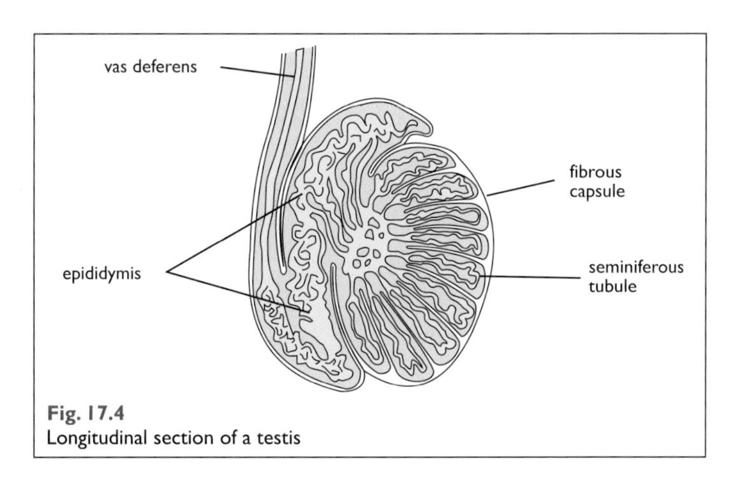

Fig. 17.4
Longitudinal section of a testis

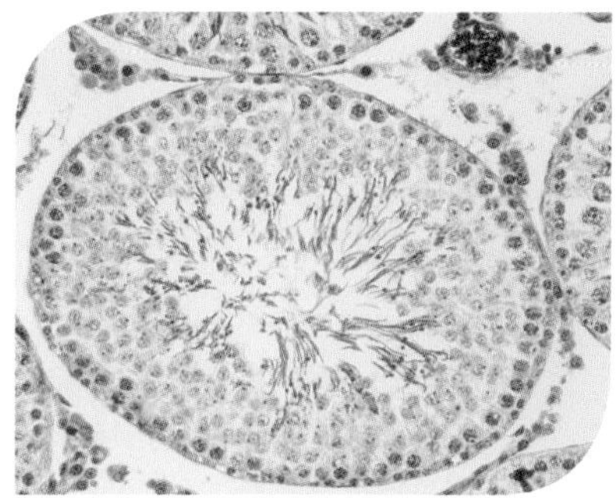
Section through a seminiferous tubule of testis of rat

scrotum. This keeps the testes at about 35°C — 2°C below body temperature — which is necessary for sperm production. If the testicular temperature falls below the optimum, the smooth muscle in the scrotum contracts, pushing up the testes closer to the body where it is warmer.

Sperm

A sperm has three parts (Fig. 17.5):

1. A head, containing the *nucleus* and tipped by an **acrosome**, a sac containing enzymes the sperm uses to penetrate the egg.
2. A *middle piece*, containing mitochondria in which ATP for movement is produced.
3. A *tail*, by which it swims.

In the wall of each seminiferous tubule, sperm are produced at a rate of about 200 million a day in a process called **spermatogenesis** (Fig. 17.6).

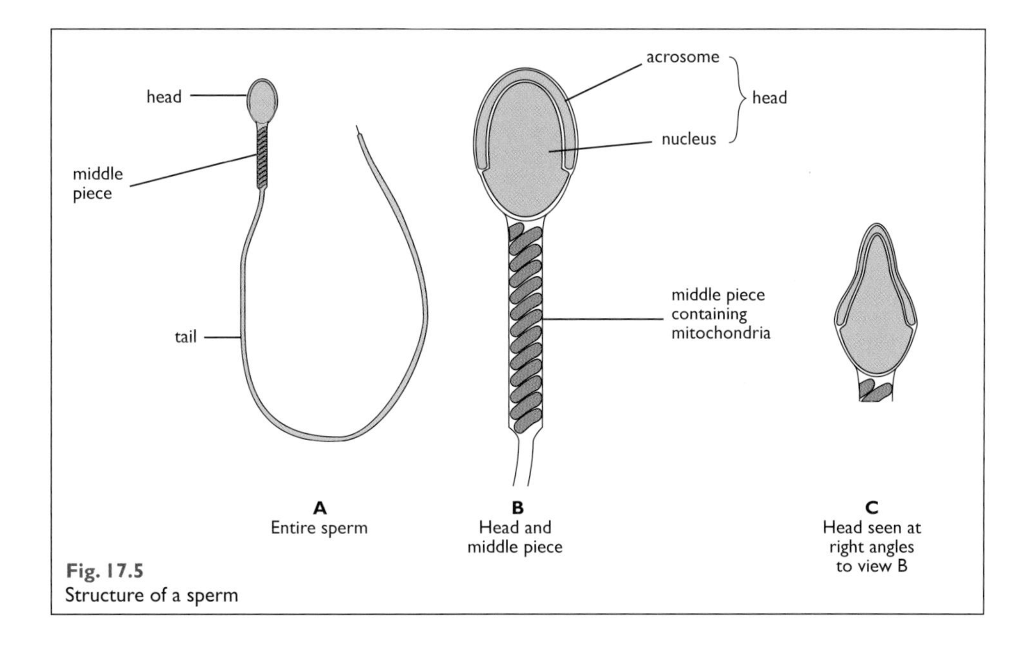

Fig. 17.5
Structure of a sperm

The 'plumbing' of the male system

Leading from the epididymis is the **vas deferens** (plural: *vasa deferentia*). This is a muscular tube carrying sperm to the base of the bladder. As shown in Fig. 17.2, it loops over the ureter. This is because as the testis descended into the scrotum before birth, it dragged the vas deferens with it, forming a loop. Three glands open into the final part of the vas deferens: The **seminal vesicles, prostate** and **Cowper's glands**. These make various contributions of the *seminal fluid*, which is a complex cocktail of chemicals. It has the following functions:

ISBN 9780170191340

- It provides a liquid in which the sperm swim.
- It contains fructose (a sugar), which provides energy for the sperm to swim.
- It is alkaline, which may help to neutralise the acidity of the vaginal fluid, which is otherwise deadly for sperm.

A common method of birth control is **vasectomy**. This is a simple surgical operation in which the vasa deferentia are cut, preventing sperm from leaving the testis. Except for the fact that the man is sterile, there is no other effect on 'masculinity', as the male hormone testosterone leaves the testis via the blood stream.

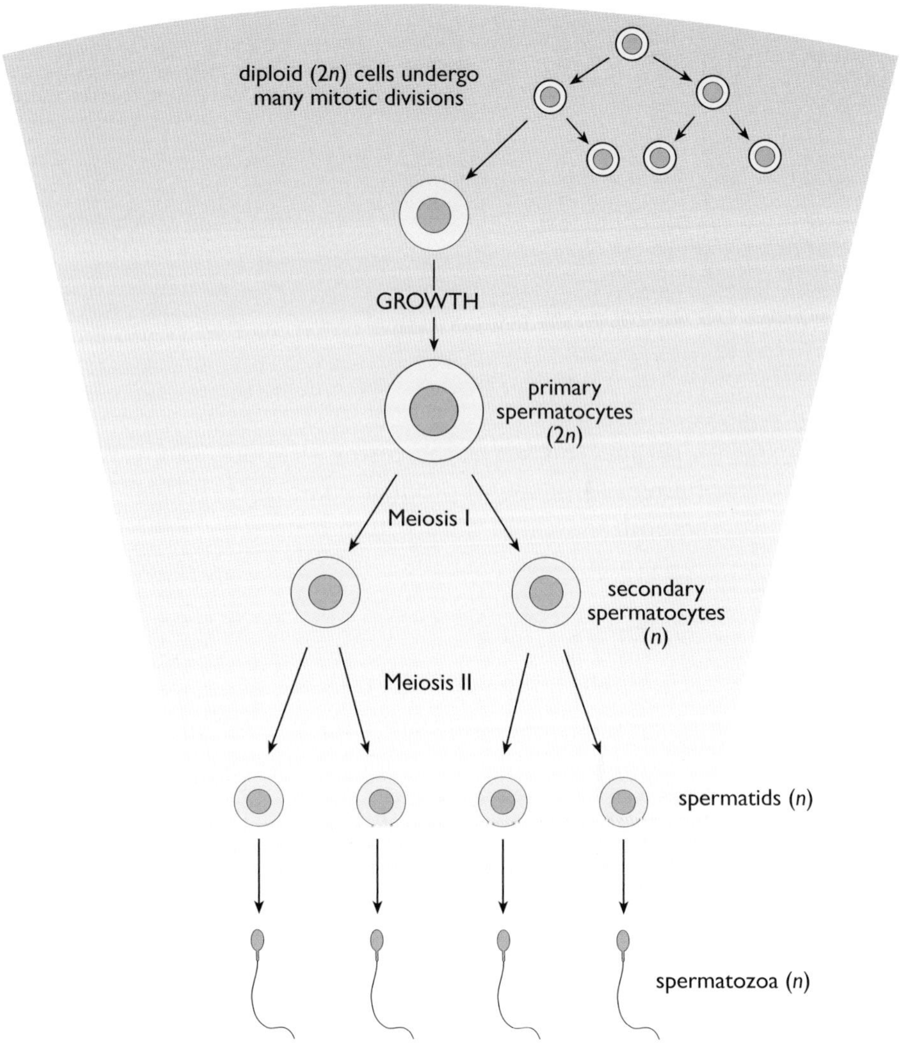

Fig. 17.6
Spermatogenesis

Before reaching the outside world, the vas deferens joins the tube leaving the bladder to form the **urethra**, which is thus a dual-purpose tube, carrying urine and sperm. The urethra runs through the **penis**, which contains three spongy blood spaces called **erectile tissue**. During sexual excitement the spaces become gorged with blood, causing the penis to become stiff and erect, enabling it to enter the vagina during intercourse. The end of the penis is covered by the *foreskin*. If too tight it can restrict urination, in which case it may be removed by a simple surgical operation called *circumcision*.

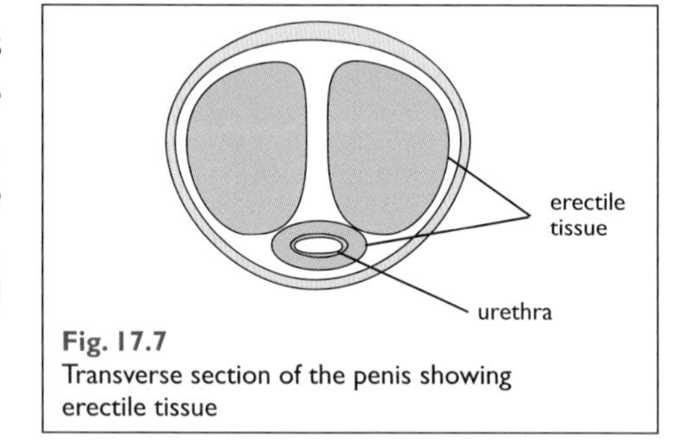

Fig. 17.7
Transverse section of the penis showing erectile tissue

THE FEMALE REPRODUCTIVE SYSTEM

The female human reproductive system is structurally simpler than that of the male, and is shown in side view (Fig. 17.8) and front view (Fig. 17.9).

The female gonads or **ovaries** are about 3 cm long and lie in the abdominal cavity. They have two functions:

1. Production of female gametes or eggs.
2. Production of female sex hormones, **oestrogen** and **progesterone**.

Close to each ovary is the opening of an **oviduct** or **Fallopian tube**, which leads to the **uterus** ('womb'), within which the baby develops. The opening of the Fallopian tube is funnel-shaped and is extended into tentacle-like **fimbriae**. The fimbriae and rest of the oviduct are lined with **cilia** which, at the time of ovulation, propel the egg towards the uterus.

ISBN 9780170191340

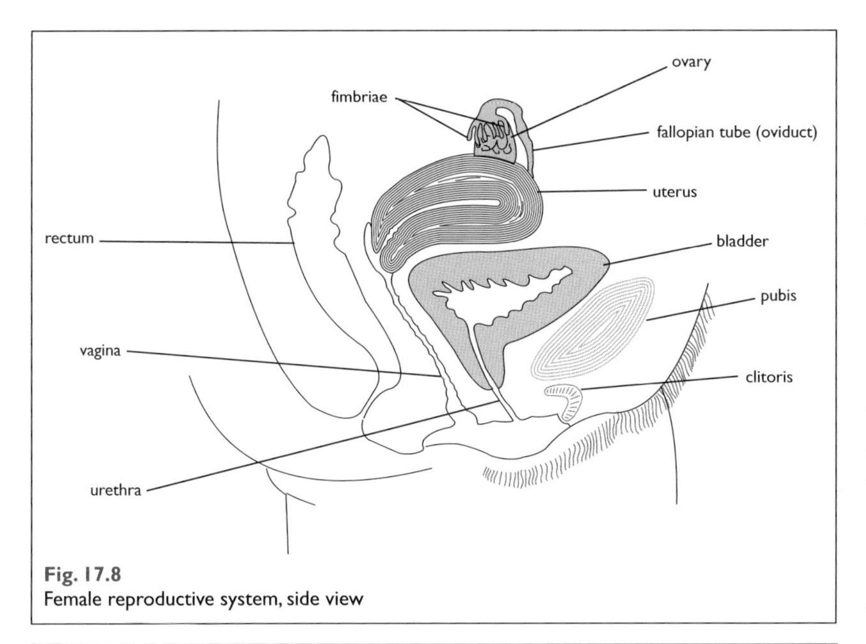

Fig. 17.8
Female reproductive system, side view

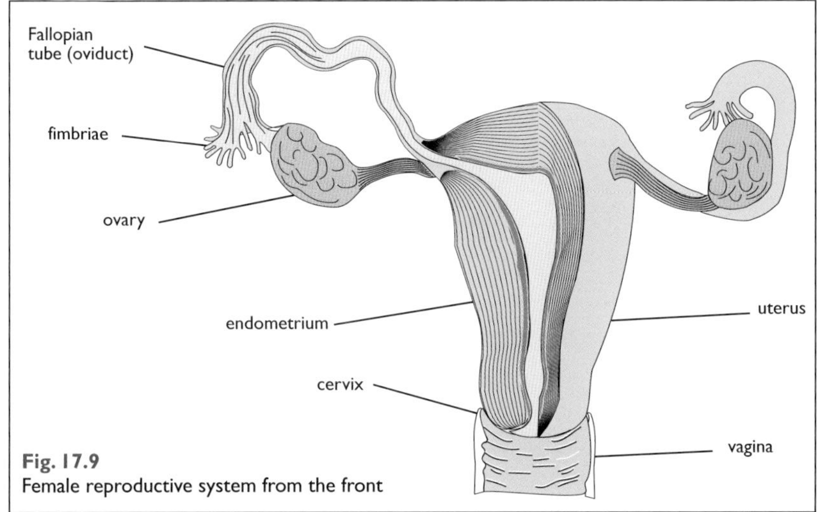

Fig. 17.9
Female reproductive system from the front

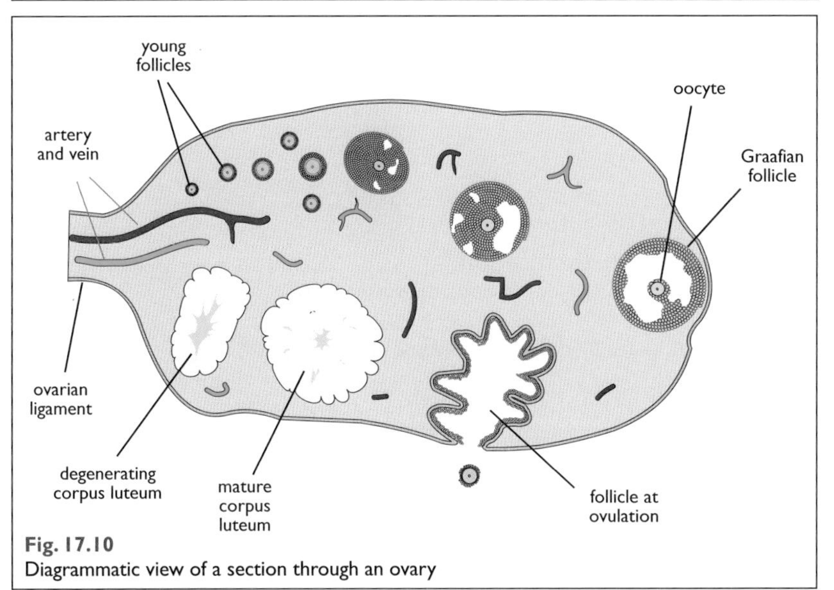

Fig. 17.10
Diagrammatic view of a section through an ovary

The wall of the uterus consists of smooth muscle, and its function is to expel the baby in childbirth. The lining of the uterus is called the **endometrium**. It contains many glands and contributes to the formation of the **placenta** during pregnancy. The placenta anchors and nourishes the baby. The neck of the uterus is the **cervix**, and leads to the exterior by the **vagina**. Near the opening of the vagina is a small area of erectile tissue called the **clitoris**, which becomes gorged with blood during sexual stimulation.

A girl is born with about two million immature eggs, but by the time she reaches puberty, all but about 400,000 have died. Each immature egg is surrounded by a cluster of cells called a **follicle** (Fig.17.10). These follicles remain 'dormant' until puberty, after which they mature, usually one at a time every 28 days or so.

The female gametes

A human egg is just visible to the naked eye, and at 0.13 mm (130 μm) across is one of the largest human cells. Eggs are produced in a process called **öogenesis** (Fig. 17.11), which differs in several ways from spermatogenesis:

- The mitotic divisions that precede meiosis cease before birth, so that a girl is born with all the potential eggs she will ever have.
- The maturation of the remaining eggs ceases at menopause.
- In meiosis the cytoplasm is distributed very unevenly between the four products, only one gamete being produced. The remaining three are tiny, functionless **polar bodies**.
- Eggs are released intermittently, normally one every 28 days.

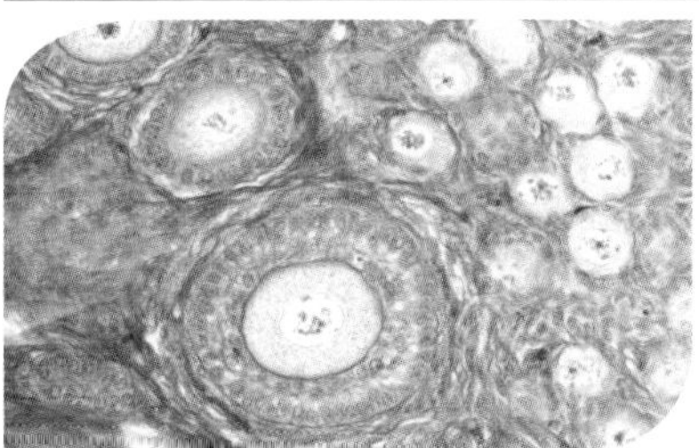
Section through ovary of rabbit, showing follicles

Ovulation

When a follicle develops, it enlarges until it reaches about the size of a pea. By this time it has developed a fluid-filled cavity and is called a **Graafian follicle** that is close to the surface of the ovary. The egg (actually a secondary öocyte) is released by rupture of the follicle at the surface of the ovary, surrounded by some of the follicle cells. This process is called **ovulation** and occurs about every 28 days

ISBN 9780170191340

from puberty to menopause at around 50 years of age. Ovulation is accompanied by a slight rise in body temperature. With careful monitoring over an extended period, ovulation can often be detected in this way.

Of the follicle cells that surrounded the egg at ovulation, some remain in the ovary and develop into a **corpus luteum** (meaning 'yellow body'). This secretes the hormone *progesterone*, which stimulates changes in the endometrium in preparation for possible arrival of a fertilised egg. About 14 days after it was formed the corpus luteum dies, so the secretion of progesterone ceases, precipitating a 'period' (see below).

At the time of ovulation and under hormonal influence, the fimbriae slowly move over the surface of the ovary. Normally the cilia sweep up a released egg and propel it down the oviduct, assisted by slow peristaltic contractions. This journey normally takes 3–5 days, during which the lining of the uterus is undergoing changes in preparation for the possible arrival of a fertilised egg.

diploid cells ($2n$ chromosomes) undergo many mitotic divisions
GROWTH
primary öocyte ($2n$)
Meiosis I
first polar body
secondary öocyte (n)
Meiosis II
second polar body
ovum

Fig. 17.11
Stages of öogenesis

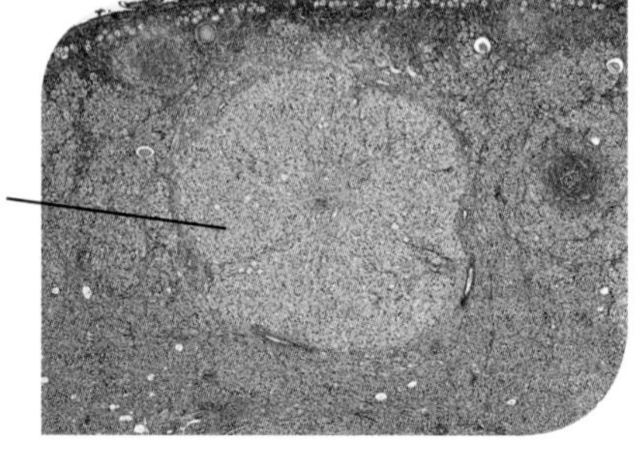

Section through the ovary of rabbit showing a corpus luteum

Puberty and the hormonal control of reproduction

We are all born with reproductive organs, but they are in an immature state. The change from a juvenile, sexually immature person to an adult is called **puberty**, and is controlled by hormones.

In males, puberty begins at about 11–13 years of age and in females a little earlier, usually 10–12 years:

- In both sexes, hair begins to thicken under the armpits and pubic region, and both begin a period of rapid growth (the 'growth spurt').
- Both males and females begin to take an interest in sex.
- In males the sperm ducts, prostate, seminal vesicles and testes enlarge and sperm production begins.
- In girls, the uterus and breasts enlarge and menstrual cycles begin.
- In boys, hair thickens on the face, chest and legs, the body becomes more muscular and the larynx enlarges, deepening the voice.

These changes take place under the direct influence of the sex hormones: *testosterone* secreted by the testes, and *oestrogen* and *progesterone* secreted by the ovaries. By the late teens these changes are complete and, physically speaking, adulthood has been achieved.

ISBN 9780170191340

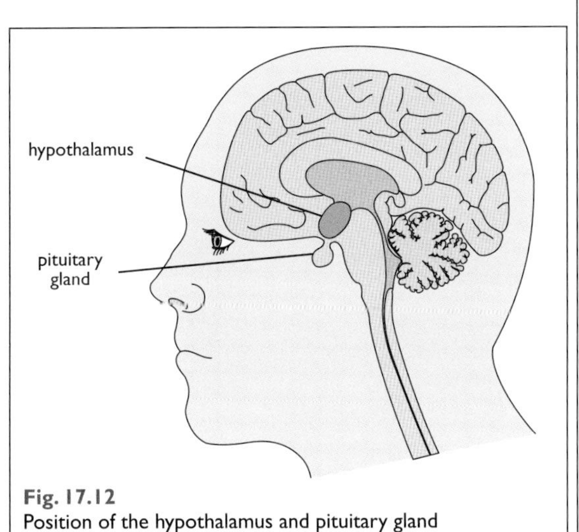

Fig. 17.12
Position of the hypothalamus and pituitary gland

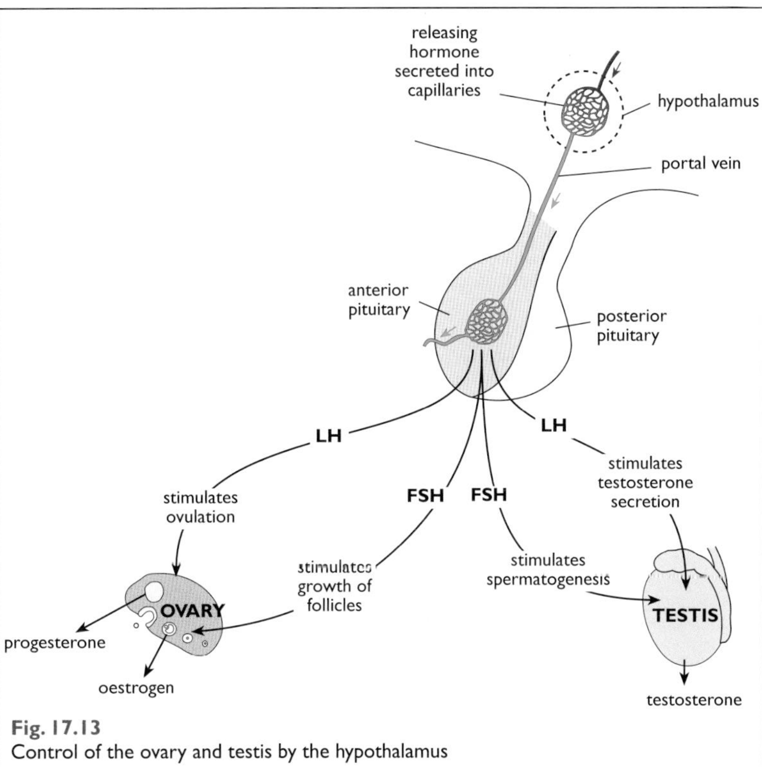

Fig. 17.13
Control of the ovary and testis by the hypothalamus

In humans, menstrual cycles and egg production cease in the late 40s or early 50s. This is the *menopause*, and one of its effects is the partial demineralisation and weakening of the bones, which can lead to *osteoporosis* (Chapter 15).

Role of the pituitary and hypothalamus

How do the ovaries and testes 'know' when puberty is due? The answer is that they do not; they are acting on signals from the anterior (front) part of the **pituitary gland** at the base of the brain (Fig. 17.12 and 17.13).

At the onset of puberty the anterior pituitary starts to secrete two hormones: **follicle stimulating hormone (FSH)** and **luteinising hormone (LH)**. Though they take their names from the effects in females, they are equally important in males.

- In females, FSH stimulates the growth of follicles in the ovary and the secretion of oestrogen by the follicle cells. In males, FSH stimulates spermatogenesis.
- In females, LH stimulates ovulation and the development of the corpus luteum. In males, it stimulates the secretion of testosterone by the testis.

The anterior pituitary gets its information from a part of the brain just above it, called the **hypothalamus**. This part of the brain controls a number of other vital functions such as body water content and body temperature.

Puberty begins to get underway when the hypothalamus starts secreting a **releasing hormone** into a tiny system of blood vessels that connect it to the anterior pituitary. The releasing hormone stimulates the anterior pituitary to produce FSH and LH.

The menstrual cycle

Approximately every 28 days, the smooth muscle in some of the arterioles in the endometrium contracts so strongly that blood flow to all but the deepest layer of the endometrium shuts down. This cuts off the oxygen supply, causing pain, and the bulk of the endometrium dies. During the next 4–5 days it comes away as blood and dead cells. In this process of **menstruation** or 'period', about 150 cm^3 of blood is lost each month.

The deepest layer of the endometrium is supplied by different arterioles, and does not die. From this deep layer a new lining grows, until a new period begins, and so on. A cycle has no beginning or end but, conventionally, the day menstruation begins is regarded as day 1.

ISBN 9780170191340

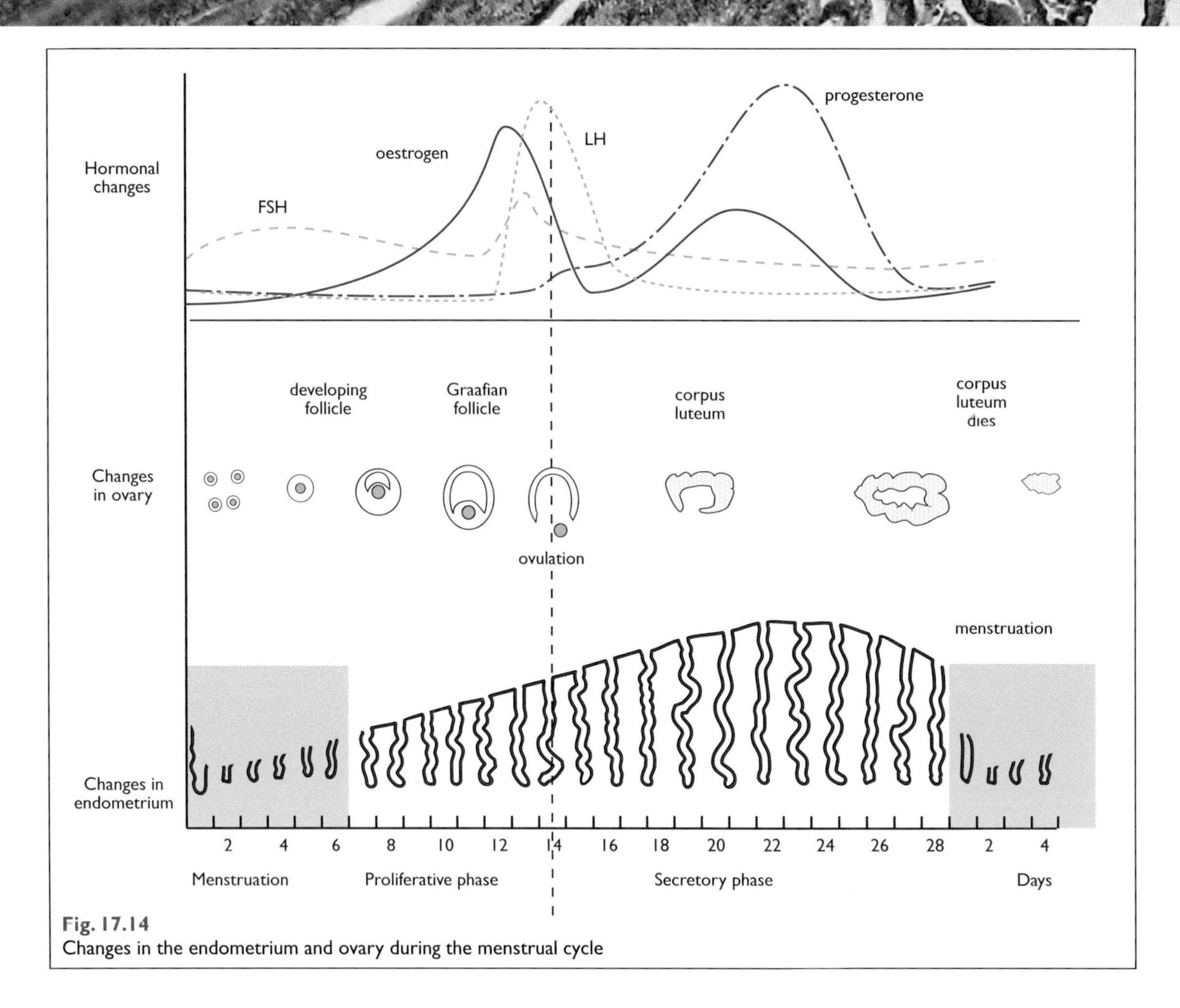

Fig. 17.14
Changes in the endometrium and ovary during the menstrual cycle

Control of the cycle

The **menstrual cycle** is so-called because the events in the uterus are its most obvious sign. Paralleling these events are changes in the ovary and activity of the pituitary (Fig. 17.14).

At about the time of menstruation the anterior pituitary begins to increase its output of FSH, stimulating a follicle to begin to mature. The young follicle secretes oestrogen, stimulating the growth and repair of the endometrium. Oestrogen also begins to inhibit the secretion of FSH by the pituitary, preventing the development of any more follicles. This is the basis for the contraceptive pill, which contains oestrogen and progesterone.

Then, around day 12 and for reasons not fully understood, there is a surge in the secretion of LH (and to a lesser extent FSH) by the pituitary. This causes ovulation on or around day 14, followed by the development of the corpus luteum and secretion of progesterone. Among the several effects of progesterone, two play an important part in the menstrual cycle:

1. It stimulates the activity of the secretory tissue of the endometrium, preparing it for possible arrival of a fertilised egg.
2. It inhibits (tends to prevent) the secretion of FSH and LH by the pituitary.

As the corpus luteum grows and its progesterone output increases, so does the inhibition of LH output by the pituitary. Eventually LH output decreases to the point that the corpus luteum dies from lack of LH. As a result progesterone secretion stops, the endometrium dies, and a 'period' begins.

With the end of progesterone secretion, the pituitary is freed from inhibition and begins to secrete FSH and LH again. A new follicle begins to mature and secrete oestrogen. This stimulates the regeneration of the endometrium that, over the next few days, increases its thickness. At about day 14, ovulation occurs and a new corpus luteum develops.

In summary, the menstrual cycle can be divided into three phases:

1. Menstruation (days 0–5).

ISBN 9780170191340

2. The *proliferative phase*, dominated by the secretion of oestrogen and regeneration (proliferation) of the endometrium.
3. The *secretory phase*, which begins with the formation of the corpus luteum and is dominated by the secretion of progesterone.

Fertilisation

Fertilisation is the joining of an egg and a sperm, and in humans it normally occurs in the oviduct. The prelude to fertilisation is intercourse, in which the penis is inserted into the vagina, and rhythmic movements culminate in **ejaculation**. The walls of the sperm ducts, seminal vesicles and prostate gland contract and semen is ejected into the upper vagina. Ejaculation is accompanied by feelings of extreme pleasure called **orgasm**. Females experience orgasm by stimulation of the clitoris.

A typical ejaculation contains about 300–500 million sperm. Semen is deposited just outside the cervix. After ovulation an egg can survive for up to 36 hours. After ejaculation, sperm can live in the female for about two days, so for fertilisation to occur therefore, intercourse must occur within a time period of about two days before to 36 hours after ovulation. In either case the sperm have to travel to the upper part of the oviduct, a distance of about 15 cm.

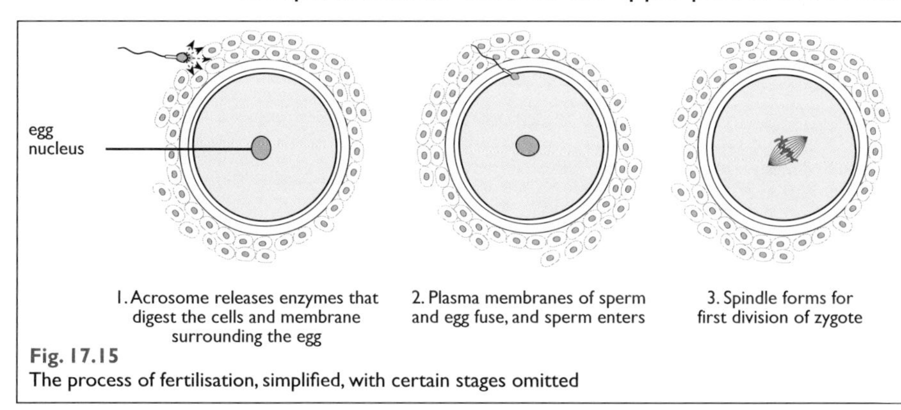
egg
nucleus
1. Acrosome releases enzymes that digest the cells and membrane surrounding the egg
2. Plasma membranes of sperm and egg fuse, and sperm enters
3. Spindle forms for first division of zygote

Fig. 17.15
The process of fertilisation, simplified, with certain stages omitted

Usually no more than a few hundred to a few thousand sperm reach the vicinity of the egg, and only one normally fertilises it. Contact between egg and sperm seems to be a matter of chance; in humans there is no evidence that sperm are attracted to the egg.

When a sperm collides with cells surrounding the ovum, the acrosome in the head of the sperm releases enzymes that digest a pathway through the cells surrounding the egg. The sperm and egg plasma membranes fuse and the sperm enters the egg (Fig. 17.15).

The entry of a sperm triggers a reaction in the egg plasma membrane which blocks the entry of other sperm. After the sperm head has entered the egg, the tail and middle piece degenerate. Since the middle piece includes the mitochondria, the mitochondria of the fertilised egg and all the body cells derived from it are derived from the mother. All your mitochondria come from your mother, and her mother, and her mother . . .

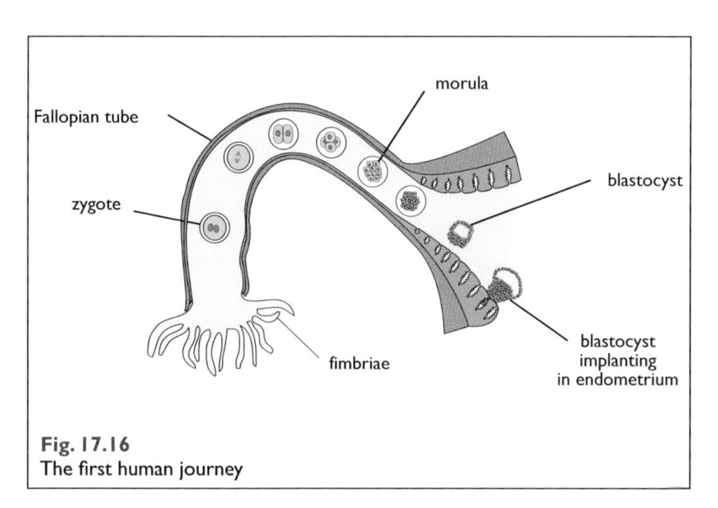
morula
Fallopian tube
zygote
blastocyst
fimbriae
blastocyst
implanting
in endometrium

Fig. 17.16
The first human journey

Cleavage and implantation

Fertilisation normally occurs in the upper third of the oviduct. The zygote divides into two, then four, eight and so on, producing a solid ball of cells called a **morula** (Fig. 17.16). The morula is slowly propelled down the oviduct by cilia, and after a journey of 3–5 days it reaches the uterus. By this time it has developed a fluid-filled cavity and is called a **blastocyst**.

On reaching the uterus the blastocyst becomes buried in the endometrium. This process is called **implantation** and marks the beginning of pregnancy.

Very occasionally a fertilised egg fails to reach the uterus and implants elsewhere usually in the oviduct, but sometimes in the abdominal cavity. The result is an *ectopic pregnancy* which, without surgery, is usually fatal.

ISBN 9780170191340

What prevents periods during pregnancy?

In the absence of pregnancy the death of the corpus luteum puts an end to the secretion of progesterone, which the endometrium needs to survive. So, how do the corpus luteum and endometrium survive during pregnancy?

The answer is that soon after implantation the blastocyst begins to secrete a hormone called **chorionic gonadotrophin**, which mimics the action of LH. Chorionic gonadotrophin can be detected in the urine and is the basis for pregnancy testing.

For the first three months, pregnancy depends on the existence of the corpus luteum; if it is removed there is an immediate miscarriage. After this, the placenta (see below) secretes large quantities of progesterone, and removal of the corpus luteum has no effect.

Early nutrition of the embryo

The blastocyst is not an embryo; this develops from an *embryonic disc* inside the blastocyst. Soon after the blastocyst has become implanted, its outer layer or **trophoblast** develops branched outgrowths called **villi**. These greatly increase the surface area for absorption of nutrients from the endometrium (Fig. 17.17). At about the same time as the villi develop, a fluid-filled **amniotic cavity** develops. Later, this plays an important part in protecting the embryo from mechanical blows.

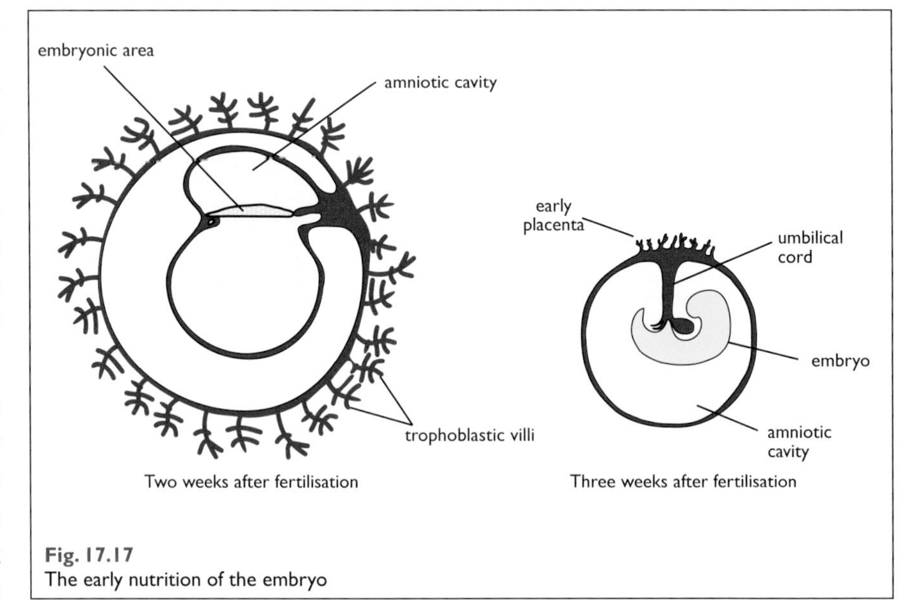

Fig. 17.17
The early nutrition of the embryo

The period of development inside the mother is called **gestation**, and in humans is approximately nine months (actually, 38 weeks, or 40 weeks after the last menstrual period). It is conventionally divided into three **trimesters**, each of three months. Changes in the appearance of the embryo and foetus are shown in Fig. 17.18.

In the first trimester all the organ systems have developed by eight weeks, after which the embryo is called a **foetus**. By the end of the first trimester the foetus is about 5 cm long and has the appearance

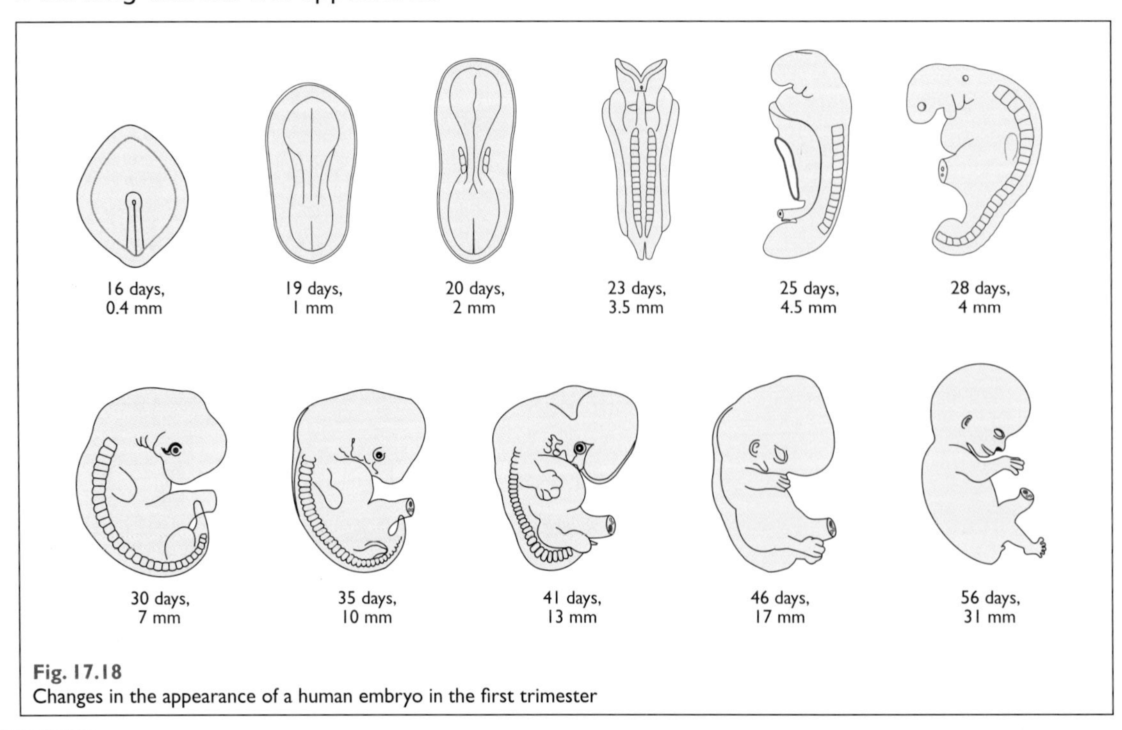

Fig. 17.18
Changes in the appearance of a human embryo in the first trimester

 ISBN 9780170191340

of a human. It is during the first trimester that the foetus is most susceptible to harmful environmental factors (see below).

For about the first 2–4 weeks the embryo obtains nutrients and oxygen directly from the endometrium, via the trophoblastic villi. By this time the embryo's needs have increased and a new structure, the **placenta**, develops.

The placenta

The *placenta* is an organ formed partly by the baby and partly from the mother and, besides anchoring the foetus, has two functions:

1. It enables the transfer of nutrients, wastes and other materials between mother and baby.
2. It produces hormones essential for the maintenance of pregnancy.

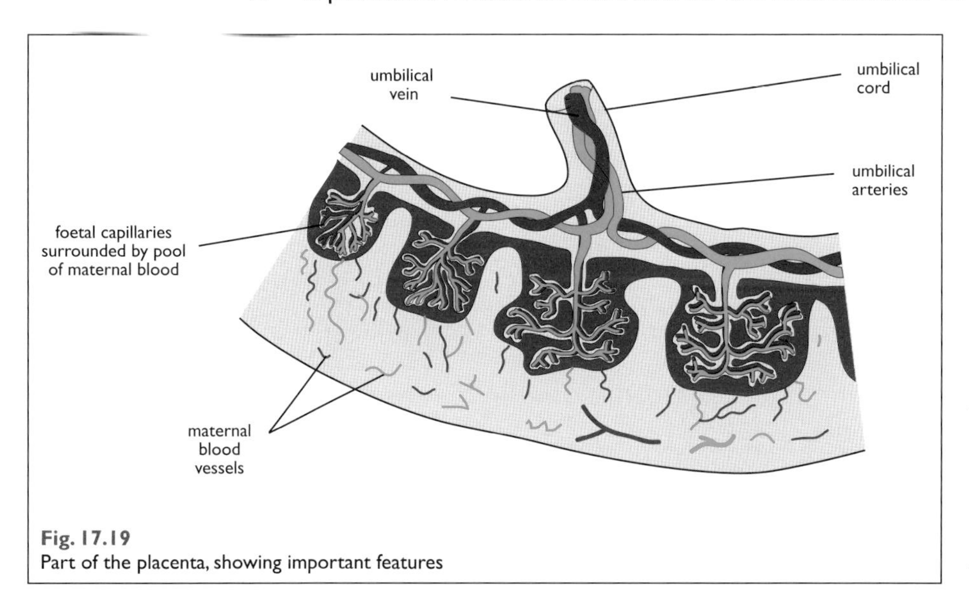

Fig. 17.19
Part of the placenta, showing important features

By the time it is fully developed it is about 4 cm thick and about 20 cm across. It is connected to the baby by the **umbilical cord**, containing two **umbilical arteries** and an **umbilical vein**. The umbilical arteries take blood from the baby to the placenta, and the umbilical vein returns it from the placenta to the baby.

In the placenta, exchange of materials between mother and baby is helped by two features:

1. The thousands of branching *villi*, each containing capillaries, have a total surface area of about 10 m^2 (Fig. 17.19).
2. The walls of the mother's blood vessels break down so that the villi are bathed in maternal blood. As a result, the blood of mother and baby are separated by a very short distance.

Transport across the placenta

Though they come very close, maternal and foetal blood *do not mix*. There are two very good reasons why this is necessary:

1. Mother and foetus may be of different blood groups.
2. Pathogens (disease-causing organisms) could cross from mother to foetus. This does sometimes happen, as in the case of viruses such as rubella ('German measles').

During its time in the uterus, the kidneys, lungs and alimentary canal of the mother look after the corresponding needs of the foetus.

There are several ways in which substances cross the placenta (Fig. 17.20). Oxygen, CO_2 and urea cross by simple *diffusion*. Glucose also diffuses across but by a mechanism called *facilitated diffusion*, and this involves special carriers in the cell membranes. Amino acids cross by *active transport*, and this requires ATP energy generated by mitochondria. Antibodies made by the mother's immune system cross by a different mechanism, called pinocytosis.

The movement of oxygen is helped by the fact that foetal haemoglobin (Hb_F) is different from that of the adult. It is slightly more 'greedy' for oxygen and can therefore take it from the mother's haemoglobin (Hb_M):

$$Hb_MO + Hb_F \rightarrow Hb_M + Hb_FO$$

ISBN 9780170191340

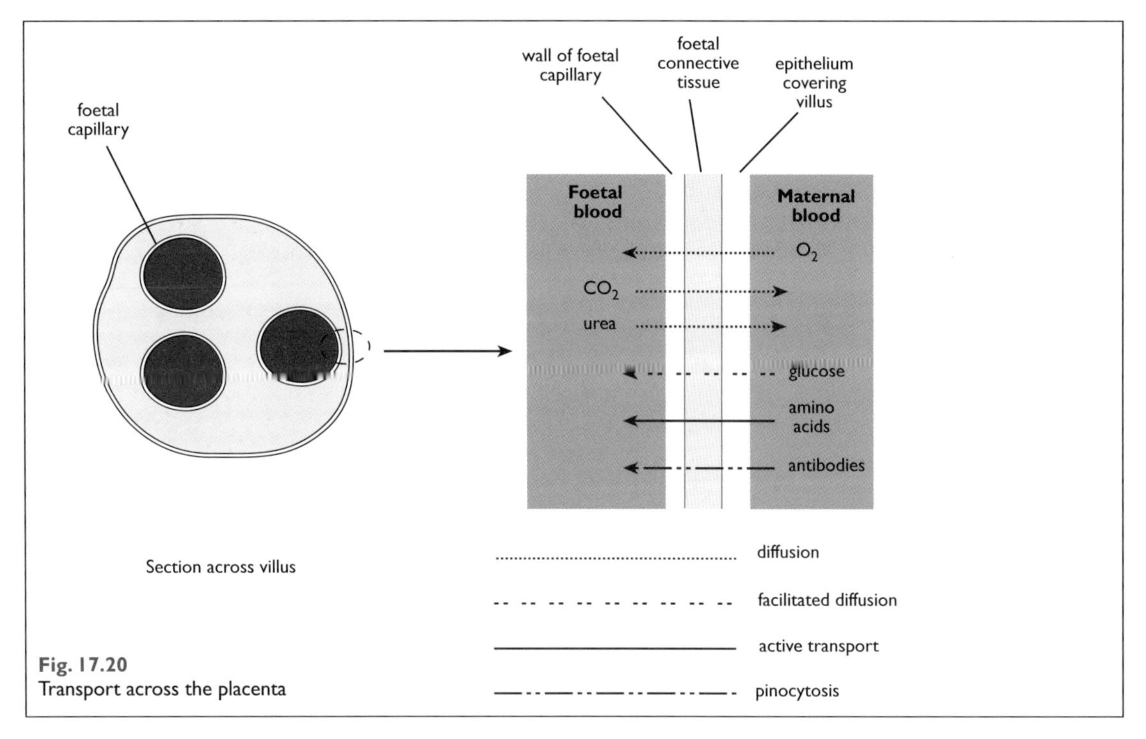

Fig. 17.20
Transport across the placenta

As soon as the baby is born, it ceases making foetal haemoglobin and starts producing adult haemoglobin.

The placenta as an endocrine organ

Besides serving as an exchange surface for nutrients, oxygen and wastes for the foetus, the placenta also secretes a number of hormones:

- Oestrogen and progesterone stimulate the growth of the uterus and the secretory and duct tissues of the breasts, but *inhibit* the secretion of milk.
- Towards the end of pregnancy it secretes **relaxin**, which causes a softening of the ligaments binding the pubic bones and the sacrum and ilium together. This allows the pelvic canal to widen in childbirth, making it easier for the baby's head to pass through.

Protection of the foetus

A human foetus needs protection against mechanical blows to the mother's abdomen. This is the function of the **amniotic fluid**, which is contained within a membrane called the **amnion** (Fig. 17.21).

Amniocentesis

During pregnancy the foetus sheds cells into the amniotic fluid. These can be removed for examination by withdrawing a small sample (10–20 cm^3) of amniotic fluid using a needle.

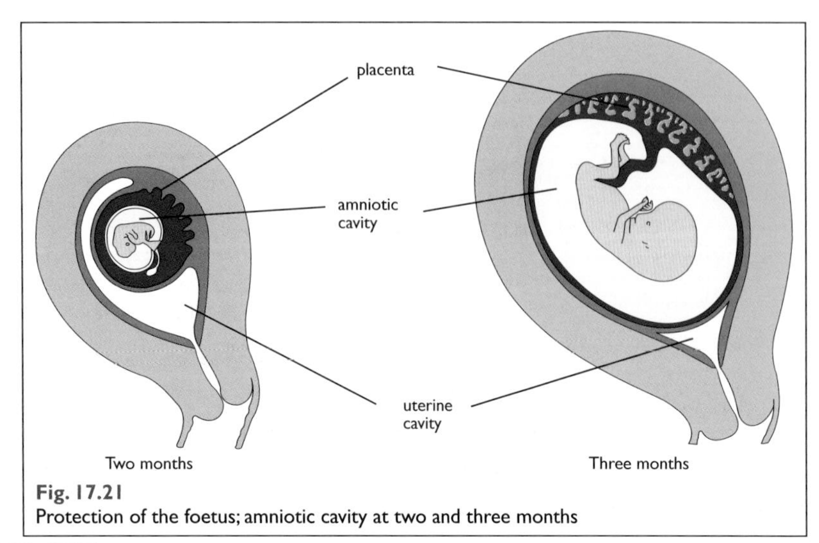

Fig. 17.21
Protection of the foetus; amniotic cavity at two and three months

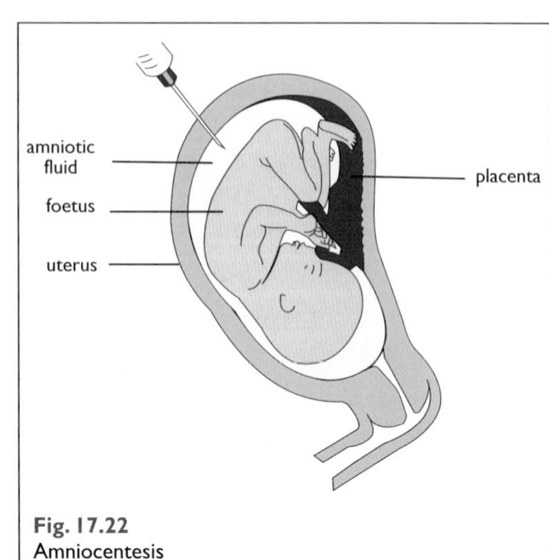

Fig. 17.22
Amniocentesis

ISBN 9780170191340

The test is done at 16–18 weeks and can reveal a variety of genetic abnormalities such as Down syndrome and haemophilia (Fig. 17.22).

Birth

By the last few weeks of pregnancy the foetus has (usually) come to lie with its head against the cervix. The exact trigger for birth is not known for certain, except that it originates in the foetus. The mother's work in the expulsion of the baby is called *labour*, and is divided into three stages (Fig. 17.23).

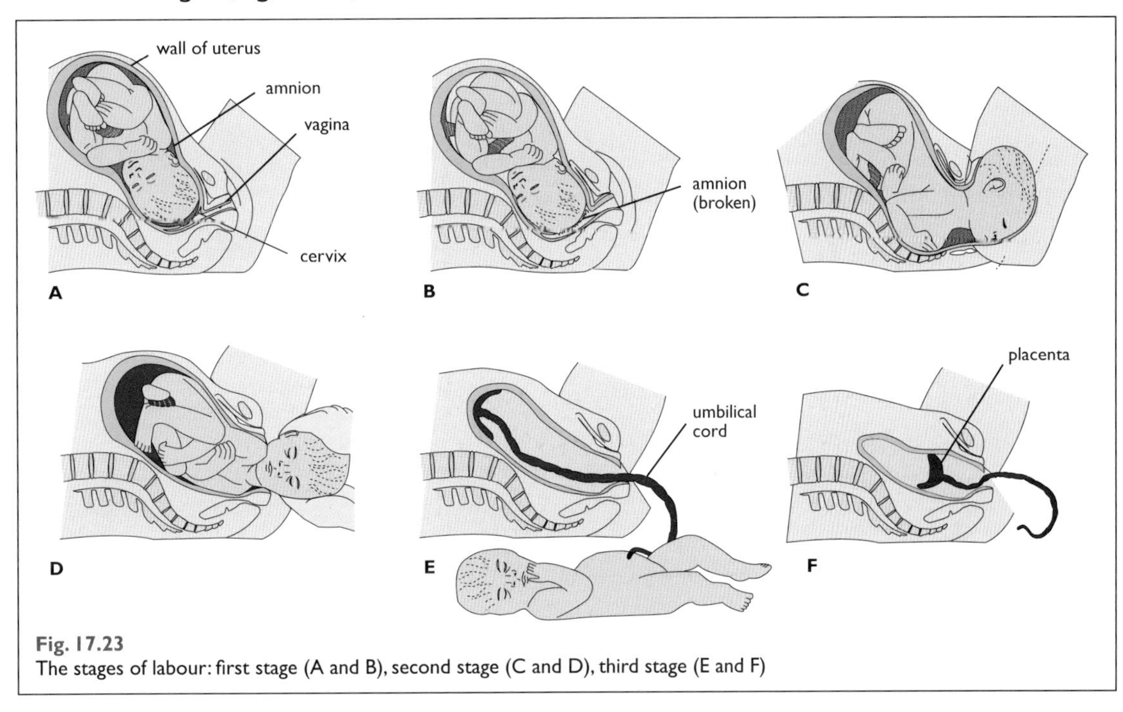

Fig. 17.23
The stages of labour: first stage (A and B), second stage (C and D), third stage (E and F)

The *first stage* is triggered by the baby. Hormonal and other chemical signals from the foetus stimulate the posterior pituitary to release **oxytocin**, a hormone that stimulates the contraction of the uterine muscle. These contractions begin about every 15 minutes, and get stronger and more frequent. Early in labour the amnion breaks, releasing the amniotic fluid. The uterine contractions slowly stretch the cervix, causing *labour pains*. This stage is by far the longest, usually lasting several hours.

Oxytocin is often given artificially to the mother to help stimulate her contractions.

When the cervix is sufficiently dilated, the mother assists the uterine contractions with contractions of her own abdominal muscles. The baby is forced out, the umbilical cord is cut and the baby takes its first breath. This is the second stage, and is quite short.

The third stage is the one they never show on TV. It consists of the expulsion of the placenta, and occurs about 15–30 minutes after the birth. It is essential that all the placenta leaves the mother, as any residue could become infected.

Extension: Changes in the circulation at birth

The foetal and adult circulations are significantly different, yet within seconds of birth the baby's circulation has to change from one adapted to obtaining oxygen from the placenta to one obtaining oxygen from the lungs (Fig. 17.24).

In the adult circulation the right side of the heart pumps deoxygenated blood through the lungs and the left side pumps oxygenated blood through the rest of the body. Since the two sides are in *series* they pump the same volumes of blood every minute, so the flow through the lungs is equal to the flow through the whole of the rest of the body.

In the foetus, on the other hand, oxygenated blood from the placenta enters the *right* side of the heart, yet the baby's brain, which has the greatest need for oxygen, gets its blood from the *left* side of the heart. Moreover, as consumers of oxygen the lungs need only a small blood supply.

ISBN 9780170191340

The problem of getting oxygenated blood entering the right side of the heart to the left side is 'solved' by means of two short cuts or *shunts*, which close immediately after birth:

1. The *foramen ovale* is a hole which allows blood to flow directly from right atrium to left atrium. It is guarded by a flap that acts as a valve, preventing blood from flowing in the opposite direction.
2. The *ductus arteriosus*, which carries blood from the pulmonary artery to the aorta.

Blood flows through these shunts because, for reasons to be explained, the pressure is higher in right atrium than in the left, and higher in the pulmonary artery than in the aorta. Immediately after birth these pressure differences are reversed.

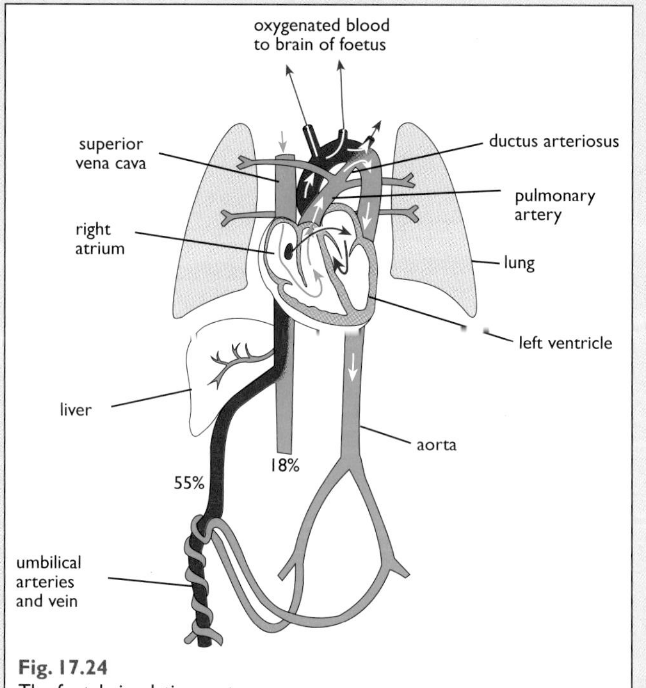

Fig. 17.24
The foetal circulation system

Blood enters the right atrium in two streams:

1. A deoxygenated stream from the head and arms. This enters the right ventricle, which pumps it through the pulmonary artery. The lungs of the foetus are unexpanded and offer high resistance to blood flow, so most of this blood goes through the ductus arteriosus to the aorta and then to the legs, abdomen and placenta. Because the flow through the lungs is so small, the flow of blood from the pulmonary veins into the left atrium is much lower than the flow entering the right atrium, and the pressure in the left atrium is correspondingly lower.
2. A partly oxygenated stream from the placenta, abdomen and legs. This passes through the foramen ovale into the left atrium and then into the left ventricle, which pumps it via the aorta to the head and arms.

When these two streams enter the right atrium, they remain largely separate.

Because of these two shunts, the left and right sides of the heart are in *parallel* and do not have to pump equal quantities of blood.

As soon as the baby is born these shunts close, separating the two sides of the heart.

The changes begin with the contraction of smooth muscle in the umbilical arteries, effectively severing the blood connection with the placenta. The CO_2 level in the baby's blood rises, stimulating the breathing centre in the brain and causing it to take its first breath. The removal of the placenta from the circulation and the opening up of the lung circuit closes off the two shunts as follows:

- The expansion of the lungs greatly reduces their resistance to blood flow, so the flow through the lungs and into the left atrium increases. At the same time the pressure in the right atrium decreases, because less blood enters the right atrium due to the loss of the placental circulation. The increase in pressure in the left atrium and the decrease in pressure in the right atrium reverses the pressure difference between left and right atria, causing the flap over the foramen ovale to close.
- The other event is the closure of the ductus arteriosus. With the shutting off of the placenta, the resistance of the systemic circuit increases, causing the pressure in the aorta to rise. Coupled with the decrease in resistance of the pulmonary artery, this reverses the blood flow through the ductus arteriosus. The blood flowing through the ductus arteriosus is now oxygenated, and this stimulates the smooth muscle in the ductus to contract, closing it off.

These changes become permanent; the flap closing the foramen ovale forms a permanent tissue junction with the atrial wall, and the ductus arteriosus slowly becomes a solid band of fibrous tissue. Occasionally the foramen ovale fails to close completely, so separation between oxygenated and deoxygenated blood is incomplete. The situation is normally corrected by surgery.

Birth also stimulates the production of adult haemoglobin, though it takes 3–4 months for all the foetal cells to be replaced.

ISBN 9780170191340

Lactation

With the loss of the placenta, milk secretion is no longer inhibited by oestrogen and progesterone. The first milk to be secreted is called **colostrum**. It is a clear yellow colour and very rich in protein. Normal milk begins to be secreted several days later.

Milk is *almost* the ideal food. It contains a sugar (lactose), fat and protein with all the essential amino acids, and also most of the essential vitamins and minerals. It is not particularly rich in vitamin C (ascorbic acid), and contains little iron, though the baby's liver stores enough iron to last several months. It is also completely lacking in fibre, which is important later.

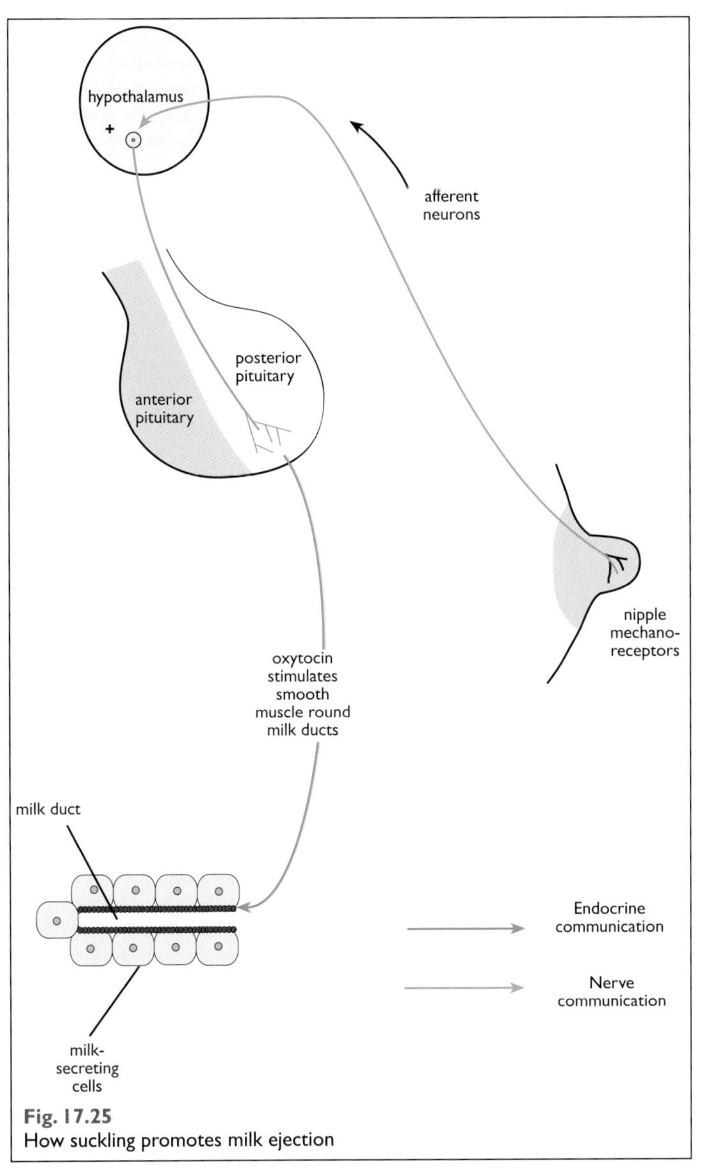

Fig. 17.25
How suckling promotes milk ejection

Breast-feeding has other advantages:

- It contains antibodies that can give temporary immunity to a number of diseases.
- It is free of bacteria (though HIV can be transmitted via breast milk).
- It promotes a strong bond between mother and baby.
- The act of suckling also *tends* to inhibit ovulation, so it can play a role in contraception (though this is not completely reliable).

The act of suckling promotes the release of oxytocin by the pituitary gland, which has two effects:

1. It stimulates contraction of smooth muscle in the milk ducts, thus helping to eject milk into the baby's mouth.
2. It also stimulates contraction of the uterine muscle, which helps it contract back to its normal size.

The ejection of milk during suckling is called 'let-down' and involves both nerves and hormones, diagrammed in Fig. 17.25. Whereas oxytocin stimulates the muscle lining the mammary ducts, **prolactin**, secreted by the anterior pituitary gland, stimulates the secretory activity of the mammary tissue.

Twins

Twins may be *similar* or *fraternal*, and are produced as shown in Fig. 17.26.

Similar ('identical') twins

Identical, or **monozygotic,** twins are produced when a blastocyst divides into two cell masses, each of which implants in the endometrium and grows into a baby. If separation is incomplete, 'Siamese' or *conjoined* twins result. Because they are produced from the same fertilised egg, they are *genetically* identical. They are not, however, phenotypically identical since no two individuals have identical environments.

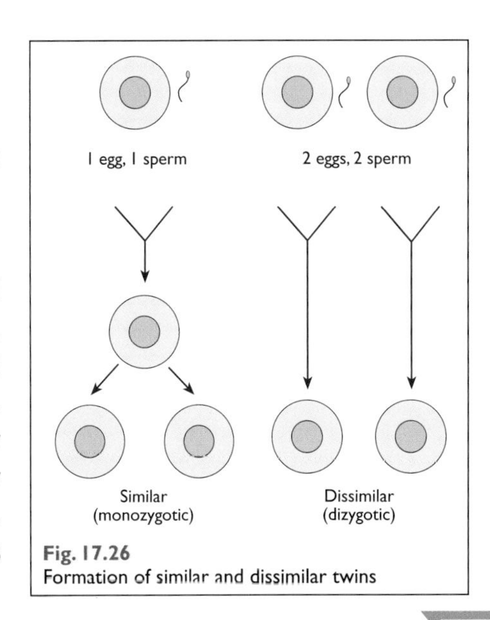

Fig. 17.26
Formation of similar and dissimilar twins

ISBN 9780170191340

Dissimilar ('fraternal') twins

Dissimilar twins are also called **dizygotic** twins because they develop from different fertilised eggs. If more than one egg is released at ovulation (probably one from each ovary), each may be fertilised by a different sperm. Fraternal twins share half their genes in common, as they do with other brothers and sisters.

Environmental factors affecting reproduction

How successful we are in reproduction depends partly — and in some cases largely — on environmental factors in pregnancy, some of which are dealt with here.

Environmental effects in pregnancy fall into two groups: Those due to *lack* of a particular nutrient, and those due to the *presence* of a harmful agent such as a chemical, virus or radiation.

Dietary deficiencies

Since a pregnant woman is 'eating for two', she needs more of all the essential nutrients. If she eats insufficient *macronutrients* (fats, protein and carbohydrates), the baby is liable to be underweight. Lack of certain micronutrients (vitamins and minerals), however, may produce serious abnormalities.

Folic acid (folate)

Otherwise known as vitamin B_9, deficiency of folate has been linked to *spina bifida*, a congenital abnormality ('birth defect'). In normal development, a flat plate of tissue in the very early embryo rolls up to form a tube, which develops into the brain and spinal cord. In spina bifida this does not happen properly.

Harmful chemicals

Among these factors are **teratogens**, which produce physical deformities. The most notorious is *thalidomide*, an anti-depressant that was at one time taken in the first trimester, and produced babies with only partially-formed limbs. Other chemicals, some of which have been around for centuries, have more subtle effect that have only been discovered in recent decades.

Alcohol

Alcohol is a known teratogen, causing a range of malformations collectively known as **Foetal Alcohol Syndrome** (FAS). Symptoms include slow growth, facial abnormalities, defective heart and other organs, and mental retardation. Less obvious are various behavioural problems such as poor attention span, over-activity and extreme nervousness.

Smoking

Mothers who smoke in pregnancy tend to have babies with lower birth weight and greater susceptibility to respiratory and other infections. There may also be a greater risk of Sudden Infant Death Syndrome (SIDS). After birth, there appear to be risks to the child of second-hand smoke.

Rubella

Otherwise known as 'German measles', rubella is a disease that causes mild symptoms in adults but during pregnancy can produce deformities in the child. Among these are malformations of the eye, ear and heart. The easiest way of avoiding these is to vaccinate girls against the disease before puberty.

Radiation

Unborn children are particularly susceptible to the effects of radiation produced by radioactivity. Among Japanese women pregnant at the time of the Hiroshima and Nagasaki atomic bombs and who survived, 28% aborted, 25% gave birth to children who died in their first year of life, and 25% of the surviving children had abnormalities of the central nervous system.

ISBN 9780170191340

Summary of key points in this chapter

- Male *gametes* are minute, mobile, and are produced in enormous numbers from puberty until old age.
- Female gametes are large, stationary and are produced (in humans) once a month until menopause.
- Both male and female *gonads* are endocrine glands as well gamete-producing organs.
- The male reproductive system consists of the *testes* and associated ducts and glands.
- The female reproductive system is anatomically much simpler than the male, but its control involves more hormones.
- In both males and females, the reproductive system is under control of hormones secreted by the *pituitary gland*, which is in turn controlled by the *hypothalamus* in the brain.
- The release of eggs is part of a complex *menstrual cycle* involving cyclical changes in the uterus, ovary and pituitary gland.
- Fertilisation normally occurs in the *oviduct*, several days before the *blastocyst* implants in the uterus.
- For the first 2–4 weeks the embryo obtains nourishment directly from the endometrium, but after this is dependent on the *placenta*, which is formed by tissues derived from both baby and mother.
- Transfer of nutrients and wastes between mother and baby is facilitated (helped) by the short distance between maternal and foetal blood and the large area between them.
- The foetus is protected from mechanical shocks by *amniotic fluid*, contained within the *amnion*.
- There are three stages of labour, the first being the most prolonged.
- Milk is an almost perfect food, containing nearly all the essential nutrients.

Test your basics

Copy and complete the following sentences.

a) Fertilisation is the joining together of two ___*___ cells, a male ___*___ or ___*___ and a female ___*___ or ___*___, to form a diploid cell called a ___*___.

b) In animals, gametes are produced in ___*___, in which the chromosome number is reduced from ___*___ to ___*___. Sperm are produced in the ___*___ and eggs are produced in the ___*___. The testes also produce male sex hormone or ___*___, and the ovaries produce the female sex hormones, ___*___ and ___*___.

c) The testes lie in a bag-like extension of the body wall, the ___*___, in which the temperature is slightly lower than body temperature. In the testes, sperm are produced in the walls of ___*___ tubules. These join together to form a long twisted tube, the ___*___, in which sperm mature. Sperm are carried from each testis by the sperm duct or vas ___*___. Opening into the last part of the vas ___*___ are three glands, the ___*___ ___*___, the ___*___ and ___*___ glands, which together make ___*___ fluid.

d) The two sperm ducts join the ___*___ which takes urine from the bladder, and pass through the penis. This contains large spongy blood spaces called ___*___ tissue.

e) In the ovary, an egg is produced inside a cluster of cells called a ___*___. When mature, this contains a fluid-filled cavity and is called a ___*___ ___*___. The egg is released by rupture of the surface of the ovary, a process called ___*___. The egg is collected by the funnel-shaped opening of the oviduct or ___*___ tube. The opening of the oviduct is fringed by tentacles which slowly move over the surface of the ovary, sweeping up the egg by the beating of millions of microscopic ___*___. The egg is slowly carried down the oviduct to the ___*___, in which a foetus will develop if fertilisation occurs. The lining of the ___*___ is called the ___*___, and its neck is called the ___*___, which opens into a tube called the ___*___.

f) The activity of the ovaries and testes is controlled by two hormones secreted by the ___*___ gland. In females, ___*___ ___*___ ___*___ (___*___) stimulates the growth of follicles and the secretion of ___*___, and ___*___ stimulates ovulation and the subsequent development

ISBN 9780170191340

of the ___*___ ___*___. In males, ___*___ stimulates the production of ___*___ and ___*___ stimulates the secretion of ___*___. In both sexes the pituitary gland is under the control of the ___*___ in the brain.

g) Approximately every 28 days the ___*___ breaks down and comes away as blood and dead cells, a process that lasts about 4–5 days. During the next 10 or so days the ___*___ regenerates (grows again) under the influence of the hormone ___*___. About 14 days after the beginning of menstruation, ___*___ occurs. The follicle cells that do not come away with the egg develop into the ___*___ ___*___, which secretes the hormone ___*___. This in turn stimulates the ___*___ to prepare for the possible arrival of a fertilised egg.

h) During intercourse, the erect penis is inserted into the vagina and rhythmic movements culminate in ejection of seminal fluid. After fertilisation, which normally occurs in the ___*___, the ___*___ divides repeatedly to form a ball of cells called a ___*___. By the time it reaches the uterus it has developed a cavity and is called a ___*___. This becomes buried in the ___*___, a process called ___*___.

i) The outer layer of the blastocyst develops branched outgrowths called ___*___, through which nutrients are absorbed. After about a month, the nutrition of the embryo is taken over by the ___*___, which is formed partly by maternal tissues and partly by the foetus. The foetus is connected to the ___*___ by the ___*___ cord, which contains two ___*___ ___*___ and an ___*___ ___*___.

j) In the ___*___, the blood of mother and baby come very close together over a large surface; this promotes the exchange of substances between the two blood systems; ___*___ and nutrients pass from mother to baby, and ___*___ ___*___ and ___*___ diffuse in the reverse direction. In addition to these functions, the ___*___ also secretes ___*___ and progesterone.

k) After about two months the embryo has all the adult organs and is called a ___*___. It is protected from mechanical shock by ___*___ fluid, within a membrane, the ___*___.

l) Birth involves three stages of labour. The first is by far the longest, and involves the slow stretching of the ___*___ as a result of powerful contractions of the wall of the ___*___. The second stage involves the expulsion of the ___*___, and the third stage, the expulsion of the ___*___.

m) With the loss of the ___*___ after birth, the mammary glands are freed from inhibition and begin to secrete ___*___.

n) Similar ('identical') or ___*___ twins are formed by division of the embryo into two, and so are ___*___ identical. Fraternal or ___*___ twins are formed from different fertilised eggs and are thus genetically ___*___.

ISBN 9780170191340

Animal Processes: Assessement Practice

In the following questions you may use labelled diagrams to support your answers.

QUESTION ONE: MOVEMENT

The diagram shows the elbow joint and two of the muscles that move it.The diagram also shows other tissues involved in movement.

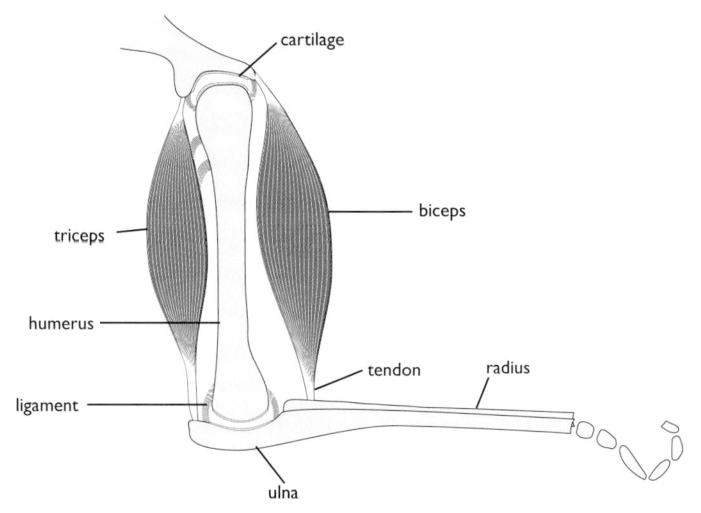

Discuss (i.e. *explain*) how the **components** of the skeletomuscular system bring about bending and straightening of the elbow joint AND consider the factors that may **reduce** a person's ability to move the elbow.

In your answer you should:

- **Describe** (i.e. *name*) the type of joint.
- **Explain** how the parts of the joint **function** during movement.
- **Identify** a **malfunction** that could affect movement of the elbow joint.
- **Discuss** the **cause** and **effect** that the **named** malfunction has on a person's ability move the elbow.

(Guidance: Your answer should be approximately 39 lines.)

QUESTION TWO: HUMAN SKELETAL SYSTEM

Strength combined with **lightness** are two features that allow the skeleton to carry out its functions.

Discuss (i.e. *explain*) how bones are produced and the features that enable them to combine lightness with strength.

Relate these features to the **functions** a skeleton.

In your answer you should:

- **Describe** at least **TWO functions** of the skeleton, **giving examples.**
- **Explain** how bone is **formed.**
- **Link** the **structure** of different bones to their **functions**.

(Guidance: Your answer should be approximately 50 lines.)

ISBN 9780170191340

QUESTION THREE: DIGESTION IN THE MOUTH

During the time food is in your mouth and during its journey to the stomach, two processes occur in the treatment of the food; one physical, the other chemical.

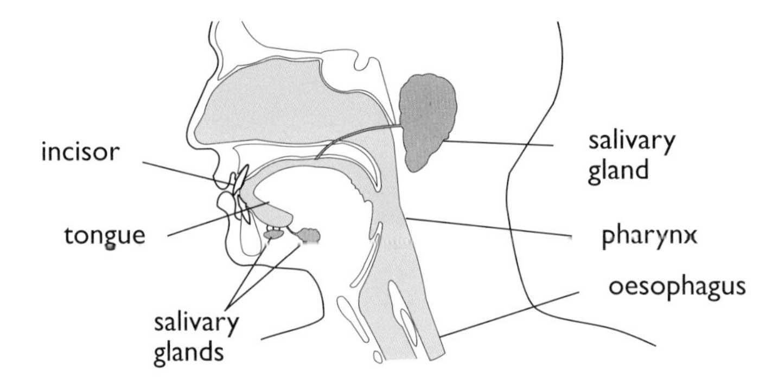

Discuss (i.e. *describe*) these processes AND the structures involved.

In your answer you should:

- **Describe** (i.e. *explain*) what digestion is.
- **Explain** (i.e. *describe*) TWO **named processes** in the treatment of the food AND **describe** the structures responsible for these processes.
- **Explain** how these **two processes** work **together** in the treatment of the food.

You may use diagrams to support your answers.

Process 1: *(Guidance: Your answer should be approximately 1 line.)*

Process 2: *(Guidance: Your answer should be approximately 1 line.)*

(Guidance: Your answer should be approximately 42 lines.)

QUESTION FOUR

(a) **Describe** the structure of the heart.

(Guidance: Your answer should be approximately 9 lines, + equal amount of blank page for your answer.)

(b) **Explain** how TWO structural features of an artery adapt it to carry out its function.

(Guidance: Your answer should be approximately 13 lines, + equal amount of blank page for your answer.)

(c) **Explain** the role of valves in the blood system.

(Guidance: Your answer should be approximately 8 lines, + equal amount of blank page for your answer.)

QUESTION FIVE

(a) **Describe** (i.e. *explain*) how air is propelled into and out of the lungs during breathing.

(Guidance: Your answer should be approximately 9 lines.)

(b) (i) **Explain** why the percentage of oxygen, carbon dioxide and water vapour differ in inhaled and exhaled air.

(Guidance: Your answer should be approximately 8 lines.)

(ii) **Explain** how TWO features of the microscopic structure of the lungs help gas exchange.

(Guidance: Your answer should be approximately 7 lines.)

(iii) **Explain** how gas exchange in the lungs is maintained.

(Guidance: Your answer should be approximately 9 lines.)

ISBN 9780170191340

QUESTION SIX

The diagrams show a longitudinal section through a human kidney (left), and a simple diagram of a nephron.

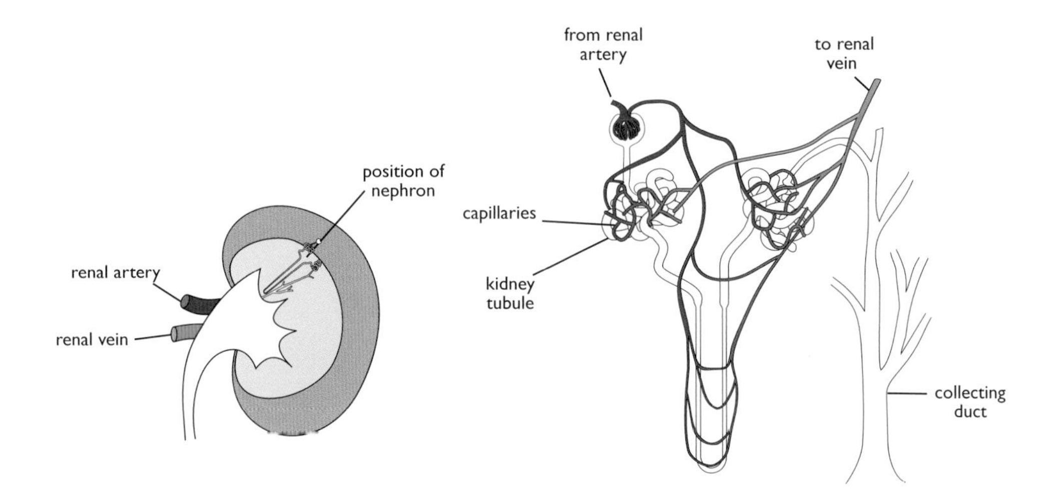

Refer to the diagrams above to answer the following questions.

(a) **Describe** the function of the renal artery and renal vein.

(Guidance: Your answer should be approximately 4 lines.)

(b) **Explain** what happens between the glomerulus and Bowman's capsule.

(Guidance: Your answer should be approximately 8 lines.)

(c) **Explain** what happens between the kidney tubule and the capillaries.

(Guidance: Your answer should be approximately 8 lines.)

QUESTION SEVEN

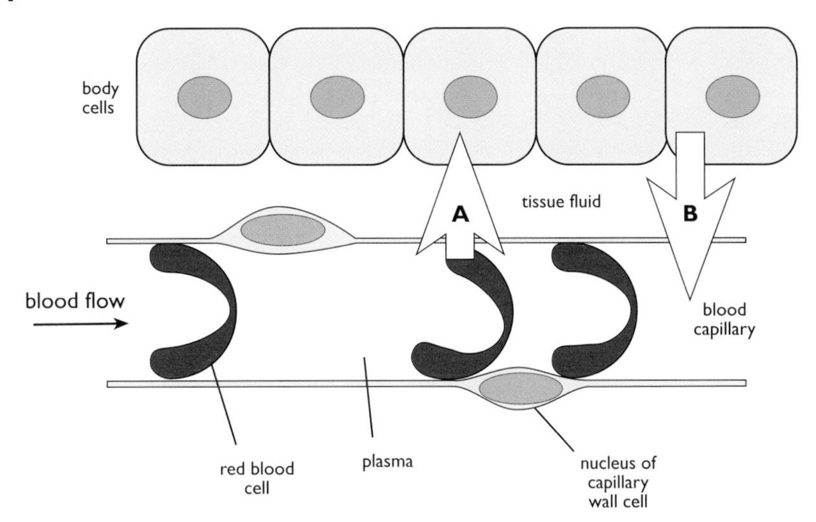

The diagram shows a capillary and some of the cells it serves, in an organ that is NOT a lung. The arrows indicate movement of substances.

(a) Name ONE substance that moves in the direction indicated by arrow A.

Name another substance that moves in the direction indicated by arrow B, and **explain how** these substances move.

(Guidance: Your answer should be approximately 6 lines.)

(b) **Explain** how a named waste product in arrow B was produced by the body cells, **and** why its movement from the body cells into the blood is beneficial to the body.

(Guidance: Your answer should be approximately 6 lines.)

ISBN 9780170191340

QUESTION EIGHT

The time when a child becomes sexually mature is called puberty. These changes are under the control of hormones secreted by certain glands.

(a) **Describe** THREE physical changes that occur in a girl as she goes through puberty (excepting armpit and pubic hair).

(Guidance: Your answer should be approximately 3 lines.)

(b) **Explain** how each of these changes you mention in (a) prepares her body for reproduction.

(Guidance: Your answer should be approximately 15 lines.)

(c) **Explain** (i.e. describe) the effects of LH, FSH and oestrogen in puberty.

(Guidance: Your answer should be approximately 15 lines.)

QUESTION NINE

(a) For sperm to reach an egg, they have to travel from testes to the oviduct, where fertilisation normally occurs. Describe the pathway the sperm make in reaching an egg.

(Guidance: Your answer should be approximately 8 lines.)

(b) **Explain** how the survival of sperm on their journey to the egg is helped by secretions from the seminal vesicles and prostate glands.

(Guidance: Your answer should be approximately 12 lines.)

(c) As sperm travel through the female body, the numbers of sperm may decrease. Describe TWO ways this can happen.

(Guidance: Your answer should be approximately 12 lines.)

QUESTION TEN

During the birth of a baby, it leaves the body of the mother. There are three stages to this process called the First, Second and Third stages of labour.

The diagrams show the three stages.

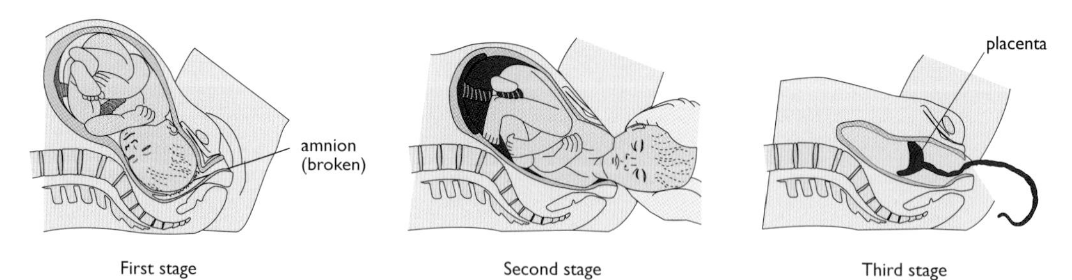

(a) **Explain** what is happening at the First Stage and why these events need to occur.

(Guidance: Your answer should be approximately 15 lines.)

(b) **Explain** what is happening at the Second Stage.

(Guidance: Your answer should be approximately 4 lines.)

(c) **Explain** why the Third Stage is necessary.

(Guidance: Your answer should be approximately 8 lines.)

(d) **Explain** how oxytocin helps the process of birth.

(Guidance: Your answer should be approximately 10 lines.)

Part Five

CARBON CYCLING

18 The carbon cycle

Until a few years ago, one only read about carbon in chemistry and biology textbooks. Nowadays it is unusual for a week to go by without some reference to carbon in connection with global warming or climate change, real or imagined, natural or human induced.

So what is the fuss all about? It is a complex subject, and before we get into details, we need to look at the way the element carbon 'circulates' in nature. This is the *carbon cycle*, and is shown in extremely simplified form in Fig. 18.1.

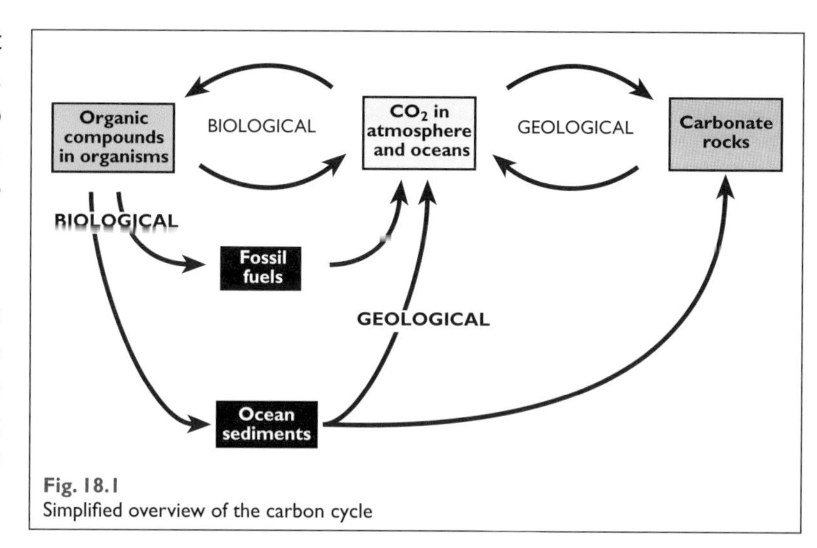

Fig. 18.1
Simplified overview of the carbon cycle

Carbon compounds

If you remove all the water from a cell, about half of what is left is carbon, in carbohydrates, proteins, fats and nucleic acids (e.g. DNA). In these compounds at least some of the carbon atoms are joined to hydrogen atoms; these compounds are *organic*. Carbon compounds in which the carbon is combined with oxygen are said to be *inorganic*, for example the calcium carbonate that makes up the skeletons of many marine (sea-living) animals, and CO_2. Inorganic carbon compounds are thus fully *oxidised* and cannot be burnt.

In nature, carbon circulates, meaning that it moves between various 'compartments' such as the atmosphere, oceans, rocks, soil, sediments and living organisms. In its 'journey', carbon forms part of a variety of compounds, which may be organic or inorganic.

The cycling of carbon actually involves two cycles; one rapid, the other extremely slow:

1. A *biological* cycle involving *photosynthesis* and *respiration*, in which carbon circulates rapidly over periods that can be as short as a few days to as long as thousands of years.
2. A *geological* cycle in which the carbon remains oxidised throughout and circulates over a time scale of millions of years.

The amount of carbon passing around the biological cycle each year is about 1000 times greater than the amount of carbon passing around the geological cycle.

The two cycles are not completely separate; as explained below, **fossil fuels** originate in once-living organisms, but their bodies are later subjected to geological processes. Taken together, they form a **biogeochemical cycle**.

Where is the carbon?

Carbon exists in various 'compartments':

- The atmosphere, mainly as CO_2 at a concentration of about 390 parts per million by volume (ppmv). There are also traces of methane (CH_4).
- The ocean, as CO_2 in physical solution, but mainly as the hydrogen carbonate (HCO_3^-) and carbonate (CO_3^{2-}) ions. The surface layers, warmed by the sun and in which photosynthesis occurs, mix only very slowly with the deeper layers.
- Land plants.
- Dead organic matter in the soil.
- Carbonate rocks (by far the largest compartment).
- Sediments at the bottom of the ocean.

ISBN 9780170191340

- Fossil fuels such as oil, coal and natural gas.

The amounts vary enormously, as indicated in Fig. 18.9.

The biological carbon cycle

The organic cycle is mainly due to two biological processes: photosynthesis and respiration.

1. Plants obtain their carbon from CO_2 in **photosynthesis**:

$$\text{light} + \text{carbon dioxide} + \text{water} \xrightarrow{\text{chlorophyll}} \text{carbohydrate} + \text{oxygen}$$

2. The carbohydrate is used to produce other biological molecules such as proteins, fats and nucleic acids. Some of the carbohydrate is oxidised in **respiration** to provide energy for the plant:

$$\text{carbohydrate} + \text{oxygen} \longrightarrow \text{carbon dioxide} + \text{water} + \text{energy}$$

Sooner or later, the plant dies and its organic compounds are used by **heterotrophs**. These are organisms (animals, fungi and most bacteria) that cannot make their own organic matter, and so depend on plants for their energy. Food moves from autotrophs along a series of heterotrophs in a **food chain**. Figure 18.2 shows a terrestrial (land) food chain, but of course there are food chains in oceans and lakes.

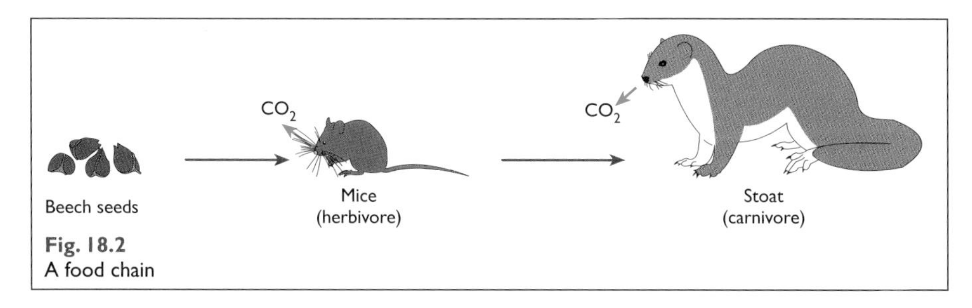

Fig. 18.2
A food chain

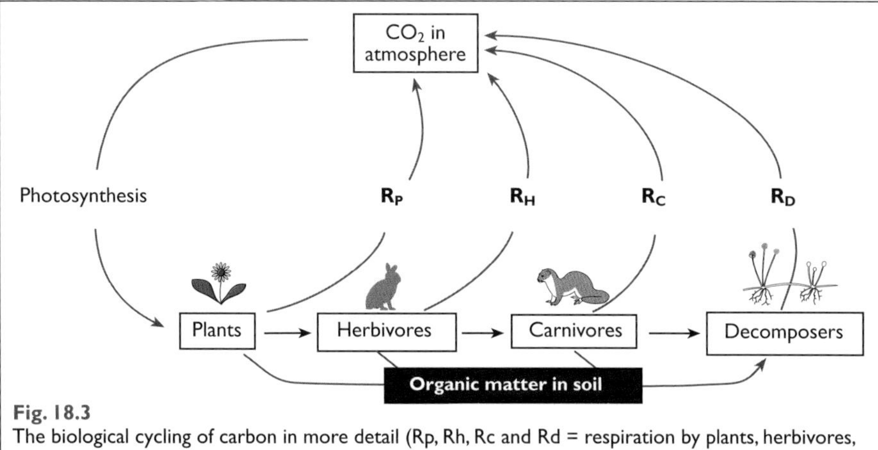

Fig. 18.3
The biological cycling of carbon in more detail (Rp, Rh, Rc and Rd = respiration by plants, herbivores, carnivores and decomposers)

At each link in the chain some of the organic matter is built up into complex organic compounds in growth, but most is converted to CO_2 and water in respiration. The result is that by the end of the food chain, practically all the carbon in the organic matter made by plants has been converted back to CO_2, to be used again in photosynthesis, as shown in more detail in Figure 18.3.

The time scale for carbon cycling can be quite short — leaves may be eaten and their carbon returned to the atmosphere as CO_2 within days of the leaves being produced. On the other hand, it can take a thousand years or more for the carbon in a tree trunk to convert back to CO_2.

Carbon cycling in the oceans differs in a number of ways from cycling on land:

- The producers on land may have huge total biomass (for example forest trees), but the producers in the ocean — the **phytoplankton** — are tiny, single-celled plant-like organisms. Even though they may be extremely numerous, their total biomass is small.
- Whereas land plants grow relatively slowly, under ideal conditions phytoplankton can divide (and thus double its mass) every 24 hours or so. Even the fastest growing terrestrial flowering plant has a generation time of several weeks. They therefore make up for their small biomass by their high growth and reproductive rates.

Sooner or later all organisms in the upper, sunlit layers die and sink to the bottom, or are eaten by other organisms that die and sink to the bottom. This constant rain of organic matter transports carbon from the surface to the ocean bed, where it forms *sediments*. Eventually, over millions of years, these sediments enter the geological carbon cycle in ways described below.

ISBN 9780170191340

This removal of carbon (as organic matter) from the surface water of the ocean is compensated by atmospheric CO_2 dissolving in the ocean. There is thus a (very slow) transport of carbon from the atmosphere to the seabed (Fig. 18.4).

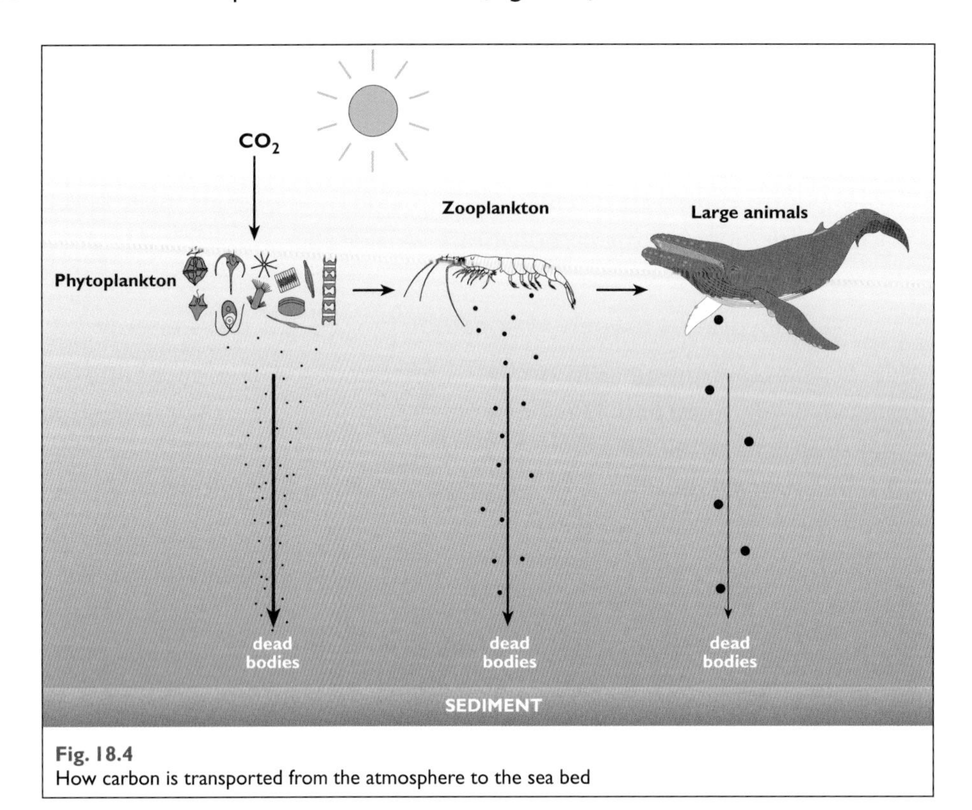

Fig. 18.4
How carbon is transported from the atmosphere to the sea bed

In most ecosystems the rates of photosynthesis and respiration are roughly equal when measured over an entire year. This is not the case over shorter intervals. At night, photosynthesis stops altogether but respiration continues, so the level of CO_2 in the atmosphere rises very slightly. During the summer, photosynthesis exceeds respiration, so the CO_2 level falls slightly. In winter, when many plants stop photosynthesising altogether, animals and other heterotrophs are feeding off what the plants have made during the previous summer. Respiration therefore exceeds photosynthesis and CO_2 level rises slightly.

Until humans began to modify the environment on a big scale, the rates of photosynthesis and respiration were approximately equal over an entire year, and hence the concentration of CO_2 in the atmosphere changed very slowly. Its concentration has been much higher than it is today even millions of years ago (see below), but as far as is known it changed much more slowly than it is changing today.

The geological carbon cycle

To understand the geological cycle, we need to know a bit about the theory of **plate tectonics** ('continental drift'). This theory (now fact) holds that the Earth's crust consists of a series of separate tectonic plates, floating on a liquid **magma**. Where two plates meet, one may slide past its neighbour in a series of jerks, each jerk being an earthquake. Alternatively, one plate may slide underneath another in a process called **subduction**. Above the subduction zone magma may from time to time burst out of the crust in a volcanic eruption. In the geological cycle, carbon can occupy various 'compartments' (Fig. 18.5), for example:

- In the *atmosphere*, as CO_2 and (as a trace) methane, CH_4.
- In the *ocean*, in solution or as hydrogen carbonate (HCO_3^-) ions.
- In *sediments* at the bottom of the sea. These may be *inorganic*, consisting of the limy skeletons of marine organisms, or they may be organic, consisting of the dead bodies of organisms (Fig. 18.5).
- In *limestone* (calcium carbonate, $CaCO_3$).

ISBN 9780170191340

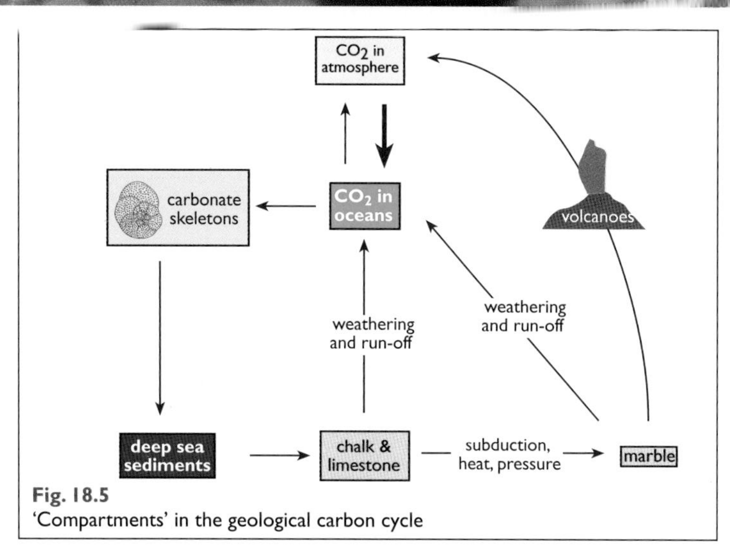

Fig. 18.5
'Compartments' in the geological carbon cycle

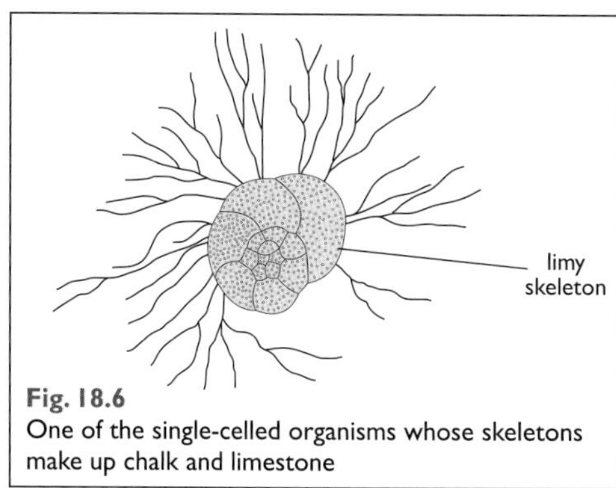

Fig. 18.6
One of the single-celled organisms whose skeletons make up chalk and limestone

The atmosphere and ocean are in contact at the ocean surface, and CO_2 can diffuse between them. At the moment the atmospheric CO_2 concentration is increasing. Much of this 'new' CO_2 dissolves in the upper layer of the ocean, where there is enough light for photosynthesis.

Sources and sinks

For the past 100–200 years CO_2 has been dissolving in the ocean from the atmosphere faster than it is leaving it. The ocean is acting as a **sink**, in which the amount of CO_2 is increasing. Gases are less soluble in warm water than in cold water, so if the ocean were to warm up substantially, CO_2 would be driven out of solution and the oceans would then be acting as a **source**. This is believed to have happened numerous times in the past, as will be explained. Another source of CO_2 is the burning of fossil fuels (see next page).

The amount of carbon in the various compartments varies hugely. For example there is about 100,000 times more carbon in limestone rock than there is the atmosphere. Many marine organisms such as corals make skeletons of lime (calcium carbonate). The most important are the untold numbers of some planktonic animals, such as those shown in Figure 18.6. When they die, their skeletons sink down and form thick layers of sediment on the ocean bed. Over millions of years these sediments become compacted and form a **sedimentary rock** such as chalk or limestone.

Over more millions of years, sedimentary rock may undergo further changes:

- As a result of movements of the Earth's crust (the outermost layer of the Earth), they may rise above the sea and even be thrust up as mountains where it is exposed to rainwater. Falling rainwater dissolves atmospheric CO_2 and becomes very dilute *carbonic acid*. Though only a weak acid, it slowly dissolves the lime to form a solution of *calcium hydrogen carbonate*. This process is called **weathering**, and is how limestone caves are formed. In some caves the calcium hydrogen carbonate solution evaporates, precipitating the limestone again as downward-pointing *stalactites* or upward-pointing *stalagmites*.
- Over millions of years the hydrogen carbonate ions washed down to the sea bed may be used to build limy animal skeletons again.
- It may be buried more deeply and, as a result of heating and great pressure, be converted to *marble*. This is a **metamorphic** rock because it has *metamorphosed* (changed) from one kind of rock to another. The marble may subsequently be uplifted above sea level and become eroded in the same way as limestone.
- Where one tectonic plate sinks below an adjacent plate it enters the Earth's mantle, where it is strongly heated and converted into molten magma. In this process the limestone reacts with other minerals and produces CO_2, which may be expelled in *volcanic eruptions*. When magma cools it forms **igneous** rock (Latin: *ignis* = 'fire').

Whichever of these three routes the carbon takes, it eventually is returned to either the atmosphere or the ocean (Fig. 18.7).

ISBN 9780170191340

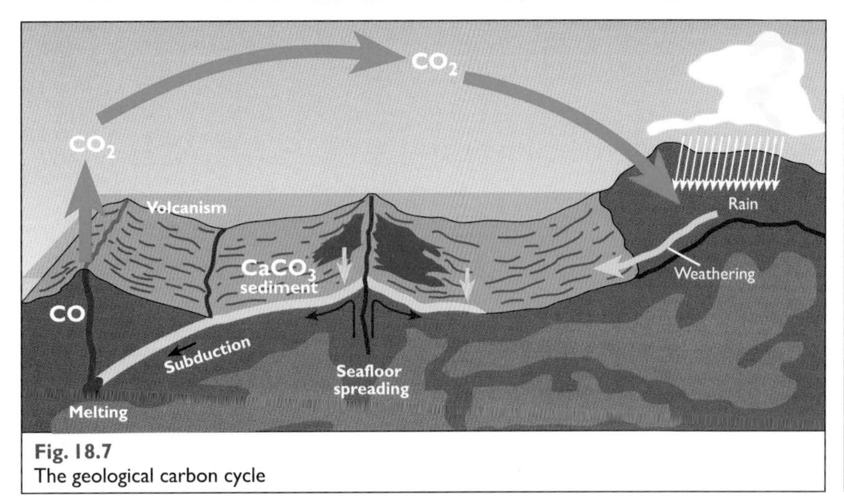

Fig. 18.7
The geological carbon cycle

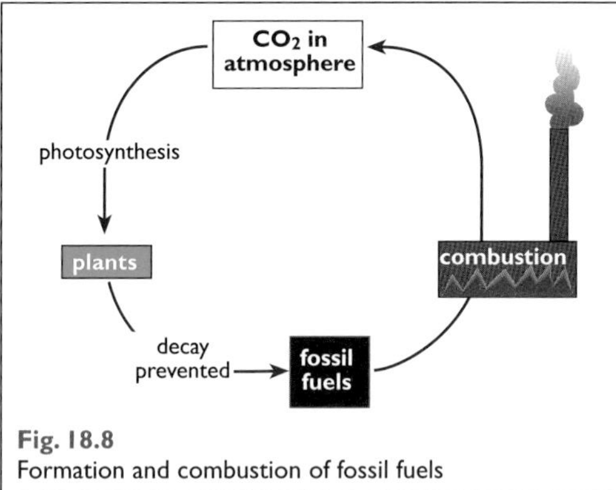

Fig. 18.8
Formation and combustion of fossil fuels

Fossil fuels

The geological and biological cycles are not completely separate, but are linked by CO_2 (in the air and dissolved in sea water), which takes part in both. Fossil fuels (coal, oil and natural gas) are another way in which they are linked; they are formed in the Earth's crust and form over geological time, but start off as living matter. The chemical energy stored in fossil fuels was originally solar (sun) energy that was trapped in photosynthesis millions of years ago.

Under certain conditions, decay of organic matter is prevented and the organic matter is slowly — over millions of years — converted to fossil fuel. In the mid 18th century, coal began to be used to provide the energy that made the industrial revolution possible. More recently oil has become dominant and, even more recently still, natural gas (Fig. 18.8).

Coal

In places such as bogs and marshes, in which conditions are anaerobic (lacking oxygen), dead plant material does not decay and is slowly converted to **peat**. If peat is buried deeply by sediment, it becomes compacted and water is squeezed out. If buried deeply enough it becomes heated and slowly converted via a series of increasingly energy-rich materials. First it forms *lignit*e (brown coal), then *bituminous coal* and finally the highest quality coal, *anthracite*. Coal formation was particularly characteristic of the Carboniferous period 360–300 million years ago.

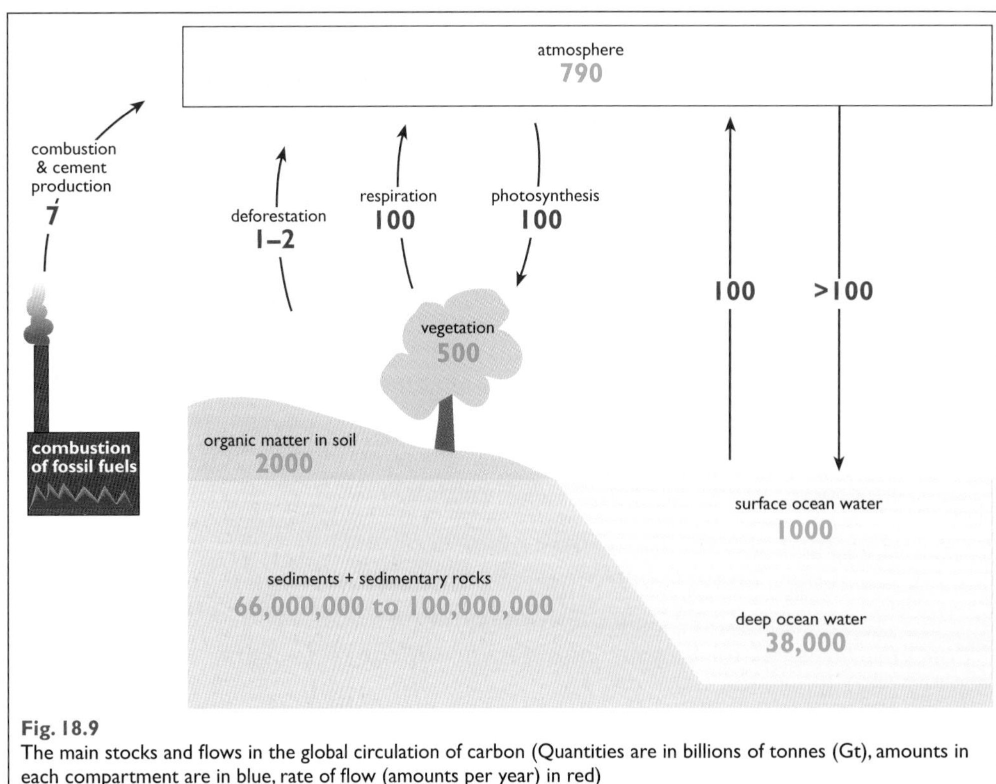

Fig. 18.9
The main stocks and flows in the global circulation of carbon (Quantities are in billions of tonnes (Gt), amounts in each compartment are in blue, rate of flow (amounts per year) in red)

ISBN 9780170191340

Oil

Oil was formed from *phytoplankton*, which consists of tiny, plant-like drifting organisms. At certain times in the past, in particular 90–150 million years ago, vast quantities of phytoplankton sank to the bottom of the sea under anaerobic conditions and were buried under sediment. The heat present at these depths slowly converted the organic matter to crude oil or *petroleum*.

Natural gas

Natural gas is formed in a similar way to oil, except that temperatures required are higher, so it tends to form deeper down. Natural gas is mainly methane (CH_4), with smaller quantities of ethane (C_2H_6) and propane (C_3H_8).

In any given year the amount of organic matter that forms fossil fuel is minute, but over millions of years the quantities are huge. Even greater amounts of organic sediments fall to the bottom of the sea and are not converted into fossil fuel.

An important feature of natural gas is that when burned, it produces 43% less CO_2 per kJ of energy released than coal, and 30% less than oil.

Summary of key points in this chapter

- *Carbon* is the key element in living matter and in all organic compounds.
- Carbon takes part in a biogeochemical cycle, in which it moves through a series of 'compartments'.
- The amount of carbon in each 'compartment' varies greatly, the largest being over a million times greater than the smallest.
- There are two cycles: A geochemical cycle in which carbon circulates over a timescale of millions of years, and a biological cycle in which carbon circulates over a timescale of days to centuries.
- In the biological cycle, autotrophs remove CO_2 from the atmosphere and sea in photosynthesis, while both autotrophs and heterotrophs add CO_2 to the atmosphere and seawater in respiration.
- Globally, over an entire year the rates of photosynthesis and respiration are approximately in balance.
- A 'compartment' in which carbon is entering faster than it is leaving is called a *sink*.
- A 'compartment' in which carbon is leaving faster than it is entering is called a *source*.
- Dead planktonic organisms with skeletons of lime (calcium carbonate) fall onto the seabed and, over thousands of years, form thick layers of carbon-containing sediment.
- By compression in the Earth's crust, sediments may be slowly converted into *sedimentary rock* such as limestone.
- Sedimentary rock that becomes deeply buried may be sufficiently heated to be changed into *metamorphic rock* such as marble.
- Sediments may slowly become buried in the Earth's crust by *subduction*, where the edge of a tectonic plate is moving beneath a neighbouring plate.
- As a result of subduction, solid crust is converted into liquid magma, which may later be ejected to the Earth's surface in volcanic eruptions. When magma cools it forms *igneous rock*.
- In some cases thick layers of dead organisms accumulate in anaerobic conditions, preventing decay. These layers may become buried and, over millions of years, converted to *fossil fuels* (coal, oil or natural gas).

ISBN 9780170191340

Test your basics

Copy and complete the following sentences.

a) Carbon is an essential constituent of all __*__ compounds, in which at least some of the carbon is bonded to __*__. Carbon also exists as __*__ compounds in which the carbon is bonded to __*__ for example CO_2 and calcium carbonate, in which the carbon is thus fully oxidised.

b) The ultimate source of all organic compounds in nature is the process of __*__, in which CO_2 is converted by plants to __*__ compounds using __*__ energy from the sun. Plants use some of this organic matter as a source of energy in __*__ by oxidising it back into CO_2. Animals and other organisms that feed off the plants, and also the decomposers, use some of the organic matter in __*__, converting more of it back to CO_2. Eventually, as the organic matter passes along a food __*__, almost all of it is oxidised back to CO_2.

c) Organisms that can make their own organic matter from CO_2 are __*__; those that cannot are __*__. These two kinds of organism are interdependent; __*__ depend on __*__ for their supply of organic matter, and __*__ depend on __*__ for their supply of CO_2 and other inorganic raw materials.

d) In certain parts of the world, some dead organisms do not decay but become buried under __*__. This has happened many times in the remote past. As the __*__ accumulated it was slowly converted to rock. Over long periods of time and under the effect of high temperatures and pressure, the organic matter was converted to __*__ fuels such as coal, __*__, and __*__ gas.

e) Over the course of a year and before the industrial revolution, the global rates of photosynthesis and respiration were almost __*__, so the concentration of __*__ in the atmosphere changed very little. Since the 18th century, fossil fuels have been burned to provide __*__ for __*__ activity. This has had the effect of __*__ the concentration of __*__ in the atmosphere.

f) The rise in atmospheric CO_2 is less than might be expected because a considerable amount has been dissolving in the ocean, causing its pH to __*__. Because CO_2 is entering the ocean faster than it is leaving, the ocean is acting as a __*__.

g) There have been times in the past when, as a result of changes in the Earth's __*__, the world's climate, and hence the oceans, have warmed. This has caused CO_2 to be driven out of solution into the atmosphere, raising its concentration. At such times the ocean has been acting as a __*__.

ISBN 9780170191340

Humans affect the cycle

'Prediction is very difficult, particularly about the future.'

Neils Bohr, Nobel prizewinning physicist

Humans have been responsible for a rapid increase in the concentration of atmospheric CO_2 in two ways:

1. By cutting down forests and oxidising the timber, either by burning it or allowing it to decay.
2. By burning fossil fuels to provide energy. The production of fossil fuel took place over millions of years, but a substantial proportion of this has been re-converted back to CO_2 in the last two centuries.

At the moment there is a huge controversy about this; some people think that the increase in atmospheric CO_2 is changing the climate, with possible catastrophic consequences for future generations. Others think that there is little cause for concern and that we can continue with 'business as usual'.

Figure 18.9 (on page 192) shows a simplified picture of the various flows of carbon into and out of the atmosphere, together with some of the main compartments involved. Exact figures are uncertain, and different sources give slightly different values. The amount present in a compartment at any one time ('stock') is shown in blue. The amount moving from one compartment to another each year ('flow') is shown in red. Flows that are only significant over millions of years are not shown.

The most important things to note are:

- The amount of CO_2 in the atmosphere is small compared with most of the other compartments.
- The amount of CO_2 added to the atmosphere by human activity is relatively small compared with the amounts added and removed by living organisms, but it is sufficient to produce a steady increase in atmospheric CO_2.
- The rate at which CO_2 enters the ocean from the atmosphere is slightly greater than the rate at which it moves in the reverse direction. The difference means that the ocean is acting as a *sink* for CO_2.
- Although the deep ocean contains far more carbon than the upper, sunlit layers, mixing between them is over a timescale of thousands of years.

The Earth's heat blanket — the 'greenhouse' effect

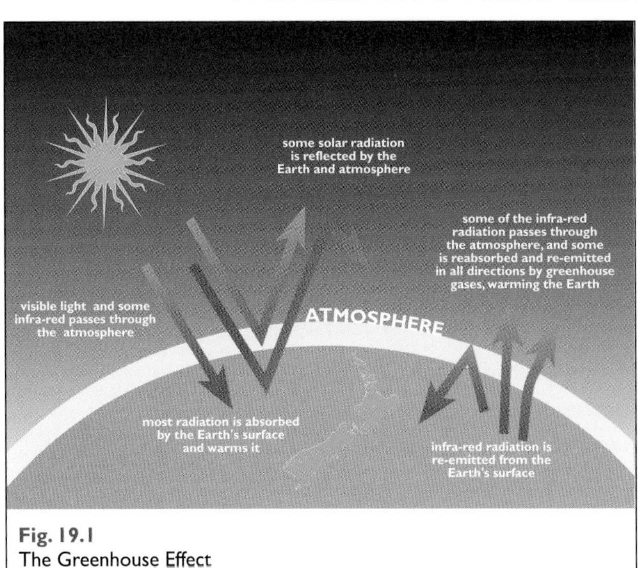

Fig. 19.1
The Greenhouse Effect

Although CO_2 accounts for less than 0.04% of the atmosphere today, it helps keep us warm by trapping **infra-red** (heat) radiation by the so-called **'greenhouse' effect**. In simple terms, what happens in an actual greenhouse is this; light passes through the glass and is absorbed by the interior, which gains energy. This extra energy is then re-radiated (given out again) as infra-red radiation. Glass is semi-opaque to heat (does not transmit it well), so the heat energy is trapped inside the glasshouse, making it warmer. (In reality, most of the warming in a glasshouse is due to the glass preventing warm air from escaping.)

The atmosphere acts in a similar way, due to the presence of **greenhouse gases** — mainly *water vapour* and *carbon dioxide*. Both these gases are transparent

ISBN 9780170191340

to light but relatively opaque to infra-red radiation, and trap outgoing heat. As a result the average temperature of the Earth's surface over a year is about 30 °C warmer than it would be without this atmospheric 'heat jacket' (Fig. 19.1). Because carbon dioxide has probably always been part of the atmosphere, the greenhouse effect is a *natural* effect.

Another greenhouse gas: Methane

Besides carbon dioxide, methane is another gas that is causing concern, particularly in New Zealand. This is because methane is produced by microorganisms living in the rumen of sheep and cattle, which are of great importance economically. World wide, methane is also produced in anaerobic respiration by microorganisms in rice paddy fields, the guts of termites, bogs and marshes and organic matter in landfills.

Although the concentration of methane in the atmosphere is only about 1.8 parts per million, each methane molecule is 30 times more potent in trapping heat than each CO_2 molecule. On the other hand, it only has an atmospheric 'shelf life' of about eight years since it is fairly rapidly oxidised to CO_2.

Methane is also present on the ocean bed as solid, ice-like **methane hydrate**. This is only stable at the cold temperatures and high pressures that exist on the ocean bed; as soon at it is brought to the surface it fizzes and breaks down to methane. The quantities of methane hydrate are believed to run into the hundreds of billions of tonnes, exceeding all the fossil fuels put together.

Atmospheric carbon dioxide is increasing rapidly

Since 1958, atmospheric CO_2 concentrations have been regularly measured at a station on Mauna Loa, a volcano on Hawaii. The station is high up (3397 m) and far from major human settlements. The increase in CO_2 is shown in Fig. 19.2. The graph shows three important features:

1. The concentration of CO_2 has steadily increased from 315 ppmv (parts per million by volume) in 1958 when measurements began, to about 387 ppmv in 2010.
2. Each summer the CO_2 level decreases slightly as photosynthesis exceeds respiration, and increases slightly each winter as respiration exceeds photosynthesis.
3. The rate of increase is increasing.

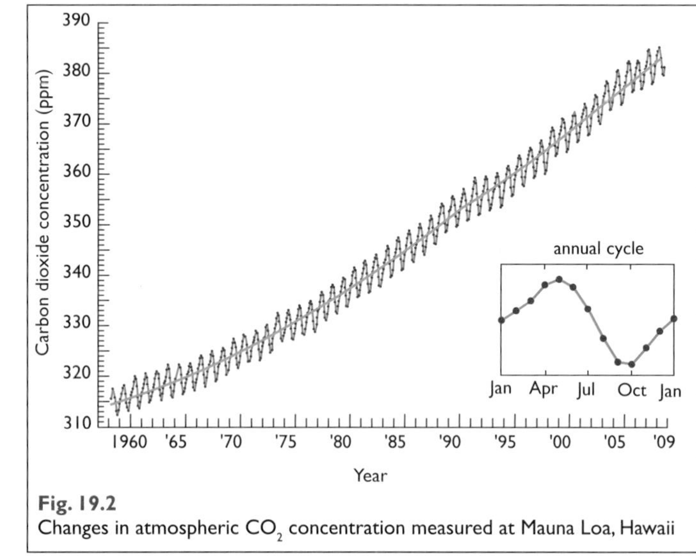

Fig. 19.2
Changes in atmospheric CO_2 concentration measured at Mauna Loa, Hawaii

Many scientists are concerned that the increase in atmospheric CO_2 is warming the Earth's climate.

The ocean as a sink for carbon dioxide

Atmospheric CO_2 is actually increasing only about half as quickly as would be expected from the known amounts of fossil fuel being burned. The difference is due to CO_2 dissolving in the oceans, which are thus a *sink* for CO_2.

One of the effects of this increasing atmospheric CO_2 is a reduction in the pH of the seas (*acidification*). CO_2 dissolves in water to form a weak acid, *carbonic acid*:

$$CO_2 + H_2O \rightarrow H_2CO_3$$

Since the beginning of the industrial revolution the pH of the oceans has decreased from 8.179 to 8.10. This seems very small, but the pH scale is *logarithmic*; it represents approximately a 25% increase in hydrogen ion concentration. Many scientists are concerned about this because of the possibility that it would interfere with the ability of marine organisms such as corals and 'shellfish' (molluscs and crustaceans) to produce their skeletons of calcium carbonate. Millions of people depend on these creatures for food.

ISBN 9780170191340

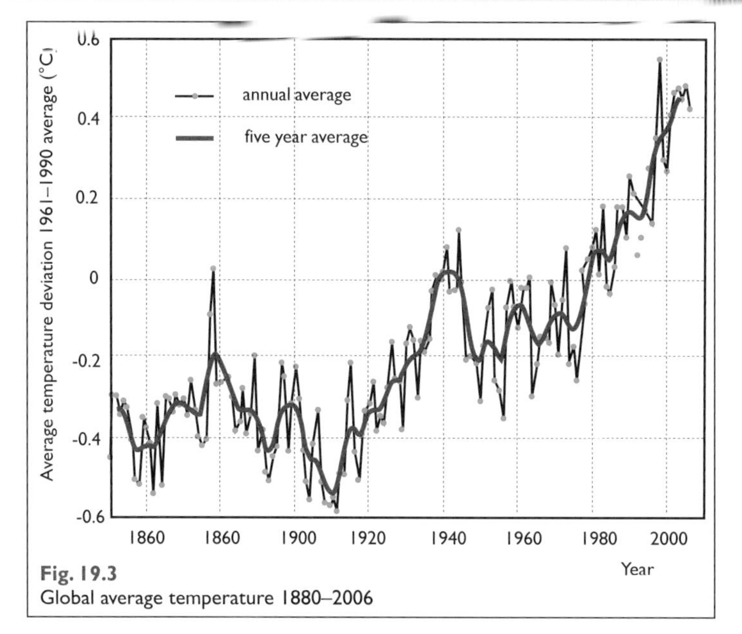

Fig. 19.3
Global average temperature 1880–2006

Are global temperatures increasing?

Figure 19.3 shows average global temperatures from 1860–2006, as presented by NASA (US National Aeronautics and Space Administration). From the mid 20th century the world has been getting warmer.

Positive feedback effects

A positive feedback effect is a 'vicious cycle'. It occurs when the result of a process acts to speed up that process. It is a *self-accelerating* process and is therefore *destabilising*. Here are some possible examples (Fig. 19.4):

- Polar ice absorbs less than 20% of solar radiation, most being reflected back into space. The ocean, on the other hand, absorbs up to 90% of solar radiation. In technical language, the **albedo** or reflectivity of ice is much higher than that of water. As a result, if the area of polar ice decreases, more heat from the sun is absorbed, causing further melting. This in turn increases the amount of energy absorbed, and so on.
- The Arctic *permafrost* is ground that has been frozen for thousands of years. If the permafrost were to melt, the massive amounts of organic matter in the soil would decay, producing CO_2, which could cause further warming, and so on.
- Warming of the oceans would reduce the solubility of CO_2 in the sea, reducing the ability of the ocean to act as a sink. The result would be to accelerate the rise in atmospheric CO_2, causing further warming. Eventually, the warming of the ocean could begin to drive off CO_2 back into the atmosphere. In this situation the oceans could begin to act as a *source* for CO_2.
- The melting of Arctic permafrost causes the decay of organic matter and releases methane. Added to that are the enormous quantities of methane hydrate on the ocean floor. A small rise in temperature could cause it to bubble off as methane gas. Methane is a much more potent greenhouse gas than CO_2, and its release would lead to further warming. Potential quantities that could be released are enormous, and some scientists are concerned that this could trigger a 'runaway' feedback cycle.

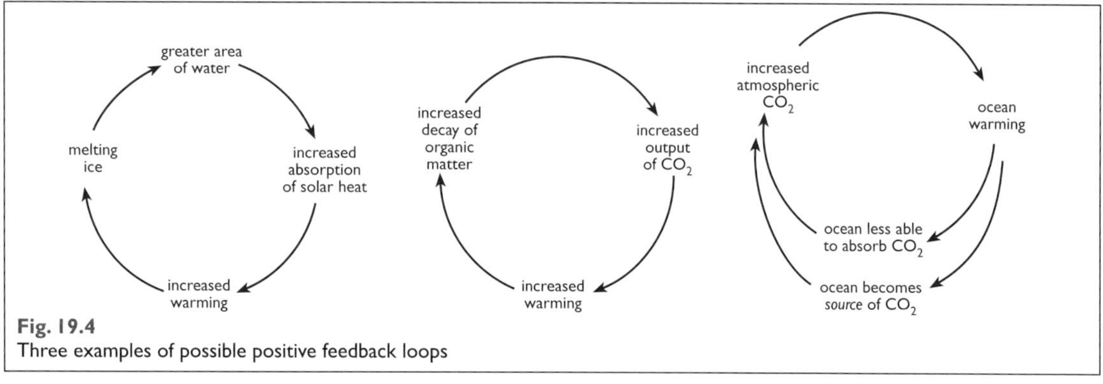

Fig. 19.4
Three examples of possible positive feedback loops

Negative feedback effects

If the climate were to get warmer, the atmosphere would hold more water vapour. In clear skies this would lead to greater greenhouse warming (positive feedback), but there could also be more clouds formed. Clouds reflect solar radiation, and in this way would reduce warming.

Effects on climate: Temperature, rainfall patterns

Many climate scientists believe that the consequences of global warming are likely to be more complicated than just an increase in average temperatures. It is considered that:

ISBN 9780170191340

- Warming is likely to be greater toward the poles than near the equator.
- Weather is likely to be more variable with both low and high temperature extremes (though the highs exceeding the lows), as well as extremes in rainfall amounts, both high and low. The distribution of rainfall is likely to change, so that some areas become wetter on average, and others drier.

If predictions for rainfall come true, the implications for agriculture would be serious.

Effects on sea levels

Over the longer term, melting of Greenland and Antarctic ice, together with the thermal expansion of the seawater, is likely to raise sea levels. The melting of the Greenland ice cap is estimated to raise sea level by about six metres, and a similar increase would result from the melting of the West Antarctic ice cap. If the East Antarctic ice cap were to melt as well, the total rise in sea level is estimated to be about 70 m. Ice that is floating, such as the Arctic ocean ice cap, would have no effect on sea level.

A sizeable proportion of the world's population lives within 100 m or so of sea level. London, Shanghai, New York and all coastal cities would be at least partly flooded.

However, the complete melting of the polar ice caps would take centuries, so should we be concerned? There are three reasons why we should be giving the matter serious thought:

1. Some countries are literally only a few metres (or less) above sea level, such as the Maldives and parts of Bangladesh. These populations would be threatened within decades rather than centuries.
2. World population is increasing at about 80 million per year (about the population of Germany), and food production is heavily dependent on climate.
3. Because of the possibility of positive feedback effects, we could reach a situation in which warming becomes unstoppable, no matter what action is taken.

The controversy

As the evidence for climate change has accumulated, the intensity of the argument has increased. Entire industries have grown up on the assumption that their waste products can be absorbed by the atmosphere and oceans without harm to the environment. The suggestion that there are limits to growth is not good news to industrialists, shareholders and politicians. As the American novelist Upton Sinclair put it:

'It is difficult to get a man to understand something if his salary depends on his not understanding it.'

In 2008, Kendall Zimmerman of the University of Illinois published the results of a survey of Earth scientists, in which they were invited to answer two questions:

1. When compared with pre-1800s levels, do you think that mean global temperatures have generally risen, fallen, or remained relatively constant?
2. Do you think human activity is a significant contributing factor in changing mean global temperatures?

The number surveyed was 10,257, of whom 3146 (30.7%) responded. The scientists were mainly from the United States; the remainder were from 22 other nations. Among the Earth scientists were geochemists (15.5%), geophysicists (12%), oceanographers (10.5%) and climatologists (5%).

Overall, 90% of the participants answered 'risen' to question 1, and 82% answered 'yes' to question 2. The results are shown in Fig. 19.5. They indicate that as the level of climate expertise increases, so does agreement with the two questions. Of the climate specialists, 96.2% answered 'risen' to question 1, and 97.4% answered 'yes' to question 2. In contrast, a recent Gallup poll indicated that only 58% of the general public would answer 'yes' to question 2.

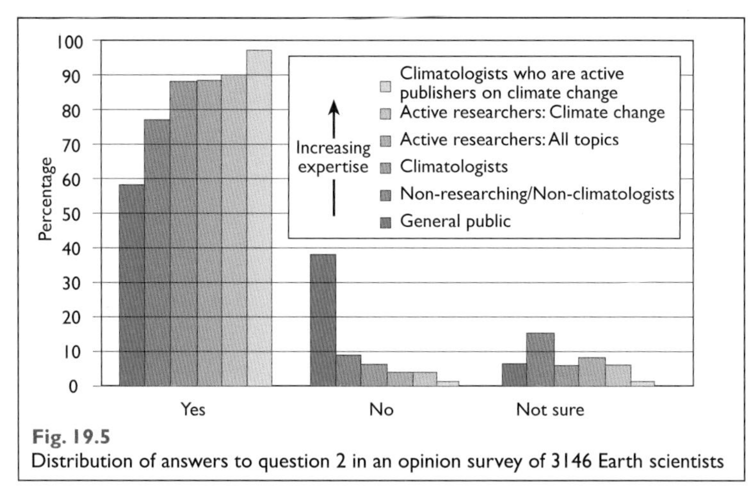

Fig. 19.5
Distribution of answers to question 2 in an opinion survey of 3146 Earth scientists

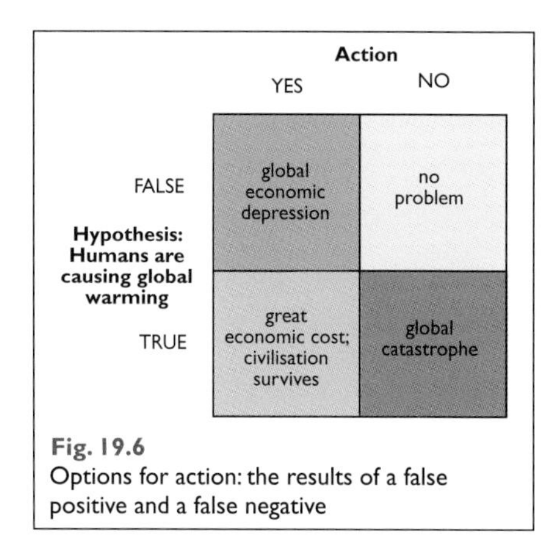

Fig. 19.6
Options for action: the results of a false positive and a false negative

What should be done?

Given that some (2.6%) of the climate scientists who responded were not convinced that human activity is changing the climate, it can be argued that there is still some doubt. What, then, should be done? We have to decide between two kinds of error:

1. We take action which turns out to be unnecessary ('false positive').
2. We take no action but it turns out that it was necessary ('false negative').

Our situation is represented in Figure 19.6. In the diagram there are two lines and two columns:

- The lines indicate two possible *situations*, we do not yet have final proof that either is true.
- The columns indicate our possible *actions*, over which we *do* have control.

Given these two possible situations, and two possible actions for each, there are four possible outcomes:

1. We take action that turns out to be unnecessary. There would be considerable economic cost, but possibly some benefits — like a reduction in air pollution that would result from greatly decreased use of fossil fuels.
2. We take no action and it was not necessary anyway. There would be neither costs nor benefit.
3. We take action and it turns out that global warming was occurring and was due to human activity. There would be considerable economic dislocation, but we would have preserved the world as a habitable place for our children and grandchildren.
4. We take no action and suffer the consequences of global warming; extreme weather such as drought, crop failures, storms and, in the longer term, flooding of New York, London, Shanghai and numerous other major cities.

It really boils down to the question of whether you would ensure your house against fire, or whether public buildings should have fire escapes. Though these measures cost money, they ensure against something infinitely worse.

This is also the way certain politicians think — on other issues. The *New York Times* (8 December 2009) reported that after 9/11 the US Vice-President Richard Cheney expressed concerns that a Pakistani scientist was offering nuclear weapons expertise to Al Qaeda. Cheney reportedly said: 'If there is a 1% chance that Pakistani scientists are helping Al Qaeda build or develop a nuclear weapon, we have to treat it as a certainty in terms of our response.' Cheney took the view that what the US was faced with was a low-probability, high-impact event.

In other words, Cheney was advocating the 'precautionary principle'. Yet when confronted with climate change, which has a probability considerably greater than 1%, the thinking changes. Why?

ISBN 9780170191340

Extension: Common arguments against human-caused climate change

A relatively small number of scientists (and a large proportion of industrialists and politicians) argue that there is no cause for concern. They put forward a number of arguments, the most frequent of which are detailed below.

'Human activities are only a minor source of atmospheric CO_2'

Indeed they are, but natural flows of CO_2 into the atmosphere are almost exactly balanced by natural flows out of the atmosphere. It only needs a small 'anthropogenic' (human-made) input to disturb this delicate balance.

'Last winter was the coldest for ten years'

This common 'argument' fails to distinguish between 'climate' and 'weather'. 'Climate' relates to conditions over the long-term, such as decades or centuries. 'Weather' relates to the short-term, such as days, weeks or months. There may be freakishly cold periods even in a period of climatic warming, and vice-versa.

'Water vapour is a more important greenhouse gas than CO_2 is'

This is a misleading half-truth. Water vapour *does* absorb infra-red radiation to a considerably greater extent than CO_2 — partly because it is present in far higher concentrations. There are, however, two further points:

1. The amount of water vapour that the atmosphere can hold is determined by temperature. If all the water were suddenly removed, within a few weeks it would be restored by evaporation. Likewise, if the water vapour concentration were suddenly increased, the surplus would condense and fall as rain or snow. At any given temperature then, there is a fixed amount of water vapour that the atmosphere can hold.
2. Carbon dioxide absorbs some wavelengths that water vapour does not. This extra absorption warms the air a bit more, enabling the atmosphere to hold more water vapour, *increasing* the warming effect of water (Fig. 19.7). The effect of water vapour is to approximately double the warming effect of CO_2 alone. Water is therefore acting as part of a *feedback* (secondary or indirect effect) mechanism rather than a *forcing* (primary effect) agent.

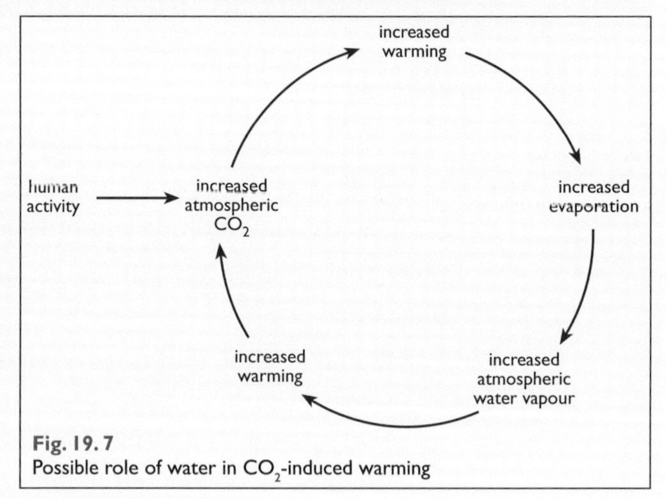

Fig. 19.7
Possible role of water in CO_2-induced warming

Although CO_2 is only responsible for 9% of the greenhouse effect (compared with 26% for water), it is the only one of the two gases that can vary in a way that *could* drive global warming.

'The climate has always been changing'

Indeed it has. We know this from analysis of gases trapped in ice cores drilled from Antarctica, a continent covered with ice that in some parts is over 4 km thick. The ice is being continually added to from above by snowfall. At the same time, the ice moves towards the sea, where it breaks off and forms icebergs.

Scientists have been able to take samples of Antarctic ice by drilling long cylindrical cores and analysing the contents. The deeper the ice, the longer ago it was formed from snow.

In addition to measuring the concentration of trapped CO_2, scientists can get information about temperature changes. They do this by measuring the proportions of two isotopes of oxygen — the normal ^{16}O and the heavier isotope ^{18}O.

Figure 19.8 shows how the temperature and atmospheric CO_2 concentrations have changed over the last 800,000 years. During this time, the Earth has undergone cyclical changes in temperature and atmospheric CO_2 concentrations, with about 100,000 years between peaks. These changes are believed to be due to cyclical changes in the Earth's orbit and rotation, which affect the amount of solar radiation reaching the planet.

Though it is not obvious from the graph, the temperature changes precede the changes in CO_2 by about 800 ± 200 years, suggesting that temperature changes have caused the changes in CO_2.

ISBN 9780170191340

One hypothesis as to how it works is as follows:

- Changes in the Earth's orbit and rotation result in an increase in solar radiation.
- The oceans begin to warm up, decreasing the solubility of CO_2 in the ocean.
- CO_2 is driven out of the ocean into the atmosphere.
- Increased atmospheric CO_2 warms the climate further, driving more CO_2 into the atmosphere in a positive feedback cycle.
- At some point, changes in the Earth's orbit and rotation reduce the incoming solar radiation. This causes cooling, reversing the positive feedback cycle; cooler oceans can now act as a CO_2 sink, producing further cooling.

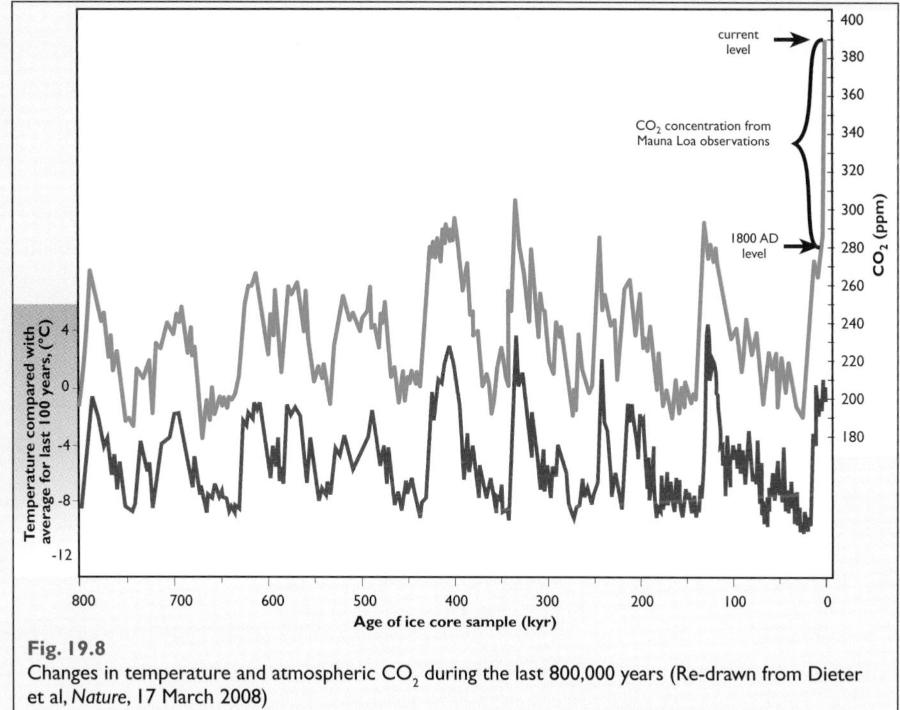

Fig. 19.8
Changes in temperature and atmospheric CO_2 during the last 800,000 years (Re-drawn from Dieter et al, *Nature*, 17 March 2008)

These positive feedback cycles may have been reinforced by other positive feedback effects mentioned earlier.

Despite these undoubtedly natural (pre-human) changes, during the last 800,000 years CO_2 levels have not exceeded 300 ppmv until the last 100 years (0.0125% of the total). This coincides with the large scale burning of fossil fuels.

The above hypothesis has led some climate scientists to argue that a similar thing is happening today; increased energy from the sun is warming the ocean, driving CO_2 from ocean to atmosphere. If this were true, the CO_2 content of the ocean would be falling and the pH of the ocean would be rising. In fact, the pH of the ocean is falling.

'The increase in atmospheric CO_2 is caused by volcanoes'

This is clearly false: If volcanoes were a significant cause of the increase, the graph showing the increase in atmospheric CO_2 would show a series of spikes, each spike corresponding to an eruption. The steady upward trend is, however, consistent with increased output from burning fossil fuels and deforestation. According to the US Geological Survey, anthropogenic (human-produced) CO_2 is about 30 billion tonnes a year — more than 130 times as much as volcanic CO_2 production.

Not only can we be certain that the increase in atmospheric CO_2 is *not* due to volcanoes, there is very strong evidence that it *is* due to human activity. This evidence comes from studies of carbon isotopes.

Like all elements, carbon exists as several different forms or *isotope*. About 99% of carbon has an atomic mass of 12 (^{12}C), but the remaining 1% consists of a slightly heavier isotope, carbon 13 (^{13}C). $^{13}CO_2$ diffuses slightly slower than normal carbon dioxide ($^{12}CO_2$). For this reason the CO_2 taken in by plants — and all the organisms that feed off them — contains slightly less than 1% ^{13}C. Since the carbon in fossil fuels was ultimately derived from photosynthesis, fossil fuels contain a slightly lower proportion of ^{13}C than carbon from non-biological sources.

Not surprisingly, the CO_2 produced when fossil fuels are burnt is slightly enriched in ^{12}C compared with carbon from inorganic sources. If fossil fuels are a significant source of increased atmospheric CO_2, we would therefore expect proportion of $^{12}CO_2$ in the atmosphere to be rising slightly — and this is exactly what scientists have found.

Another piece of isotopic evidence comes from the radioactive isotope, carbon ^{14}C. This forms about 1 part per billion of carbon in atmospheric CO_2 and in living matter, but half of it decays to nitrogen every 5730 years. Over geological time it disappears completely, so fossil fuels and the CO_2 produced when they are burnt contain no carbon-14. As a result we would therefore expect the proportion of ^{14}C in the atmosphere to have been decreasing since the Industrial Revolution — and this is exactly what has been happening.

ISBN 9780170191340

'Both atmospheric CO_2 levels and temperatures have been much higher in the past than they are today'

This is true. Experiments have shown that plants grown in atmospheres artificially enriched with CO_2 produce leaves with fewer stomata (pores in the epidermis through which gas exchange occurs) per mm^2, and plants grown in artificially low CO_2 levels produce more stomata per mm^2. Studies of the surfaces of fossil leaves have suggested that at certain times in the past, CO_2 levels have been up to ten times higher than they are today.

One such time was at the end of the Permian period, 250 million years ago. It has been suggested that massive volcanic action raised the CO_2 levels to a point sufficient to trigger positive feedback effects, causing 'runaway' warming.

Whatever the cause, it coincided with the greatest mass extinction of all time. In this 'great dying', 95% of all marine species and 75% of all terrestrial (land-living) species ceased to exist. This was a much greater extinction than the one that took the dinosaurs 65 million years ago.

So, although there have been much higher CO_2 levels in the past, in at least one case it seems the effect was catastrophic for nearly all life on the planet.

A WILD CARD: PEAK OIL

One aspect of the burning of fossil fuels that is only just beginning to receive attention is 'Peak Oil', or the peaking of oil extraction (what oil companies like to call 'production').

Oil is uniquely useful. It is easily transported and is extremely *energy-dense*, meaning that each kilogram contains enormous energy. A litre of petrol (obtained by refining oil) will take an average car 15 km, at a cost of less than NZ$2 (2010 prices). That litre of petrol contains the energy equivalent of 130 hours of hard human work. At the minimum 2010 wage of $12.50 per hour, this single litre of petrol can be valued at 130 x $12.50 = $1625!

Fossil fuel energy is so cheap that it is not surprising that it has become the lifeblood of industrial civilisation. In fact, the Industrial Revolution could just as easily be called the Fossil Fuel Revolution. Everything we make, eat or move depends on oil. The average US household (and New Zealanders are not very different) uses the energy equivalent to having 50 slaves working full time.

To give but one example: To bring 1 kJ of food energy to your plate takes, on average, 8–10 kJ of fossil fuel energy, mainly from oil:

- Soil is cultivated using machinery that used oil in its manufacture and needs oil to run.
- Pesticides and fertilisers are made from oil (nitrate fertiliser is made from natural gas, another fossil fuel).
- Water is pumped for irrigation using energy from oil.
- Food is packaged using plastic made from oil.
- Food is transported using diesel refined from oil.

To produce the food than an average American eats each year requires 1500 litres of oil. If human energy were used, this is the equivalent of 100 humans working a 40 hour week.

Tourism, because of its dependence on the aircraft industry, is entirely dependent on cheap oil. With only a very limited rail system, New Zealand's entire transport system (and our economy) depends on oil.

Oil was produced millions of years ago, and is a one-off gift from nature, yet the economic planning of all governments — including successive New Zealand governments — has always assumed that oil will continue to be available at a reasonable price for the foreseeable future.

How well-placed is this confidence? Figure 19.9 shows one aspect of the situation. Across the world as a whole, oil discoveries peaked in the mid-1960s, before many of your parents and teachers were born. Oil consumption, on the other hand, has risen steadily. In 1980 consumption began to exceed discovery, and the world is now using oil several times faster

ISBN 9780170191340

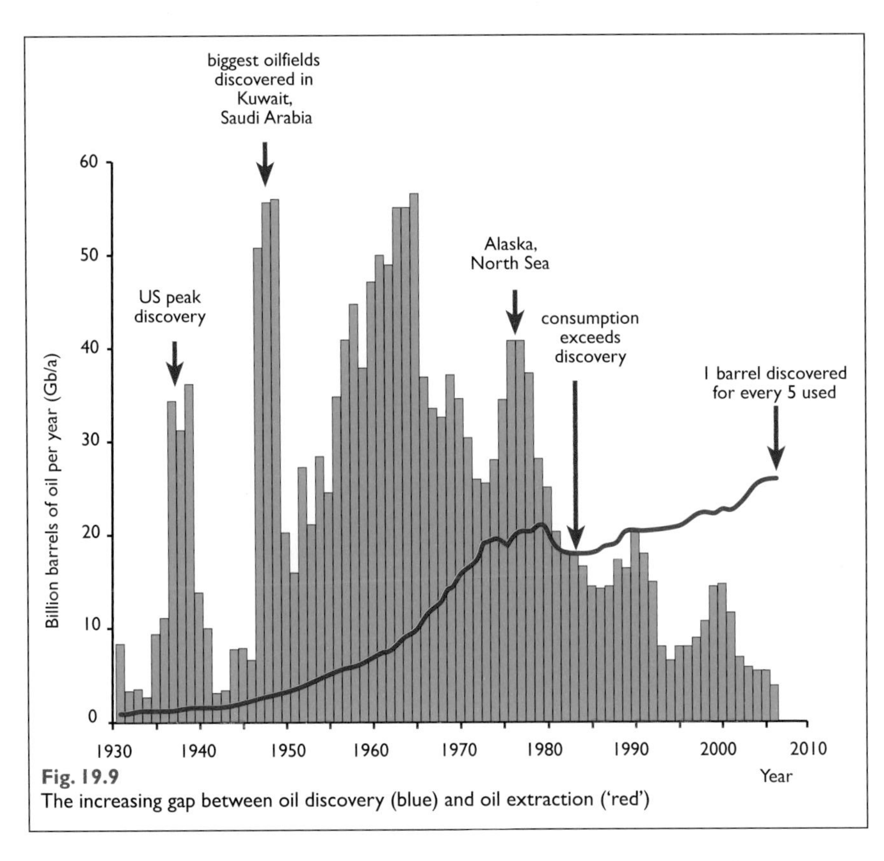

Fig. 19.9
The increasing gap between oil discovery (blue) and oil extraction ('red')

than it is being found. Oil extraction in over 75% of oil exporting countries is in decline. For these countries, 'peak oil' is not a theory (as many economists insist), but historical fact.

We are told that there are enormous amounts of oil in the ground. This is true, but it misses the point; what matters is not the *amount*, but how quickly it can be extracted, and at what energy cost. In the 1930s, you could extract 100 barrels of oil for every barrel of oil you spent on exploration and drilling. In other words, the energy return/energy invested (or ERoEI) was about 100:1. In 1970 the ERoEI was about 25:1, and in the 1990s it was between 18:1 and 10:1. Now, recent oil finds are returning a ratio of about 5:1 to 3:1 because they tend to be small and in very difficult-to-get-at places, such as 10,000 metres beneath the surface of the sea. The 2010 pollution disaster in the Gulf of Mexico shows all too clearly that the difficult-to-extract oil is also the riskiest to extract.

The tar sands in Alberta, Canada are believed to contain reserves comparable to the entire Middle East — yet it costs enormous amounts of energy (in the form of natural gas) to extract, with an ERoEI of about 5:1 or less. Once you get to 1:1, it ceases to be a resource — no matter how much is in the ground.

What about renewables? Methanol, which can be made from biomass, has an ERoEI of about 3:1, and biodiesel has an ERoEI of about 2:1. Corn-based ethanol produced in the US has an ERoEI of about 1 — in other words, it costs about as much energy to produce it as you get when you burn it. On top of that, it uses land that otherwise could be used for producing food.

So what has 'Peak Oil' got to do with climate change? It could have a great deal; as oil gets more expensive, there may be an increasing reliance on coal, which produces considerably more CO_2 for each unit of energy released.

Politicians are reluctant to think about dangers further away than the next election. Young people can play an important part in educating them.

ISBN 9780170191340

Summary of key points in this chapter

- Together with water vapour and methane, carbon dioxide is a 'greenhouse' gas, so-called because it helps to trap the sun's heat and warms the Earth.
- Before the Industrial Revolution the atmospheric concentration of CO_2 changed only slowly, but in the last century it has been increasing rapidly due to the burning of fossil fuels and the cutting down and burning of forests.
- About 40% of the CO_2 pumped into the atmosphere by industry is absorbed by the surface layers of the ocean.
- There is widespread agreement among climate scientists that the increase in atmospheric CO_2 may be warming the Earth's climate.
- The results of climatic warming are predicted to be more extreme weather, with changes in rainfall patterns, and in the longer term, raising sea levels.
- The situation is made more complex and more difficult to predict because of *feedback effects.*
- A complicating factor is the impending shortage of oil that may lead to increasing emphasis on coal, which produces more CO_2 than other fossil fuels.

Test your basics

Copy and complete the following sentences.

a) Carbon dioxide and other atmospheric gases are transparent to __*__ radiation but absorb longer wavelength (_*_-_*_) radiation. Light from the sun is absorbed by the Earth's surface that re-radiates the energy as __*__, which is absorbed by CO_2. In acting as a kind of 'heat jacket', CO_2 is said to be a '__*__' gas. Several other gases act in a similar way, most notably __*__ vapour and __*__.

b) Molecule for molecule, methane is a more potent greenhouse gas than CO_2. It is produced in __*__ respiration by microorganisms living in the __*__ of cattle and sheep, and also in bogs, marshes and rice fields. Methane is present in huge quantities as methane __*__ on the sea bed, and is only stable at the low __*__ and high __*__ that occur in these locations.

c) One effect of increasing CO_2 concentrations is to raise its concentration in the oceans, making it more __*__. This is likely to have adverse effects on animals with skeletons of __*__.

d) Predicting the effects of increased atmospheric CO_2 is complicated by feedback effects, which are of two kinds. In positive feedback, the results of a process act to __*__ that process. In negative feedback, the results of a process act to __*__ it.

e) Polar ice reflects over 80% of solar radiation back into space, but the ocean absorbs most incoming radiation. If the area of polar ice were to shrink, its reflectivity or __*__ would decrease, tending to __*__ the Earth's temperature. This would be an example of __*__ feedback.

f) If a rise in CO_2 causes the ocean to warm, this would __*__ the solubility of CO_2 and therefore the ability of the ocean to absorb further CO_2. This would tend to speed up a rise in atmospheric CO_2, and would be an example of __*__ feedback.

g) Clouds __*__ solar radiation, so that if atmospheric water vapour were to increase as a result of climatic warming, this would increase cloud cover. This in turn would reflect more radiation, tending to reduce further temperature increase. This would be an example of __*__ feedback.

h) If the ice floating on the Arctic ocean were to melt, the effect on sea level would be __*__. If polar ice on land were to melt, the effect would be to __*__ sea level. Sea level would also tend to rise as a result of thermal __*__ of sea water.

ISBN 9780170191340

Carbon Cycling: Assessement Practice

QUESTION ONE

The diagram shows a simplified representation of certain parts of the carbon cycle. In the cycle, carbon can exist in various states, represented by numbered boxes.

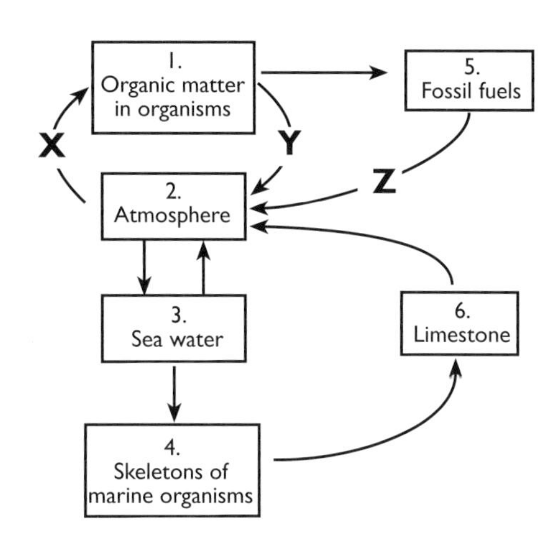

(a) **Name** TWO kinds of chemical compound that contain carbon and which are constituents of all organisms.

(Guidance: Your answer should be approximately 2 lines.)

(b) **Name** the processes represented by the arrows labelled X, Y and Z.

(Guidance: Your answer should be approximately 3 lines.)

(c) Using Boxes 2, 3 and 5 as examples, **explain** what is meant by the terms *source* and *sink*.

(Guidance: Your answer should be approximately 6 lines.)

(d) The amount of carbon dioxide in the atmosphere is increasing.

(i) **Explain** the reasons for this increase.

(ii) **Discuss** possible consequences.

(Guidance: Your answer should be approximately 30 lines.)

ISBN 9780170191340

QUESTION TWO

Since 1958, the atmospheric carbon dioxide concentration has been measured at various stations around the world. The diagram below shows results from a research station at Baring Head, near Wellington, 1970–2005.

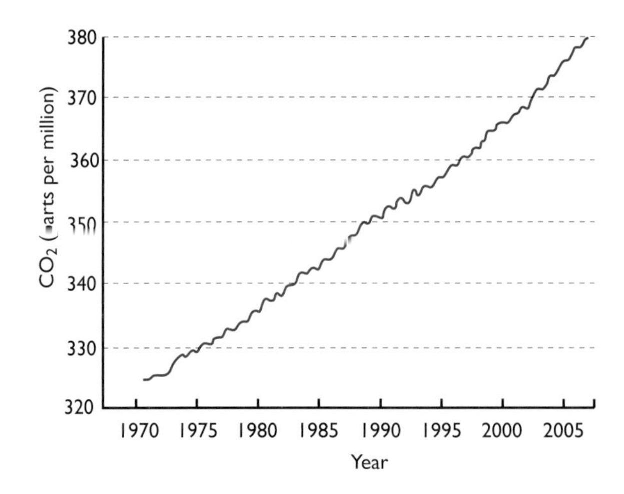

(a) Approximately what has been the annual increase in parts per million between 1995 and 2005?

(Guidance: Your answer should be approximately 1 line.)

(b) Does the rate of increase appear to be increasing or decreasing? Justify your answer.

(Guidance: Your answer should be approximately 1 line.)

(c) **Explain** the reason for the small 'steps' in the graph.

(Guidance: Your answer should be approximately 4 lines.)

(d) It is sometimes argued that the main cause of the increase in CO_2 concentration has been volcanic eruptions.

Explain how the graph shows that this cannot be true.

(Guidance: Your answer should be approximately 3 lines.)

QUESTION THREE

(a) **Explain** what is meant by the term 'greenhouse gas'.

(Guidance: Your answer should be approximately 4 lines.)

(b) **Explain** what would be the immediate effect if all the CO_2 in the atmosphere were to be suddenly removed.

(Guidance: Your answer should be approximately 4 lines.)

(c) What would be the effect of the change you mention in your answer to (b) on:

(i) The concentration of water vapour in the atmosphere.

(ii) The temperature of the Earth?

(Guidance: Your answer should be approximately 8 lines.)

(d) At certain times in the past, atmospheric CO_2 concentration has been much higher than it is today. It is believed that *feedback effects* have played an important part in past increases in atmospheric CO_2.

With reference to both **positive feedback** and **negative feedback**, discuss possible mechanisms that could have affected atmospheric CO_2 concentration in the past.

(Guidance: Your answer should be approximately 20 lines.)

ISBN 9780170191340

Key Terms

Absorption	The movement of digested food through the lining of the alimentary canal into the blood
Acrosome	Sac in sperm tip containing digestive enzymes used to penetrate the egg
Active transport	The movement of molecules or ions across a cell membrane with the expenditure of energy, usually against a concentration gradient
ADH	Antidiuretic hormone; promotes water conservation by concentrating the urine
Adrenaline	Hormone that prepares the body for urgent action, secreted by the adrenal glands
Aerobe	Organism that grows in the presence of oxygen
Aerobic	In the presence of oxygen
Afferent arteriole	Blood vessel supplying the kidney glomerulus
Agar	Jelly-like material on which bacteria and other microorganisms are grown
Albedo	The extent to which a material on the Earth's surface reflects solar radiation
Alimentary canal	Tube extending from mouth to anus, within which food is digested
Allele	One of two or more alternative forms a gene can take; often one is dominant to the other
Alveoli	Minute, thin-walled, bag-like endings of the bronchial tree; and the site of gas exchange
Amino acid	A building unit of a protein
Ammonia	Product of deamination of amino acids
Amnion	Membrane round embryo/foetus containing amniotic fluid
Amniotic cavity	Cavity surrounding foetus and containing protective fluid
Amniotic fluid	Liquid that protects the foetus against mechanical shock
Ampulla	Swelling on the end of a semicircular canal, containing receptors that detect rotation of the head
Amylase	Enzyme that breaks down starch into maltose
Anaerobe	Organism that can grow in the absence of oxygen
Anaerobic respiration	Oxidation of organic matter by cells to release useful energy, using substances other than oxygen (e.g. nitrate)
Annual	A plant that completes its life cycle in one year or less
Annual ring	The secondary xylem that is produced in an entire year, sharply distinct from that of previous and subsequent years
Antagonistic muscle	Muscle that has the opposite effect to another muscle
Anther	The part of a stamen that produces pollen
Antibiotic	Substance produced by a microorganism that kills or inhibits growth of bacteria or other microorganisms
Antibody	Protein produced in response to presence of 'foreign' substance, usually a protein or complex carbohydrate
Antigen	Substance that can recognised by the immune system as 'foreign', stimulating it to produce an antibody
Antitoxin	Antibody that neutralises a toxin
Aorta	Artery taking blood from the left side of the heart
Appendix	Blind-ending outgrowth of large intestine; vestigial in humans
Aqueous humour	Liquid in front part of the eye which circulates, bringing nutrients to the lens and cornea
Arteriole	Extremely small artery
Artery	Vessel taking blood away from the heart

ISBN 9780170191340

Asexual reproduction	Reproduction that does not involve meiosis and fertilisation, so there is no genetic change. The offspring are genetically identical to the parent
Assimilation	The absorption of digested food by cells
Asthma	Condition in which the bronchioles constrict as a result of excessive contraction of smooth muscle in their walls, accompanied by an over-secretion of mucus
ATP	Adenosine triphosphate, an immediate source of energy in cells
Atrio-ventricular valve	Heart valve preventing blood flowing back from the ventricle to the atrium
Atrium	Heart chamber that receives blood from veins
Autoclave	Pressure-cooker used for sterilising laboratory equipment
Autosome	A chromosome not involved in the determination of sex
Autotroph	Organism that can produce its own organic matter from CO_2 and water
Axon	A long threadlike extension of a neuron which transmits impulses
Bacteriophage	Virus that attacks bacteria
Ball and socket joint	Joint allowing movement in all planes
Bark	Outer, protective layer of stem or root, produced during secondary growth
Basilar membrane	Membrane upon which the hearing organ (organ of Corti) is located
Bicuspid valve	Valve between the left atrium and left ventricle
Biennial	Plant that takes two years to complete its life cycle
Bile	Yellow-green juice secreted by the liver; emulsifies fats
Binary fission	Reproduction by dividing in two
Biogeochemical cycle	Cycle of elements involving both biological and geological processes
Biosynthesis	Making large molecules from smaller ones, such as glycogen from glucose
Bladder	Muscular sac in which urine is temporarily stored
Blastocyst	Hollow ball of cells produced by cleavage of the fertilised egg of a mammal
Blind spot	Part of the retina where there are no photoreceptors
Blood	Liquid in which materials are transported around the body of an animal
Bowman's capsule	Blind-ending part of a nephron, where the blood is filtered.
Bronchiole	The smallest tubes in the bronchial tree, lacking cartilaginous support
Bronchitis	Inflammation of the bronchi
Bronchus	Tube in lungs that transport air; supported by cartilage
Bulb	Perennating structure consisting of a tiny, vertically growing underground stem surrounded by thick, scale-like bases of foliage leaves
Caecum	Blind-ending part of large intestine
Calcium phosphate	Main mineral constituent of bone
Callus	Mass of cells produced after wounding a plant, especially if stimulated by a plant hormone
Cambium	Meristem responsible for secondary growth (growth in thickness) of a stem or root
Canine	Single-rooted tooth between incisors and premolars
Capillary	Microscopic blood vessel where substances enter and leave the blood
Capsid	The protein coat of a virus, enclosing the genetic material
Carbon dioxide	Gaseous waste product of respiration
Carbon monoxide	Colourless, odourless gas that combines with haemoglobin and prevents it carrying oxygen
Cardiac muscle	Muscle of the heart
Cardiac vein	Vein returning blood from the heart muscle to the right side of the heart
Carnivore	Animal that feeds on other animals
Carpel	Female unit of a flower; each flower may have one or several, in which case they may be separate or joined
Carrier	Person who is infected by a pathogen but shows no symptoms

ISBN 9780170191340

Cartilage	Flexible skeletal material on the ends of long bones and certain other places
Cellulose	Complex carbohydrate that is the main constituent of plant cell walls
Cement	Bone-like substance covering the root of a tooth
Central nervous system	Brain and spinal cord
Centromere	The part of a chromosome that attaches to the spindle in cell division
Cervix	The neck of the uterus
Chemosynthesis	Use of chemical energy to convert CO_2 to carbohydrate. Nitrifying bacteria are chemosynthesisers
Chitin	Material forming the cell walls of fungi, insects and other arthropods
Chlorosis	Deficiency symptom in which there is lack of chlorophyll, resulting from lack of magnesium or nitrogen
Chorionic gonadotrophin	Hormone produced by the blastocyst after implantation, and which maintains the corpus luteum
Choroid	The middle layer of the wall of the eye, containing blood vessels
Chromatid	One of two identical products of replication of a chromosome
Chromosome	Threadlike structure consisting of a 'backbone' of DNA, plus proteins
Chyme	Creamy-white liquid produced in stomach as result of partial digestion of food
Cilia	Hair-like extensions of the plasma membrane, which beat and create current in the fluid outside the cell
Ciliary body	Ring of muscle used in adjusting the curvature of the lens in accommodation
Clitoris	Small area of erectile tissue near the entrance of the vagina
Cochlea	Part of the inner ear concerned with hearing
Collagen	Main protein present in bone, tendon and ligament
Collecting duct	Branching tube in the medulla of the kidney, in which urine may become concentrated
Colon	Largest part of the large intestine, where water is absorbed and faeces formed
Colony	Cluster of bacteria or fungal hyphae that has grown from a single spore
Colostrum	Special protein-rich milk secreted for a few days after birth
Compensation point	The light intensity at which photosynthesis equals respiration
Complementary base pairing	Pairing between specific bases in DNA (adenine with thymine, and guanine with cytosine)
Cones	Photoreceptors used in bright light, giving detailed, colour vision
Conidium	A fungal spore produced from the tip of an erect hypha
Conidiophore	A vertically-growing fungal hypha that produces spores at its tip
Conjunctiva	Layer containing blood vessels, covering the front of the sclera and lining the eyelid
Cork	Outer, dead, protective layer of older secondary stems and roots; the main constituent of bark
Cork cambium	Meristem whose activity produces cork, a major constituent of bark
Corm	Perennating structure consisting of a vertically growing underground stem protected by thin, scale-like bases of foliage leaves
Cornea	The transparent front of the outer layer of the eye, which acts as a lens
Coronary arteries	The first two branches of the aorta, supplying the heart muscle
Corpus luteum	'Yellow body' formed from follicle cells after ovulation
Cortex	Outer layer of the kidney
Cotyledon	A young leaf of a plant embryo
Cowper's gland	One of the glands that produces seminal fluid
Crista	Sense organ in each semicircular canal in the inner ear involved in the sense of balance
Cupola	Pad of jelly in each semicircular canal in which sensory 'hairs' are embedded
Cuticle	Waxy, water-resistant layer covering the epidermis of plants and insects

ISBN 9780170191340

Deamination	Process by which the nitrogen-containing amino group of an amino acid is removed as ammonia; the organic acid residue may then be used in respiration
Decay	Process by which microorganisms convert complex organic substances to simple inorganic substances
Decibel	Unit of sound intensity; an increase of 3dB doubles the sound intensity
Decomposer	Organism that brings about decay by breaking down complex organic materials to simpler, inorganic materials
Dehiscent fruit	Fruit that opens to release the seeds
Denaturation	The permanent inactivation of an enzyme or other protein, usually by heat
Dendrite	A thin, threadlike extension of a neuron that collects information from the environment or from other neurons
Denitrification	Conversion of nitrate to nitrogen gas by anaerobic bacteria
Dentine	Substance making up the bulk of a tooth
Deoxyribose	The 5-carbon sugar that is a constituent of DNA
Diabetes	Condition in which blood glucose level is high; a major cause of blindness
Diabetes insipidus	Condition arising from the inability to secrete ADH, resulting in massive urine output and perpetual thirst
Diaphragm	Dome-shaped sheet of muscle separating the chest and abdomen
Diffusion	The movement of a substance from a higher to a lower concentration by random movement of its particles
Digestion	The breaking down of complex food molecules into smaller molecules
Diploid	Having two sets of chromosomes
Discontinuous variation	Variation in which individuals can be placed in discrete categories, with no intermediates
Distal tubule	The last part of a nephron
Dizygotic twins	Twins formed by fertilisation of two eggs by two sperm
DNA	Deoxyribonucleic acid, the genetic material
DNA bases	Adenine, guanine, cytosine and thymine
Dominant	A trait that is expressed in both homozygotes and heterozygotes
Duodenum	The first part of the small intestine
Ear drum	Thin membrane that collects sound from the environment
Effector	A cell that carries out a response to a stimulus
Efferent arteriole	Vessel taking blood from a glomerulus in the kidney
Egestion	The getting rid of undigested food via the anus
Egg	A female gamete, stationary and large due to stored energy reserves
Ejaculation	Emission of seminal fluid in orgasm
Embryo	A very young plant, contained within a seed
Emphysema	Condition caused by smoking in which the alveoli become greatly reduced in number but increased in size, thereby decreasing the area for gas exchange
Enamel	Layer covering the crown of a tooth, and the hardest substance in the body
Endocrine gland	Gland that secretes a product (hormone) directly into the blood; a ductless gland
Endocytosis	Process by which an animal cell takes in viruses and other large particles
Endometrium	Lining of the uterus
Endoskeleton	An internal skeleton, characteristic of vertebrate animals
Endosperm	Energy storage tissue present in all developing seeds and some mature seeds
Endospore	Highly resistant spore produced by some bacteria
Endothermic	Able to maintain a body temperature above that of the environment by heat production in metabolism
Enzyme	Protein catalyst that speeds up chemical reactions

Ephemeral	Plant with a very short life cycle (weeks)
Epidemic	Sudden outbreak of a disease
Epidemiology	Study of the pattern of occurrence of diseases
Epidermis	The outer, protective layer of the skin or outermost tissue on a leaf, young stem or young root
Epididymis	Long tube in which sperm mature, prior to ejaculation
Epiglottis	Flap-like structure protecting the opening to the windpipe and preventing food going down the 'wrong way'
Erectile tissue	Tissue containing spaces that can become gorged and dilated with blood during sexual stimulation
Etiolation	Characteristics of a shoot that has grown in darkness; long spindly stems, small, yellow leaves
Eukaryote	Organism whose cells contain distinct nuclei, separated from the cytoplasm by a nuclear envelope
Eustachian tube	Tube connecting the middle ear and the pharynx, enabling pressure in middle ear and environment to be equalised
Excretion	The getting rid of metabolic waste
Exocytosis	Process by which material that has been enclosed by a membrane is 'exported' from a cell
Exoskeleton	Skeleton that forms the outer layer of the body, as in insects and other arthropods
Expiration	Breathing out, or exhalation
Exponential growth	Increase in numbers by a constant percentage in each given time interval
External ear	Part of the ear that conveys sound waves in the air to the ear drum
Extracellular digestion	Digestion outside cells; occurs in fungi and bacteria
F_1 generation	Hybrid produced by crossing two pure-breeding varieties
F_2 generation	Offspring produced by inbreeding an F_1 generation
Facultative anaerobe	Organism that can grow without oxygen, but grows better in its presence
Faeces	Indigestible material plus bacteria that is egested via the anus
Fallopian tube	Tube connecting the ovary to the uterus; also known as the oviduct
Feedback effect	Process in which the results act to speed up (positive feedback) or slow down (negative feedback) a process
Fermentation	Process enabling cells to obtain useful energy from organic matter without net oxidation
Fertilisation	The joining together of two haploid cells (gametes) to form a zygote
Fimbriae	Tentacle-like extensions of the opening of the oviduct, which collect eggs after ovulation
Flagellum	Whip-like structure in many bacteria used for propulsion (the 'flagella' of eukaryotes are quite different)
Foetal alcohol syndrome	Group of effects on the foetus resulting from alcohol consumption by the mother during pregnancy
Foetus	A human baby that has reached the stage at which its organ systems have developed
Follicle	Cluster of cells surrounding a developing egg in the ovary of a mammal
Follicle stimulating hormone	Hormone that stimulates the growth and development of eggs and surrounding cells in the ovary
Food chain	A sequence of organisms through which energy is transferred
Food spoilage	Process by which food is rendered inedible by the activity of microorganisms
Fossil fuels	Organic materials such as coal, oil and natural gas, formed in the Earth's crust from thick deposits of dead organisms
Fovea	Small area at the back of the retina that is particularly sensitive to detail and colour

ISBN 9780170191340

Fruit	A ripened ovary of a flowering plant
Gallbladder	Small bag-like structure that temporarily stores bile
Gamete	A haploid cell that can only develop further when it joins with another such cell to form a zygote
Ganglion	A swelling on a nerve formed by a collection of nerve cell bodies
Gas exchange	The diffusion of oxygen and CO_2 in opposite directions across a surface
Gene	A length of DNA coding for a polypeptide; a unit of heredity that controls production of a protein
Generation time	Time between successive generations; in bacteria, the time for the population of cells to double
Genome	The complete set of genes of an organism
Genotype	An organism's genetic makeup; the information it inherits from its parents
Gestation	Period of foetal development within the uterus
Glaucoma	Form of blindness caused by build-up of pressure in the eyeball
Glomerular filtrate	Liquid produced by filtering the blood from the glomerulus into Bowman's capsule
Glomerulus	'Little knot' of capillaries dipping into Bowman's capsule; the site of filtration
Glucagon	Hormone secreted by the Islets of Langerhans in the pancreas, and antagonistic to insulin.
Glycogen	Complex carbohydrate resembling starch, stored in liver and skeletal muscle
Glycolysis	Conversion of glucose to pyruvic acid, yielding a small quantity of ATP
Gonad	General name for an ovary or a testis
Graafian follicle	A mature follicle
Gravitropism	Growth of a plant organ in a direction that relates to the direction of gravity
Greenhouse gas	Gas that traps infra-red radiation and acts as a 'heat blanket' around the Earth
Grey matter	Central area of the spinal cord containing nerve cell bodies
Guard cell	One of two such cells surrounding a stoma, whose movements regulate the size of the stomatal opening
Haemoglobin	Oxygen-carrying protein in the red blood cells
Hair cells	The receptors in the inner ear involved in hearing and balance
Haploid	Having one set of chromosomes
Haversian canal	Channel in compact bone through which a capillary runs
Heart	Muscular pump that drives blood round the body
Hepatic portal vein	Large vein taking digested food from the gut to the liver
Herbaceous plant	Plant in which the aerial parts do not survive for more than one year, so the stems never have more than one year's secondary growth
Herbivore	Animal that eats plants
Heterotroph	Organism that cannot produce organic compounds from CO_2
Heterozygous	Having two different alleles for a given trait
Hinge joint	Joint allowing movement in one plane
Homeostasis	Process by which the internal conditions of the body are regulated at near-constant levels
Homeothermic	Maintaining a constant body temperature
Homologue	One of a pair of chromosomes
Homozygous	Having two of the same allele
Hormone	'Chemical messenger' carried in the blood, produced by an endocrine gland and which has a specific effect on the metabolism of the target cells
Host	Organism from which a parasite obtains its food
Hydrostatic skeleton	Skeleton consisting of fluid
Hypha	One of many threadlike structures forming the body of a fungus

ISBN 9780170191340

Hypothalamus	The part of the brain that monitors the state of hydration of the blood, among other functions
Ileum	The major part of the small intestine
Implantation	Process in which a blastocyst becomes buried within the uterine lining
Impulse	Electrical change travelling along an axon of a nerve cell
Inbreeding	Mating between close relatives
Incisor	Chisel-shaped tooth at the front of the mouth
Incubation period	Period after infection during which the person shows no symptoms of disease
Inflorescence	Cluster of flowers
Infra-red	Heat radiation
Ingestion	The taking in of food at the mouth
Inner ear	The location of the hearing and balance organs
Inspiration	Breathing in, or inhalation
Insulin	Hormone secreted by the pancreas that acts to lower blood glucose level, stimulating the liver to convert it to glycogen
Integuments	Thin coats around an ovule from which the testa (seed coat) develops
Intercostal muscles	Muscles that move the ribs
Interneuron	A neuron connecting sensory and motor neurons
Iris	Ring of muscle controlling the intensity of light on the retina
Islets of Langerhans	Microscopic clusters of cells in the pancreas, which secrete the hormone insulin
Karyogram	Photograph or diagram of the chromosomes of a species arranged in order of size
Karyotype	The chromosomal characteristics of an organism
Lacteal	Blind-ending 'twig' of the lymphatic system in a villus of the small intestine
Lag phase	Period immediately after introduction of a microorganism into a new environment during which numbers do not increase
Leguminous plant	Plant belonging to the legume family, e.g. clover, gorse, lupin, bean
Lens	Elastic body in the eye used in adjusting the amount of refraction when viewing objects at different distances
Lenticel	Small cluster of powdery cells with air spaces between, through which gases can diffuse through cork or bark
Ligament	Fibrous tissue connecting bones at a joint
Lignin	Substance that stiffens the walls of xylem vessels and supporting cells such as fibres
Limiting factor	The environmental factor that determines the rate of a process because it is in shortest supply
Lipase	Pancreatic enzyme which breaks down fat into fatty acids and glycerol
Liver	Large organ that secretes bile and stores glycogen (among many other functions)
Locus	The location of a gene on a chromosome
Loop of Henle	Region of a kidney nephron responsible for building up a high solute concentration needed to concentrate the urine
Lung	Deep, air-filled intucking of the body surface in which gas exchange occurs
Luteinising hormone	Hormone that promotes ovulation and formation of the corpus luteum
Lymph	Liquid in lymphatic vessels
Lymph node	Swelling on lymphatic vessel containing large numbers of white blood cells
Lymphatic system	System that returns surplus tissue fluid to the blood
Lymphocyte	Blood cell that can develop into a cell that makes antibodies
Lysozyme	Enzyme that breaks down the cell walls of many bacteria; present in tears and other body fluids

ISBN 9780170191340

Macronutrient	In plants, a mineral nutrient required in relatively large amounts e.g. nitrogen, phosphorus, potassium, magnesium, calcium
Magma	Liquid material lying below the Earth's crust
Medulla	The inner tissue of the kidney containing collecting ducts and loops of Henle
Meiosis	Two successive divisions of the nucleus to produce four haploid, genetically different nuclei
Melanin	Black pigment produced in cells at the back of the retina, also in certain cells of the epidermis
Menstrual cycle	Cycle of events in the uterus and ovary, in which menstruation and ovulation alternate
Meristem	A part of a plant in which cell division is occurring, such as a shoot tip or root tip
Messenger RNA (mRNA)	Type of RNA used in the transfer of information from DNA to the ribosomes (which make proteins)
Metabolism	Collective term for all the chemical reactions in cells
Metamorphic rock	Rock that has been changed as a result of heat and pressure in the Earth's crust
Methane hydrate	Solid formed on the sea bed by combination between methane and water at high pressure and low temperature
Micrometre	A millionth of a metre
Micronutrient	Mineral nutrient required in extremely small amounts
Micropyle	Minute hole in the testa of an ovule through which the pollen tube normally grows prior to fertilisation
Microvilli	Finger-like extensions of the plasma membrane of cells lining the small intestine; a site of active transport
Middle ear	Air-filled cavity containing three bones that transfer sound from air to the inner ear
Mitochondrion	Organelle that is the site of respiration; the 'powerhouse of the cell'
Mitosis	Division of a nucleus into two genetically identical nuclei
Molar	Adult tooth that does not replace an earlier one. Molars are at the rear of the jaw and have several roots
Monozygotic twins	Twins (formed by splitting of cell mass) that have developed from a single fertilised egg to form two genetically identical embryos
Morula	Solid ball of cells produced by division of the fertilised egg
Motor neuron	A neuron that carries impulses to an effector
Mucus	Slimy substance that traps dust particles and bacteria in the breathing passages
Mutation	Change in the genetic material that persists over many generations
Mutualist	Organism that lives in a relationship with another organism, both obtaining benefit
Mycelium	The body of a fungus, consisting of many threadlike hyphae
Mycorrhiza	Mutualistic relationship between a flowering plant and a fungus that lives in its roots
Myelin sheath	Insulating layer of fatty material surrounding an axon
Nanometre	A billionth of a metre
Nectar	Sugary solution secreted by many insect-pollinated flowers, serving as a 'reward'
Nectary	Patch of glandular tissue that secretes nectar, a sugar solution, which attracts insects
Negative feedback	Situation in which the results of a disturbance act to correct the disturbance; the greater the disturbance the greater the tendency to correct it
Nephron	One of about a million functional units of the kidney, consisting of Bowman's capsule, proximal and distal tubules, and loop of Henle
Neuron	A nerve cell
NIHL	Noise-induced hearing loss
Nitrate	The form in which plants absorb most of their nitrogen

ISBN 9780170191340

Nitrification	Conversion of ammonia to nitrate (via nitrite) by bacteria
Nitrogen fixation	Conversion of nitrogen gas to a combined form such as ammonia
Notifiable disease	Disease that must be reported by a doctor to the authorities
Nucleotide	One of the 4 kinds of sub-unit in a nucleic acid (DNA or RNA)
Nucleus	Part of a cell that contains the chromosomes, bounded by a nuclear envelope
Obligate aerobe	Organism that grows only in the presence of oxygen
Obligate anaerobe	Organism that cannot grow in the presence of oxygen
Oesophagus	Muscular tube connecting pharynx with the stomach
Oestrogen	Hormone secreted by follicle cells during first phase of the menstrual cycle
Omnivore	Animal that eats both animal and plant food
Öocyte	A diploid (primary öocyte) or haploid (secondary öocyte) immature egg
Öogenesis	Production of eggs in ovary
Organ	Group of tissues working together to carry out a particular function
Organ of Corti	The part of the cochlea containing the sound receptors
Organelle	Structure within a cell with a distinct function, e.g. the mitochondria (site of respiration)
Orgasm	Period of intense sensory pleasure during sexual stimulation
Osmoreceptor	Cell that detects changes in solute concentration
Osmoregulation	The regulation of the osmotic (water) concentration of the blood
Osmosis	The movement of water through a partially permeable membrane
Osteocyte	Bone cell
Osteomalacia	State in which the bones have become weakened through lack of dietary vitamin D or lack of calcium
Osteoporosis	State in which the bones have become weakened through loss of calcium phosphate in old age
Outbreeding	Mating between non-relatives
Ovary	Female part of a flower containing one or more ovules, from which seeds develop; in animals, the female gamete-producing organ
Ovulation	The release of an egg by the ovary
Ovule	Structure that, after fertilisation has occurred, develops into a seed
Ovum	A mature female gamete in animals
Oxytocin	Hormone that stimulates contraction of the smooth muscle in the uterus and ducts of the mammary gland
Palisade mesophyll	The main photosynthetic tissue in a leaf, consisting of cells elongated at right angles to the epidermis
Pancreas	Banana-shaped organ that secretes a number of digestive enzymes, which helps control blood sugar level
Pandemic	World-wide epidemic
Parasite	Organism that feeds on another organism usually without killing it
Parathyroids	Patches of tissue in the thyroid gland that secrete a hormone regulating blood calcium level
Partially permeable membrane	A membrane that allows the passage of small molecules such as water, but not passage of larger molecules
Pasteurisation	Brief heat treatment of milk or other drink, killing most bacteria
Pathogen	Organism that causes disease
Peat	First stage in the conversion of plant material into coal
Pedigree	Family tree showing the expression of a trait over several generations
Penis	The male organ that introduces semen into the vagina
Pepsin	Protein-digesting enzyme secreted by the stomach
Perennation	Survival of a plant over successive years

ISBN 9780170191340

Perennial	Plant that survives from year to year
Pericardium	Membrane around the heart, containing fluid that lubricates its surface
Pericarp	The wall of a fruit
Periodontal ligament	Tissue that binds tooth to the socket in the jaw
Peristaslis	Muscular waves that propel liquid along a tube, such as the ureter and intestine
Petal	Modified leaf serving to make a flower conspicuous to animal pollinators
Petri dish	Dish in which microorganisms are grown
Phagocyte	White blood cell that eats bacteria
Pharynx	Cavity at the back of the mouth and nose passages through which food and air travel
Phenotype	The characteristics of an organism that can be detected by examination
Phloem	Tissue specialised for transporting sugars and amino acids in plants
Photoperiod	The length of the day; an important factor influencing the time of year many plants flower
Photoreceptor	Light-sensitive cell in the retina
Photosynthesis	Process by which plants (and some bacteria) convert CO_2 to organic matter using light energy
Phototropism	Growth of a plant organ in a direction that relates to the direction of light
Phytoplankton	Microscopic photosynthetic organisms that float in the upper layers of the ocean and lakes
Pituitary gland	Gland beneath the brain that secretes numerous hormones
Placenta	Organ connecting baby and mother, formed from both
Plasma	Liquid component of the blood
Plasma membrane	Membrane forming the outermost layer of the cytoplasm
Plate tectonics	Idea that the Earth's crust consists of a series of plates floating on semi-liquid magma below
Pleural cavity	Narrow space around the lungs containing lubricating fluid
Plumule	The young shoot of a plant embryo
Polar body	Tiny, functionless cell produced in meiosis in female animals
Pollen	Single cells produced by meiosis in stamens of a flower, which produce male gametes (by mitosis) inside them
Pollen sac	Cavity in anther where pollen is produced
Pollen tube	Long, threadlike structure that grows from a pollen grain and conveys the male gametes to the female gamete
Pollination	Transfer of pollen from an anther to a stigma of a flower of the same species
Polynucleotide	A chain of nucleotides joined together
Polypeptide	A chain of amino acids
Posterior pituitary	Gland at the base of the brain that secretes ADH and oxytocin
Premolar	Adult tooth that replaces an earlier 'milk molar'
Primary growth	Growth in length of a shoot or root
Progesterone	Hormone secreted by corpus luteum and placenta, essential in pregnancy
Prokaryote	Organism in which the cells have no clearly-defined nucleus
Prolactin	Hormone stimulating milk secretion, produced by anterior pituitary gland
Prostate	Gland around the urethra that secretes a component of seminal fluid
Proximal tubule	The first part of a renal tubule
Puberty	Change from sexually immature juvenile to mature adult
Pulmonary artery	Vessel taking blood from the right side of the heart to the lungs
Pulmonary vein	Vessel taking blood from the lungs to the left side of the heart
Pulp cavity	Central part of a tooth containing nerves and blood vessels

ISBN 9780170191340

Punnett square	Diagrammatic representation of the various ways in which different kinds of gamete can combine, and their proportions
Pupil	Opening in the iris through which light passes to the retina
Pure-breeding	When mated with its own type over successive generations, continues to produce offspring like the parents. Alternate name for homozygous
Radicle	The young root of a plant embryo
Receptacle	The tip of a flower stalk to which all the other parts are attached
Receptor	A cell or nerve ending that detects stimuli
Recessive	A trait that is only expressed in homozygotes
Red cells (erythrocytes)	Oxygen-carrying cells in the blood
Reflex action	An automatic response to a stimulus
Relaxin	Hormone stimulating the softening of pelvic ligaments in the last weeks of pregnancy
Releasing hormone	One of several hormones, produced by the hypothalamus, each of which stimulates the anterior pituitary to secrete a particular hormone
Renal artery	Vessel taking blood to the kidney
Renal tubule	Long tube in kidney where filtered plasma is slowly converted to urine
Renal vein	Vessel taking blood from the kidney
Respiration	Oxidation of organic matter by cells to release useful energy
Rhizome	Perennating structure consisting of a horizontally-growing underground stem
Rhodopsin	Photosensitive pigment in the rods of the retina
Ribosome	Minute granules of RNA and protein, used in the synthesis of proteins
Rickets	Malformation of bones in childhood caused by vitamin D deficiency
RNA (ribonucleic acid)	A nucleic acid that is involved in protein synthesis (and, in certain viruses, is the genetic material)
Rods	Photoreceptors in the retina that are used in dim light
Root hair	Long threadlike extension of an epidermal cell of a root
Root system	The part of a plant specialised for absorbing water and minerals, and also for anchorage
Runner	A stem that grows from a parent plant over the ground surface and gives rise to a 'daughter' plant
Salivary gland	Gland that secretes saliva into the mouth
Saprobe	Organism that feeds on dead matter, bringing about decomposition
Scion	Part of a plant (stem or bud) grafted onto another plant (the stock)
Sclera	Tough, fibrous, outer layer ('white') of the eye
Scrotum	Bag-like extension of the body wall within which the testes lie
Secondary growth	Growth in thickness of a stem or root, resulting in the production of secondary xylem and secondary phloem
Sedimentary rock	Rock formed by slow compaction of sediments at the bottom of the sea
Seed	A very young plant (embryo), surrounded by a coat of parental tissue (testa), adapted for dispersal
Segregation	The movement of homologous chromosomes or of alleles to different daughter nuclei in meiosis
Semicircular canals	Fluid-filled tubes in the inner ear involved in detecting rotation of the head
Semiconservative replication (of DNA)	Replication in which the product is half new and half original
Semilunar valve	Valve at base of aorta and pulmonary artery
Seminal vesicle	Gland that produces a component of the seminal fluid
Seminiferous tubule	Tubule within testis where sperm are produced
Sensory neuron	A neuron that carries impulses from a receptor
Sepal	Modified leaf serving to protect the rest of the flower in bud

ISBN 9780170191340

Sexual reproduction	Process involving re-shuffling of genes in meiosis and fertilisation, resulting in genetic variation
Shoot system	The part of a plant that is typically above ground and which absorbs light and CO_2
Sink	A stage in a geological or biological cycle in which material enters faster than it leaves
Skeletal muscle	Muscle that is typically attached to bone
Source	A stage in a geological or biological cycle in which material leaves faster than it enters
Sperm	A male gamete
Spermatogenesis	The production of sperm in animals
Sphincter	Ring of muscle around a tube (e.g. the exit to the stomach) controlling movement of materials through it
Spindle	Barrel-shaped system of protein fibres that move the chromosomes in cell division
Spongy mesophyll	Tissue below the palisade tissue in a leaf, consisting of a network of cells orientated in many directions
Sporangium	Sac-like structure in which fungal spores are produced asexually
Sporangiophore	Vertically-growing fungal hypha that produces a sporangium at the tip
Spore	Single-celled reproductive body that can give rise to a new organism
Stamen	One of a number of male reproductive structures in a flower
Stem tuber	Perennating structure consisting of the end of an underground stem, swollen with energy reserves
Stigma	Part of a flower upon which pollen normally germinates
Stimulus	A change in the environment to which an organism can respond
Stock	Part of a plant on which another plant part (scion) is grafted
Stoma	Microscopic pore in epidermis of leaf or stem through which gas exchange occurs
Stomach	Muscular sac in which food undergoes partial digestion and is liquefied
Style	Stalk-like structure connecting the stigma to the ovary of a flower
Subduction	Process by which the edge of a tectonic plate slides below another
Substrate	The food material in which a fungus or bacterium grows
Summation	Process in retina in which stimuli detected by different photoreceptors are added together, giving greater sensitivity in dim light
Suspensory ligament	Fibres linking the eye lens to the cilary body
Synapse	Narrow gap separating adjacent neurons on a nerve pathway
Synovial fluid	Lubricating fluid in joints
Synovial membrane	Membrane that secretes synovial fluid
Tectorial membrane	Ribbon of jelly-like material in which the 'hairs' of sound receptors are embedded
Tendon	Fibrous tissue connecting muscle to bone
Teratogen	Substance that interferes with normal embryonic development, producing abnormalities
Testa	The outer layer of a seed
Testcross	A cross between an organism of unknown genotype with an organism that is homozygous recessive, in order to determine its genotype
Testis	Gamete-producing organ in a male animal
Testosterone	The male hormone, stimulating development of male characteristics in mammals
Tissue	A group of similar cells working together to carry out a particular function
Tissue fluid	Liquid that seeps out of capillaries and bathes the cells
Toxin	Poisonous substance produced by bacteria

ISBN 9780170191340

Toxoid	Toxin that has been chemically treated to make it harmless, without affecting its ability to stimulate an immune response
Trachea	Tube carrying air to and from the lungs
Trait	A characteristic that varies discontinuously, without intermediates
Transcription	Copying (of DNA) to make RNA
Translation	The process of using the base sequence in messenger RNA to join amino acids together in the correct sequence in a polypeptide
Translocation	Transport of solutes such as minerals, sugar and amino acids around the plant
Transpiration	The evaporation of water from the leaves and other aerial parts of a plant
Transpiration pull	Tension that develops in the xylem resulting from the evaporation of water from leaves
Tricuspid valve	Valve between the right atrium and right ventricle
Trimester	One of three 3-month stages of human development
Trophoblast	The outer layer of a blastocyst, through which nutrients are absorbed
Umbilical artery	Vessel taking blood from foetus to placenta
Umbilical cord	Cord containing two umbilical arteries and an umbilical vein
Umbilical vein	Vessel taking blood from placenta to foetus
Uracil	One of the four kinds of base in RNA, and not present in DNA
Urea	Nitrogenous waste product of humans and other mammals
Ureter	Tube taking urine away from the kidney
Urethra	Tube carrying urine from bladder to the exterior, and in male mammals also semen
Urine	Excretory liquid produced kidneys
Uterus	Muscular organ in which baby develops
Utricle	Part of the inner ear connecting the three semicircular canals; contains gravity receptors
Vacuole	Large fluid-filled cavity in a plant cell, surrounded by cytoplasm
Vagina	Tube connecting uterus to exterior
Vas deferens	Tube carrying sperm from testis to urethra
Vascular bundle	Strand of xylem and phloem cells in leaves and young stems of plants
Vasectomy	Surgical procedure in which vas deferens is cut to render a man infertile
Vector	Organism that spreads a disease
Vegetative propagation	A kind of asexual reproduction in which a lateral shoot of a plant becomes separate from, and independent of, the main shoot
Vein	Vessel carrying blood toward the heart
Vena cava	Vein bringing blood into the right side of the heart
Ventricle	Heart chamber that pumps blood out of the heart
Venule	Very small vein
Vertebra	One of the individual bones of the vertebral column or 'backbone'
Vessel	Water-conducting tube formed by joining of dead, lignified cells (vessel members) with perforated end walls
Villi	Finger-like extensions of the lining of the small intestine; in human development, tiny finger-like process through which nutrients are absorbed by the developing baby
Virion	A virus particle
Vitamin D	Vitamin necessary for the absorption of calcium from the gut and its deposition in the bones
Vitreous humour	Jelly-like material occupying rear of the eye
Water potential	A measure of the tendency of water to move out of a solution through a partially permeable membrane

ISBN 9780170191340

Weathering	Process by which rock is slowly eroded by chemical action of water and materials dissolved in it
White cells (leucocytes)	Defence cells carried in the blood
White matter	Outer layer of spinal cord, containing nerve fibres
Woody plant	Plant in which secondary growth continues year after year
Xylem	Water-conducting tissue in plants
Zygomorphic	(Of a flower) Bilaterally symmetrical, having left and right halves
Zygote	Diploid product of joining together two haploid cells (gametes)

ISBN 9780170191340

Index

ISBN 9780170191340

ISBN 9780170191340

ISBN 9780170191340